LINDSAY AND BLAKISTON

PUBLISH

A MANUAL OF SACRED HISTORY;

OR,

A GUIDE TO THE UNDERSTANDING

Of the Divine Plan of Salvation

ACCORDING TO ITS HISTORICAL DEVELOPMENT.

BY

JOHN HENRY KURTZ, D.D.,

PROFESSOR OF CHURCH HISTORY IN THE UNIVERSITY OF DORPAT, ETC.

TRANSLATED FROM THE SIXTH GERMAN EDITION,

BY

CHARLES F. SCHAEFFER, D.D.,

OPINIONS OF THE PRESS.

"A very comprehensive, accurate, and methodical digest of the Sacred History — done with genuine thoroughness and scholarship. There is nothing among our manuals of Biblical History that corresponds with this. It is simple in style, and orthodox in sentiment."—*N. Y. Evangelist.*

"The Observations (introduced by the author) are replete with the results of extensive research—meeting objections and cavils, solving difficulties, explaining obscure passages, reconciling apparent discrepancies, pointing out connections, exposing and rectifying errors, unfolding the nature and design of sacred institutions and ordinances, and showing the relation of events, persons, institutions and prophecies, to the great central fact and theme of Scripture, man's redemption through the incarnate Son." — *Evangelical Review, April,* 1855.

"This is the best book of the kind we have ever examined, and one of the best translations from German into English we have ever seen. The author makes no parade of learning in his book, but his exegetical statements are evidently founded on the most careful, thorough, and extensive study, and can generally be relied upon as among the best results, the most surely ascertained conclusions of modern philological investigation. We by no means hold ourselves responsible for every sentiment in the book, but we cordially recommend it to every minister, to every Sunday school teacher, to every parent, and to every intelligent layman, as a safe and exceedingly instructive guide, through the entire Bible history, the Old Testament and the New. It is a book which actually accomplishes more than its title promises," &c. &c.—*(Andover) Bibliotheca Sacra, April,* 1855.

Notices by the Press of Kurtz's Sacred History,

PUBLISHED BY LINDSAY AND BLAKISTON, PHILADELPHIA.

Dr. J. H. Kurtz's Manual of Sacred History is the production of a very able and pious divine of our church in Europe. The author is particularly distinguished for his learning, his orthodoxy, his liberality, his piety, and his originality. He writes with great clearness and condensation, and presents in a brief compass a large amount of matter. His various works, and particularly his Histories, have received the highest endorsement abroad in their popularity and multiplied editions, and are commended in the strongest terms by the most eminent divines. Guericke, Bruno Lindner, and Rudelbach, laud his Histories in the strongest terms, and the Evangelical Review,* in the United States, has furnished evidence of his great merits from authentic sources. The admirable Manual of Sacred History, translated by Dr. Schaeffer, (and, having examined some parts of the translation, we may say *well* translated,) will constitute a rich contribution to our theological literature. Having encouraged the translator to undertake the work, we are the more free to express our high opinion of it, and the fidelity with which it has been executed. We hope this will be the forerunner of other translations of works of the author.

C. P. KRAUTH,

Professor of Sac. Phil. Church Hist. and Past. Theol.. Gettysburg, Pa.

Sept. 16, 1854.

The *Sacred History* of Dr. J. H. Kurtz, does not belong to the ordinary class of historic Manuals, with which the literature of Germany abounds. On the contrary, after considerable acquaintance with it, we hesitate not to pronounce it a production of very superior merit in its department, possessed of high literary and theological excellence. Its style is pure and perspicuous, its divisions are natural and appropriate, and the grouping of events felicitous and impressive. Without assenting to every sentiment of the author, we cordially recommend his work to the patronage of the Christian public, and consider Dr Schaeffer as entitled to the gratitude of the church, for presenting this Manual to the English public in so accurate aud excellent a translation.

S. S. SCHMUCKER,

Professor of Didactic, Polemic and Homiletic Theology, in Theol. Sem. of Gettysburg.

Sept. 17, 1854.

I know of no work in the English or German language which gives, in so short a compass, so full and clear an account of the gradual development of the divine plan of salvation, from the fall of man to the resurrection of Christ and the founding of the apostolic church, and which is, at the same time, so sound in sentiment, so evangelical in tone, and, without being superficial, so well adapted for popular use, as the "Manual of Sacred History," by Dr. J. H. Kurtz. The translation of the Rev. Dr. Charles F. Schaeffer seems to me, as far as I have examined it, to do full justice to the German original, as well as to the English idiom.

PHILIP SCHAFF,

Prof. of Ch. Hist., &c.

Mercersburg, Pa., Jan. 31, 1855.

* July, 1853, p. 138.

THE BIBLE AND ASTRONOMY;

AN

Exposition of the Biblical Cosmology,

AND ITS

RELATIONS TO NATURAL SCIENCE.

BY

JOHN HENRY KURTZ, D.D.,

PROFESSOR OF CHURCH HISTORY IN THE UNIVERSITY OF DORPAT,
AUTHOR OF "MANUAL OF SACRED HISTORY," ETC.

Translated

FROM THE THIRD AND IMPROVED GERMAN EDITION

BY

T. D. SIMONTON.

PHILADELPHIA:
LINDSAY & BLAKISTON.
1857.

STEREOTYPED BY J. FAGAN. . . . . PRINTED BY C. SHERMAN & SON.

TRANSLATOR'S PREFACE.

The work here presented to the public in English dress, first appeared some fifteen years ago, in the form of a quite small volume. Meeting with a favorable reception, both from its vigorous treatment of the vital questions which called it forth, and from a native interest belonging to the higher themes upon which it touches — an interest to which the human mind is ever alive — the author was twice led to enlarge and improve the treatise, until in the third edition it has reached its present size.

In the work of translation I have endeavored fairly and faithfully to present the sentiments and views of the author, without omission or accommodation — all responsibility for their character of course resting with himself. The polemical cast of some portions of the work, particularly of the notes, will be accounted for by the circumstance that the author in this edition takes occasion to reply to numerous objections urged against his views as presented in former editions of the work. Keeping in mind the remark of the author, that "we by no means design to give *instruction* in regard to matters of science in the present volume," and also the obvious fact, that general and established principles, rather than more rapidly accumulating individual facts, are wanted for the discussion before us, I have refrained from attempting much addition to the scientific portions of the work. A few facts of recent discovery in the sphere of Astronomy, evidently calling for mention, as well as some results of much interest, and serving to carry out more fully the design of the author in placing together, in a general way, such facts and views as may present to the mind with sufficient distinctness, the astronomical theory of the world,—a few such matters have been introduced in the form of unpretending notes and additions in the fifth chapter, whilst an occasional note has been added here and there throughout the work at large.

It may be well to state that a short treatise upon Geology and the Bible, together with several appendices of kindred matter, is to be found in the volume from which I translate, but which in no measure affect the unity and completeness of the work here presented. As the references to authorities are almost universally to German works, the indications of page, volume, &c., refer to the original, though translations of the works may have appeared in this country, except it be otherwise distinctly stated.

T. D. S.

HARRISBURG, May 1st, 1857.

AUTHOR'S PREFACE

TO

THE THIRD EDITION.

THE present or *third* edition of this work has assumed a new form in many respects, both in its theological and scientific portions. Whilst in respect to Astronomy I have found it necessary only to add or incorporate the results due to the rapid progress of this science, I have been compelled, on the other hand, to wholly recast many sections of the work which more particularly involve questions of theology. Since the preceding edition was given to the public, I have become conscious of the erroneousness and inadmissibility of several fundamental views as therein promulgated, materially affecting the whole cast of the work, which with their far-reaching consequences must now be avoided. I may mention in this connection, particularly, the material change in my apprehension of the Hexæmeron, and the no less important alteration in my view of the Incarnation, which I now, in harmony with the old divines of our Church, acknowledge to have been conditioned alone by the sin of man. Besides, I have felt myself called upon to defend my views against the attacks of several recent writers, who have not only referred to my work, but also earnestly contested many of my positions. I refer more particularly to J. P. Lange (*posit. Dogmatik*), Ebrard (*Abhandlung über Bibel und Naturwissenschaft*), Hofmann (*Schriftbeweis*), and Delitzsch (*Erklärung der Genesis*). Especially have the two last-mentioned works, from which I frankly and gratefully acknowledge myself to have derived much benefit, both in the form of information and suggestion, and by which I have willingly suffered myself to be corrected in several points connected with the subject before us—especially have these works frequently called upon me to enter upon a somewhat lengthened defence of my own views in opposition to those presented against them. This has been *so often* and *so strikingly* the case in regard to the spirited production of *Delitzsch*, that it might almost appear to the uninitiated as though my theological sympathies lay in a wholly different direction from his, whilst I am joyfully conscious of standing upon the same ground of Christian faith and doctrine, and of theological science, with my esteemed friend, the honored author. The more frequently, therefore, I am compelled to disagree with the learned writer in the present volume — confessedly, however, only in points not vitally affecting the grounds of Christian faith and doctrine — so much the more do I rejoice that I shall soon in another place, have occasion to show how highly I prize and regard the late work of this author, and to testify to the advantage and stimulus I have derived from its perusal, as well as to show how closely my theological opinions coincide with his own.

The present edition of this work has demanded also, in those portions not requiring to be again wholly elaborated, manifold improvements and enlargements, and sometimes, no less, abridgements, just as the more matured taste and judgment of the author has dictated. May the many alterations and additions made, be found to indeed improve and enrich the volume, and may it in its new form meet with the same cheerful and appreciative reception as in the former editions!

THE AUTHOR.

DORPAT, August, 1852.

CONTENTS.

CHAPTER I.

CHAPTER II.

CHAPTER III.

CHAPTER IV.

CHAPTER V.

CHAPTER VI.

"Jehova, unser Herr!
Wie herrlich ist Dein Name auf der ganzen Erde,
Der Du mit Deiner Pracht den Himmel gekrönet!
Aus dem Munde der Kinder und Säuglinge
Bereitest Du Dir eine Macht,
Um zu schwichtigen Feind'und Rachgierige.
Wenn ich sehe Deinen Himmel, das Werk Deiner Finger,
Den Mond und die Sterne, die Du gegründet hast:
Was ist der Mensch, dass Du sein gedenkest,
Und der Menschensohn, dass Du ihn besuchest?
Wenig unter göttlichen Stand erniedrigst Du ihn,
Krönest ihn mit Ehre und Herrlichkeit.
Du lässest ihn herrschen über das Werk Deiner Hände,
Allest legtest Du unter seiner Füsse,
Schafe und Rinder allzumal,
Sammt den Thieren des Feldes,
Den Vögeln des Himmels und den Fischen des Meeres,
Was nur durchwandert Pfade des Meeres.
Jehova, unser Herr!
Wie herrlich ist Dein Name auf der ganzen Erde!"

The Bible and Astronomy.

CHAPTER FIRST.

THEOLOGY AND NATURAL SCIENCE.

We have also a more sure word of prophecy, whereunto ye do well that ye take heed, as unto a light that shineth in a dark place, until the day dawn, and the day-star arise in your hearts.[1]

Truly the Word of God, as spoken unto us by holy men of old, moved by the Holy Ghost, is a sure word; for though heaven and earth should pass away, no jot or tittle of it shall fail: it is a precious word, full of the energies of a divine life, a lamp unto our feet and a light unto our path.

But nature also, to him who has learned to read therein, is a divine book laid open; for *the invisible things of Him, from the creation of the world, are clearly seen, being understood by the things that are made, even his eternal power and Godhead.*[2] *The heavens*, also, *declare the glory of God, and the firmament showeth forth his handiwork.*[3] All that the starry heavens reveal in lines of light; all that the seas, the depths and mountains of the earth,

[1] 2 Peter 1 : 19. [2] Rom. 1 : 20. [3] Psalm 19 : 2.

proclaim,—the cheerful day and the stormy night, the glorious bloom of spring, with the hail which crushes and the frost which blasts its beauty; the lily of the field, the sparrow on the roof, the hemlock in the meadow, the serpent in the grass; yea, even a mote in the sunbeam, or a grain of sand, are, when carefully read and correctly understood, a Word of God; testifying of former days, declaring His wisdom and His power, but also His holiness; revealing His creative love, but also His avenging justice.

The yearning and earnest expectation of the creature[1] are no less a sermon from God, fraught with the deepest lessons of wisdom and knowledge, testifying of blessings and curses, of death and the resurrection, of sin and redemption.[2]

"Although," says one who has devoted his whole life[3] to the study of this divine writing—"although the book of Nature in comparison with the holy book of Revelation, appears but like an obelisk covered with hieroglyphics, standing amid the ruins of an overthrown city; whose characters have in part become unintelligible to the present race of men, and in part defaced and obliterated by a hostile hand; yet have we good grounds upon which to maintain an agreement between the contents of these hieroglyphic tracings, which were originally also a

---

[1] Rom. 8 : 19–21.

[2] Compare, for example, the interesting remarks of *G. H. v. Schubert, Ansichten von der Nachtseite der Naturwissenschaft*, 4th ed., Dresd. 1840, p. 259 seq.

[3] *G. H. v. Schubert, Symbolik des Traums*, 3d ed., Leipzig, 1840, p. 44 seq.

revelation of God to man, and the contents of the Holy Scriptures. Yea, nature also, with unmistakeable clearness, bears witness of Him from whom and through whom are all things; and in the present age of the world, when man is perversely inclined more to investigate and delight in physical and intellectual truths, in which he would fain find a full supply of his wants, than in an examination of Holy Writ, it is perhaps not wholly unnecessary to call attention to the solemn testimony of nature, and the harmony of its teachings with those of the Sacred record.

True, the written *Word of God* contains all that is necessary for our welfare: true, the Sacred Oracles speak to us more clearly, intelligibly, and unmistakeably, than the characters of the obelisk: they speak just as clearly to the learned as to the unlearned, to the rude and unlettered as to the talented and refined. For they are like "a stream of varying depth, in which the elephant may swim and the lamb wade," and whosoever hopes, with the book of Nature, to dispense with the book of Revelation, his eyes are blinded no less to the witness of the one than the other, to the being and works of God. Yet still must we also give heed to that voice, *whose sound goes out through all the earth, and its words to the end of the world,*[1] and learn from it what is revealed to us through the creative word of God; and this the rather, since nature—originally a message from God *for* us — may yet become a witness *against* us, for it is written, *so that they are without excuse.*[2]

Therefore, let the theologian, and indeed all *Chris-*

---

[1] Ps. 19 : 4. [2] Rom. 1 : 20.

tians, deign to learn of the student of nature: let the student of revelation give honor to whom honor is due: let him cheerfully permit the masters of science to disclose to his view a new world of wonders, the product of his Father's hand. Let him frankly acknowledge the truth, and strive to appreciate the bold and laborious research by which fresh treasures are brought to light from the deep and hidden mines of science, and cast into current coin.

But, in like manner, let the *man of science* give honor to whom honor is due, the master become the disciple, the teacher the pupil. Let him sit in the humble and teachable posture of a second Mary of Bethany, at the feet of a higher Master, and there learn the words of eternal life, and a wisdom which dates not its origin in time—there learn what neither his microscope nor telescope can reveal, and yet what alone can lend to his wisdom a true sacred character. Let him not forget that if nature be a book full of Divine lessons and teachings, yet is the Bible the lexicon and grammar, whereby alone the etymology and syntax of its sacred language, the form and history, the sense and signification, of the single words, may be learned—that it alone is the teacher of that criticism, hermeneutics, æsthetics, and logic, whereby the "disjecta membra poetæ" are to be arranged, explained, and understood.

But what if the Bible and Nature, instead of explaining, amplifying, and completing, should contradict each other?

The Bible and Nature, since both are the work of God, must agree. Where this does not appear to be the case, the exegesis either of the theologian or the student of nature must be at fault. And not merely the latter, but also the former is, alas! too often the case, and has begotten incalculable difficulty in the question with regard to the harmony of nature and the Scriptures.

Wherever honest *doubt*, desirous only of reliable and incontestable truth, or hostile *unbelief*, delighting ever to disgrace the cause of Bible truth in the eyes of the world, have brought forward pretended —or apparent—contradictions not capable of reconciliation, between the teachings of Scripture and the results of science, they have generally referred to the Biblical history of the creation; and not only divines, but perhaps more frequently men of science, have enlisted all their learning and sagacity to do away with these pretended contradictions, and bring out in all its beauty and symmetry, the agreement between the Bible and science.

And behold! just here, where the conflict would fain be the most unmistakable, and the contradictions most numerous—just here it is, that, with an adequate idea of Divine Revelation, and a proper understanding of the Divine record, a contradiction is wholly *impossible*. And for this reason, that the Bible neither reveals nor was designed to reveal what is attainable by scientific investigation; and conversely, that no knowledge to be gained by scientific research, comes within the province of revelation:—because these two sources of knowledge do not *encroach* in their teachings *upon each other*, but lie *side by side*, and hence of course cannot contradict and supplant, but only (the correctness of their teachings in other respects granted) complete each other.

The Mosaic history of the creation,[1] as the Bible in general, was by no means designed to give instruction in regard to natural science. Nothing was more foreign to its object. The efforts of the human mind after secular culture, after art and science, were never designed to be mere tributaries to, and dependent upon, special Divine revelation. As man was to gain by the sweat of his brow, his daily bread, for the support of his *physical* life, from the earth he inhabits; so also must he acquire from nature *in*, *around*, *below*, and *above* him, by wearisome effort and diligent research, science and knowledge for the support and culture of his *mental* being. In no case whatever has either mathematical, physical, or medical science, been communicated to him by Divine revelation. None of the prophets of the old dispensation, no apostle of the new, gained scientific knowledge through revelation. No one of them was raised by Divine illumination in this respect, beyond the stage of knowledge and culture belonging to their own age. All that a *Moses* knew in the several spheres of Astronomy, Geology, Natural History and Medicine; in regard to the constitution of the starry heavens, the structure of the earth's crust, the signs of clean and unclean animals, the course or treatment of leprosy,

[1] We cannot here anticipate the detailed explanations of the subsequent chapters of this work, and must hence for the present confine ourselves somewhat to generals.

the economy of the sexes, etc., he had learned under the tuition of the Egyptian Magi,[1] or had acquired from personal observation and study during the forty years he spent in the wilderness. But Divine wisdom knew well how to avail itself of knowledge thus acquired, by natural means, and to consecrate it as the vehicle of imperishable ideas of grace and justice, of sin and redemption. All that a *Solomon*, whose wisdom attracted the Queen of the South, spoke or sung,[2] in his three thousand Proverbs, or in his one thousand and five songs; in regard to trees, from the Cedar of Lebanon to the hissop upon the wall; in regard to beasts and birds, creeping things and fishes; was the fruit of his own deep contemplations of nature; but it was also a channel through which Divine wisdom might be conveyed to the minds of men.

Yea, we go even further; we boldly maintain, and with the fullest assurance of not in the least compromising the Divine character of the sacred books, that holy men of God, both of the Old and New dispensations, who, under the influence of the Spirit were moved to divine words or deeds, may very easily have been involved, as far as scientific know-

[1] Hence, the circumstances which brought Moses into connection with these wise men, must be regarded as having been specially under the divine direction. He who was to give to Israel the law and divine service, and with them fresh treasures of divine revelation, must also be learned in all the wisdom of the Egyptians; in order, thereby, to attain to the highest preparation of his natural gifts and talents, and also sufficiently comprehensive knowledge, for the fulfilment of his Divine mission.

[2] 1 Kings, 4 : 32, 33.

ledge is concerned, in the common and prevailing errors of their age. Such errors did not in the least detract from the religious truths they were called upon to announce, and impress upon the hearts of men. If it be true, for example, that in the time of *Joshua* the common opinion prevailed, that the sun, together with the whole starry heavens, revolved around the earth in 24 hours, certainly *Joshua* himself was not raised above this error; and it doubtless lay at the foundation of that command, evincing such signal faith and so often commented upon: "Sun, stand thou still upon Gibeon, and thou moon in the valley of Ajalon."[1] Joshua spoke the command of faith *as he* understood the matter, but the Divine hearing of the command was carried out *as God* understood it.[2] Nor should our faith be any more estranged from the Scriptures, on finding that the geocentric view underlies their teachings in other passages.[3] Moses also may have had very

---

[1] Jos. 10 : 12, seq.

[2] The desire of Joshua was, to see the light of day remain, and the darkness of night prevented, until he had secured his object in the pursuit of his enemy. And *this* desire he gained through his extraordinary faith. It was a matter of no moment to the faith of Joshua centuries ago, nor is it now, to the *faith* of the reader, by what natural means such a supernatural effect should be produced.

[3] All attempts, therefore, to prevent the inspiration of the Bible from suffering in respect to matters of human science, by proving that though the Scriptures may indeed speak geocentrically, the heliocentric view, nevertheless, underlies their teachings, must be regarded as having mistaken their object, and as tending to error. This is nothing more than the opposite pole of that perverse and mistaken spirit, which sought the rejection of

many physically erroneous views touching the nature of the starry heavens, or the structure of the earth, as he in the spirit of prophecy conceived the history of the creation of the heavens and the earth, without

the Copernican system, because a few passages of Scripture involve the geocentric view. Such was the error committed by the worthy G. Fr. v. Meyer, who, in his *Blättern für höhere Wahrheit*, viii, 342 seq., tries to defend the formal proposition: "The Bible, in thought, takes the heliocentric view." In order to effect this, he adduces particularly, James 1st, 17th: "Every good gift and every perfect gift is from above, and cometh down from the Father of lights, with whom is no variableness, neither shadow of turning." In explanation, he adds: "Here allusion is made to the earth, in contrast to the lights of the firmament, and there is attributed to the former, not as an accidental, but as an inherent quality, what is denied as an essential or inherent property with God, namely, a variableness and darkness produced by turning (or rotation). Were reference here made to a revolution of the heavens about the earth, in connection with which the sun produces the alternation of day and night, this motion would be something external to, and not belonging to the earth, which would not be in harmony with the import of the contrast. We concede that the hint is a very slight one, not rising to the character of a direct animadversion; and we regard the passage as a proof, as delicate as it is clear, that no less according to the wisdom of the Spirit of God, than the teachings of modern science, the earth rotates upon its axis; yea, that it holds, as is scarcely to be separated from its axial rotation, its annual course around the sun, and hence, of course, that the heliocentric system must underlie the views of the Bible." *G. F. G. Goltz* (in his otherwise not unworthy small treatise: Die Stillstehende Sonne zu Gibeon, Berlin, 1833, p. 35) concurs in this course of reasoning, and adduces as an additional argument, Gen. 1 : 5: "Then from evening and morning arose the first day," [as Luther renders the passage]. By this he would have us understand: "Then came the day from evening (the west) towards morning (the east)." And thus does the passage contain evidence that the

its at all being necessary that his mind should have been thereby disabused of these errors; for the Mosaic history of the creation was by no means intended to give instruction in physics, but its design was wholly to impart religious knowledge.

True, however, it is also conceivable that a physical truth may be interwoven with the revelation of religious truths, either as the necessary vehicle of the latter, or perhaps as the accidental attendant, or in a measure, the exponent of such truths. Doubtless, the religious or ethical position of something in nature, which is the object of revelation, may be so conditioned by its physical constitution, which is matter of scientific inquiry, that an erroneous apprehension of the latter would give to the former a false character and an erroneous tendency. Thus, for example, the physical constitution of the universe, the adjustment and connection of the different bodies composing it, their mutual relations, and the like, may also have a religious significance; which, as such, might be matter of revelation, so far as a more profound knowledge of it might lend to our minds a more comprehensive or a clearer insight into the Divine plan of the world, in which we are also included, and so deeply interested. But even in such cases, physical instruction cannot be so connected with revelation, either as its direct object, nor yet as its consequence, that the human mind

---

earth rotates from the west to the east; for, were the Scriptures to speak geocentrically, they should say, in accordance with the apparent course of the sun: "Then from the morning and the evening, arose the first day."

given up to the influence of the Spirit, should thereby be occasioned or forced to give up erroneous views in the sphere of physics; or rendered capable of so anticipating the future developments of human science, that its knowledge thus acquired, should stand in open conflict with the stage of development belonging to that age: for both these alike would directly conflict with the character of Divine revelation. Revelation in such cases refrains from communicating knowledge, as indeed it does not design to reveal at once, all and everything that is of religious significance. Nay, rather, it is like a governess, who does not at once impart to the child committed to her care, all she knows; but merely at each time what is immediately required for the advancement of her pupil, or what its previous culture has fitted it to receive and apply. The Holy Scriptures in such cases, evince their *Divine* character in this; that they leave room for all the future progress and results of science, that they never are found in error, and that no new science may approach them with the remark, "Had ye but been silent." But we may rest in the confident assurance that hereafter —in eternal life—a revelation of a vastly higher and more comprehensive kind, will rectify the errors of our human science, fill up its chasms, and disclose to our eager minds its higher religious import.

The error on the part of the student of the Bible, when contradictions seem to arise on a comparison of the facts of revelation with the results of science, consists frequently in this: that he expects to find in the Bible, information that is wholly foreign to it, and which it would have no motive in communicating; since for the time being it still lies wholly without the sphere of its objects. Deluded by the ignis fatuus of this false expectation, the Scriptures are examined; and, thus examined, naturally appear in a false light and are wrongly apprehended.

But none the less may the pretended or supposed contradiction rest upon an erroneous interpretation on the part of the student of nature, in that he too may approach the Book of nature with unwarranted pre-suppositions, and there read from its pages what he himself put into them. As the scientific and physical point in regard to any matter, is not the subject of revelation, so, conversely, the specifically religious point does not fall within the sphere of empirical science. Whether, for example, the world was created in time, and from nothing, through the will of a personal God, as the Bible teaches; or whether, as erring belief of both ancient and modern times teaches, it be itself God and eternal, so that the origin of new forms of life in it manifest only its proper self-development; is a question which no man of science is able to determine, from the results of his observations and investigations; for the instrument of such knowledge is faith alone. It ever eludes the grasp of empirical science, and the more conscientious and faithful such science is, so much the more carefully will it abstain from all such erroneous and presumptuous expectations. It were the gravest possible self-delusion for the student of nature, or any one else, to imagine that the results of his empirical investigations required him to deny the Biblical doctrine of the creation of the world. Not science, but speculation (for error may exist in the magnet or compass no less readily than faith or truth), is to blame for such vain assumptions.

As with the Biblical doctrine of the creation, so is it with the other fundamental doctrines of the Scriptures, which are regarded as being incompatible with the results of science. There is drawn or extorted from these results — true and false—a theory of the world, in which the Biblical doctrines of angels and spiritual beings, of original sin and the incarnation; of redemption and the end of the world, of the judgment, resurrection, and future state, no longer find room or recognition. And here again it is not natural science that is to blame; but unbridled speculation, or rather an already existing tendency of thought or imagination, which carries speculation with it, and thus does violence to the results of scientific investigation, in order to force them to say what is most pleasing to the unbelieving ear.

When we take into consideration, in addition to the arbitrary course of philosophical speculation, also the unreliable character of those scientific results which it makes use of as starting-points, it at once becomes clear how little weight is to be ascribed to its deductions, in contrast with the teachings of revelation. For it is a notorious fact, that the more deeply science attempts to press into the secrets of the origin and existence of all things, to solve the problems of time and space, so much the more unreliable do its results become. It is open to error in precisely the same ratio that difficulties presented by the sea and the mountains, the depths of the earth and the heights of heaven, grow in magnitude. It is the less liable to take appearance for reality and reality for appearance, just in the same degree that it confines itself to the surface of things, without being able to pierce to their centre. Its results are the more inadmissible, the greater the difficulty our abstract age has to surmount, in unriddling its hieroglyphics and picture-writing; and in the same ratio does a sagacious criticism become necessary, to distinguish what is genuine from what is false or interpolated, in the great Book of Nature. For let us not attempt to conceal the fact that nature no longer offers us the pure hand-writing of God: it is, in many respects, a *palimpsest*, a "codex rescriptus:" an enemy's hand has passed over it, and obliterated or rendered indistinct many a precious character, and introduced or superscribed many a word which did not originally belong to it.

Hence it is clear that the Scriptural theologian has little to fear from the pretended antagonistic character of the results of science. But if a conflict nevertheless arises, let him search the Scriptures with all diligence, and test with the greatest sacrifice of opinion his apprehension of the written word, forgetting all assumption of personal sagacity and knowledge,—let *this* be his *only* presupposition; that the word of God must stand and *will* stand, immovably firm, in opposition to all the assaults and confusion of the opinions of the age, and of human wisdom, as well as all the pretensions of human science. If he do not thus succeed in solving the supposed contradiction, let him securely remain in the fortress of the Word, under the cheerful conviction that the contradiction is either merely an apparent one — none at all — or that the error lies upon the side of science. Let him rely upon science, or rather, upon the living God, whose potent word in spite of all human pretensions, supports even science itself, that it (science), in the purifying fires of its own development, will separate from the true substance, all wood, hay, and stubble. To swear by the words of a master, may in all other cases indicate a dependent and timid mind; but there is a Master by whose words we may swear, and at the same time evince true freedom and independency: *The Lord and his Spirit*, for it is written: "*neither be ye called masters, for one is your Master, even Christ.*"[1]

Fortunately, however, the case is still propitious. True science, and particularly natural science, has in all ages of the world willingly taken a position at the feet of revelation; and has ever carefully avoided any attempts to bring its results into hostile antagonism with the teachings of Scripture. We might bring forward a not insignificant array of witnesses to the truth of Christianity, from the ranks of science; we might recount a host of celebrated names, from *Albertus Magnus* down to *Schubert* and *Cuvier*, which, at the same time, are of note in the Christian world. We might bring forward a host of witnesses, who shine as stars of the first magnitude in the firmament of science, who all

[1] Matt. 23 : 10, comp. John 8 : 32, 36.

have cheerfully asserted their unshaken faith in Holy Writ, and the harmony of its teachings with the results of their scientific investigations — if this did not lead us too far from our present object, if it had not already been frequently and satisfactorily done.

The more astonishing and greater the progress of the natural sciences collectively in our day, the more day by day our knowledge of nature increases in depth and expands in compass, and the results and views obtained become the highly prized, if not indeed sometimes over-valued common property of all cultivated minds,—so much the less is the Christian, and particularly the divine, called upon to ignore them; nay rather, so much the more is it necessary that he should bring these results and views into fruitful connection with Holy Writ, in order that the two might be united and mutually complete each other. Endeavors to consummate such a union have, indeed, often been misconstrued, and the call for them even often denied, both on the part of science and also religion. Nor have contemptuous insinuations been wholly wanting. These, however, we may well leave unnoticed; but objections *sincerely* urged, in the name of religion or science, we duly recognize; but none the less on this account do we consider the difficulties upon which they are based, as without due foundation.

Nay rather, we maintain that science can but reap advantage from a vital union with faith; nor can faith, on the other hand, derive less benefit from a similar union. Science receives its true consecration from faith, its eternal significance, outreaching far the bounds of time; faith derives from science, warmth and vivacity, vigor and fullness. Science is the younger brother of faith; both are the offspring of the *same* parent, that eternal Wisdom through which all things were made, and by which they are supported and preserved. Should not a common love to a common parent, by whom they are both sustained, be also a bond of sympathy between them? Can it be that one of them should wholly disregard the welfare of the other? Can the elder, without the guilt of fratricide, repeat the impious speech: "Am I my brother's keeper?" Or shall the younger be allowed boldly to vaunt himself, and say: I have no need of you?

But, says the advocate of science, shall the science of physics, which has but lately burst the fetters imposed upon it by a narrow-minded religion, be bound anew? Shall the more free and happy course of development, which it has but lately entered upon, be obstructed anew by the dogmas of the church? Would not its freedom of vision, in the most favorable cases, be thus obscured or obstructed by prejudices or pre-conceived notions? Were natural science again to be vitally united with faith, it would certainly lose its sovereignty and independence, and then might we expect a return of those dark ages, in which an Albertus Magnus and a Roger Bacon were decried as magicians, and a Gallileo persecuted and imprisoned.—Not so! this *should* not be the case, this *will* not be the case! Science is not to become the mere servant of faith; it is only to strive with it toward a common object, in free and unconstrained alliance, viz: the glorifying of God through a knowledge of his wisdom and power, his grace and holiness; and the advancement of man to the image of God, through a knowledge of his own calling, his position and mission in the world. Science is not to be cheated of its freedom; but is to become a co-heir of the riches laid up in the Father's house: the sphere of its calling is not to be closed against it. Nay rather, let it descend into the deepest depths, and climb to the highest heights; let it take the wings of the morning, and fly to the uttermost parts of the earth, or with the swift footsteps of light, hasten to the giddy heights of the heavens; but let it not forget the Father's house, nor fail frequently to return thither, casting its acquired treasures humbly at the feet of the everlasting Parent, in order that they may be thoroughly purified in the refining fire of eternal wisdom. Its efforts and zeal are not to be lightly esteemed or despised by faith; but, at the same time, let it not be too haughty to heed the counsel and admonitions of its elder brother, and avail itself of his wisdom and experience.

But on the opposite side we see a similar array of difficulties. Thus objects the partisan of faith: the Bible furnishes us with wholly reliable, objective truth; natural science, with unsatisfactory or inadequate perceptions, subjective views or notions. In the case of the latter, what to-day passes for indubitable, impregnable truth, is to-morrow resigned as error, only to be followed by a new view, in all probability destined for a similar fate. Are we justified in bringing together, uniting and mingling, what is divinely communicated with what is obtained by mere human means or investigations? Must we not rather ever carefully draw a line of distinction between the two, lest the objective enter the sphere of subjectiveness, the absolutely true and reliable the sphere of error, and thus itself become the subjective, the unreliable, the unsafe?

However worthy of regard sentiments expressing themselves in such fears may be under the circumstances, still there underlies them a certain weakness of faith, and also a slight consciousness of this weakness. For assuredly that is not the true faith, that overcometh the world, which fears science and dares not look it full in the face. Such a faith is far from that stability and assurance, that holy boldness which characterizes the champions of faith. True, it is not to be concealed that faith, should it enter into such a proposed alliance with human science, would be exposed to many dangers, which otherwise it might perhaps avoid. It would be forced to self-distrust, and to relinquish to some extent its self-complacent claims of directness and infallibility; it must leave the safe harbor of repose, and entrust its precious bark to unknown seas. It is not impossible that the vessel might be engulphed in the wild waves of doubt, shattered upon the cliffs of knowledge, or stranded on the sand-bars of speculation Nevertheless, faith is ever endowed with a divine energy; the bark, however weak it may appear, is possessed of an anchor capable of withstanding the most tempestuous surges, a compass which never varies, and there is besides One within the vessel who rebukes the wind and says unto the sea, "*Peace, be still.*"[1] He also calls to us and says: "*Why are ye so fearful? How is it that ye have no faith?*[2] Danger exists only where we avail not ourselves of this Divine strength or succor. If we, like the unprofitable servant, lest we lose the talent committed to us, hide it in the earth, and then in the day of reckoning like him, boldly approach our Lord with the words: "*Lo, there thou hast that is thine,*"[3] let us beware lest a like sentence be also pronounced upon us.

And wherefore this over-wrought fear lest the objective, the Divine, derive from man a subjective character or coloring? Is then the subjective, in itself, necessarily erroneous and unreliable? Has not also subjectivity its sacred, its inalienable rights? And is there in general any human knowledge, desire or emotion, which does not proceed from or pass through the sphere of the subjective? Was not the objective truth of the Scriptures subjectively imparted, passing through and deriving a coloring from the peculiarities of the human mind? Does not the preaching or writing of a Paul or a James, of a Peter or a John, bear the clear impress of subjectivity—true, of a sublime and exalted, of a holy and sanctified, of a strengthened and vigorous subjectivity? True enough, the Bible furnishes us with the objective truth in all its fulness; but this, not in its separate sentences and divisions, but in its whole scope or conception, in the unity of its prismatico-

[1] Mark 4 : 39. [2] Ibid, v. 40. [3] Matt. 25 : 25.

subjective rays. Take away subjectivity from the Bible, and you either put the Bible far away from yourself, or stand aloof from it,—your faith becomes a mere blind faith, a dead, worthless thing. It is no longer a Divine power, purifying, restoring and thoroughly strengthening you.

Far be it from *us* to seek for the subjective, a boundless sway in the sphere of religion. For by so doing we should blindly turn our weapons against ourselves, and the holy kingdom of God, to guard, protect and serve which, we deem our most sacred duty. We acknowledge and revere the *Scriptures* as the only infallible source of all religious truth, as the canon of judgment in all matters of religion; yea, further, we acknowledge and revere the *Church* as a firm barrier against all arbitrary explanation or interpretation of the written word; as a basis *upon* which we are called to build a store-house of true and sacred knowledge, increasing in extent and purity with the progress of each century. There is, indeed, much truth in the favorite proverb: "the history of the world is the judgment of the world," but much more comprehensive and incontestable is the truth, that the history of the Church of Christ, in its developments, is a judgment upon private interpretation or views which have spread abroad in the Church. It is a purifying fire in which all the dross of erroneous apprehensions, to which the most honest and truth-loving subjectivity is inevitably exposed, shall be fully separated, so that at last only the refined gold of pure doctrine may remain as a basis for future development; for we believe, in accordance with the Divine promise, in a powerful presence and influence of the Holy Spirit in the Church, which is ever victorious. But at the same time it is not to be forgotten, that all the objective, truths of the Church have proceeded from efforts and investigations in the sphere of the subjective from human knowledge, Divinely blessed. For example, those were primarily subjective views which noble and highly gifted men, such as Athanasius, Augustine, and the like, spoke, impelled by the deepest personal necessity of their life and knowledge; but it was the victorious power of indwelling truth, and the invisible influence of the spirit of God, which imparted to their views objective validity and worth in respect to the Church.

But those parts of religious knowledge which here concern us, are, in part, precisely such as the Church in her wisdom has not heretofore given a place in the sphere of the objective, and are most likely also, in part, such as must forever be excluded therefrom. There is a "*dark side of Natural Science*," of which a sage of the present day has given us profound and interesting "*views;*" there is also, no less, a *dark side of the science of revelation*, which in part coincides with the former. To the dark side of the latter pertain all those matters which are not lighted up by the rays of a clear mid-day sun, but mysteriously appear only slightly unveiled beneath the glimmering of distant stars; but which are destined, when night with her enshrouding shadows departs, and day draws on apace, to stand forth in the clearest and most satisfactory light. Only for such parts of religious knowledge do we claim the right of subjective *views*, only in such cases do we allow ourselves to seek the lights of natural science, in order to see if thus the subject may not stand forth more clearly, in more definite outline, just as, conversely, we would apply the lights of revelation to the dark side of natural science.

The results of scientific research have arrived at this point—they have acquired for themselves a position and acknowledgment by the side of religious truths. But the human mind is no abstract, dead framework or receptacle, in which the human and the Divine, each by itself, may be laid up and preserved, so that they should not mutually touch or pervade each other, and be united. The mind is an indivisible whole, a living unity; every effort or action demands the whole mind; all new knowledge which it receives thoroughly pervades it, and most intimately unites itself with the system of knowledge which is already possessed; and wherever this union is not possible, previous acquirements must either be yielded up, or the new must be denied admission. "It is the mission of faith to unite all with itself and pervade it with its own spirit; to lend to all a pure, religious and sacred character, and especially to mould, as it were, all science into Theology. And, on the other hand, it belongs to the nature of a firm scientific conviction, that it cannot abide the presence of any religious conviction or belief, unless it be fully conscious of a true and full agreement between the two."[1] This is especially the case, however, in

[1] *Lange*, Das Land der Herlichkeit, p. 6.

regard to the inductive sciences. There is inherent in them, as experience has sufficiently taught us, a power of conviction which may even endanger the authoritative claims of faith. It was upon *this* power that *Dr. David Strauss* relied, and assuredly he did not mistake as to the power *itself*, when he dared in his "*Glaubenslehre*," to oppose to revelation, a handful of sorry invalid troops, enlisted from the lazar-houses of Astronomy, Geology and Anthropology, and with incomparable assurance, attempted to persuade the public that they were a youthful, vigorous and insurmountable host.—Shall *we* look upon such scenes and fold our hands in inglorious ease? Shall science offer us in vain its blooming, youthful energies, for our protection and defence? Has not the saying: *The children of this world are wiser in their generation than the children of light*,[1] been long enough written for us to have learned the lesson it would teach?

Let us not be misunderstood. We by no means desire to see Theology beg its daily bread at the door of Natural Science, while it possesses at home the richest supplies of heavenly manna: nor do we desire to see it, as the Athenian, ever eager to see or hear some new thing, whilst the Bible freely offers its precious treasures of wisdom and knowledge—treasures ever fresh and new, and which can never be fathomed or exhausted. Neither do we wish it to imitate the example of the children of Israel in following after the gods of the Canaanites, burning

[1] Luke 16 : 8.

incense in every grove, and sacrificing upon every hill; whilst the quiet, delightful service of the God of Abraham, beckoned them to come to His courts; whilst the wondrous love and grace of Jehovah claimed their praise and adoration. But we desire that it should freely adapt to itself and apply to the glory of God, as well as to the advancement of the human mind, destined to find its end in Him, all that true science has acquired in its eager efforts after truth, in its untiring search for knowledge. We rely upon the unfailing power and life inherent in Divine truth, to separate and cast off, just as does the living organism, foreign and unappropriable matter, all error or falsehood which may insinuate itself along with truth. We desire that the vision of Theology should be rendered clear by the lamp of the Divine Word, so that it may discern gifts and spirits,[1] and not inconsiderately admit whatever is obtruded upon its attention.

---

[1] 1 Cor. 12 : 10.

CHAPTER SECOND.

THE DEISTIC AND PANTHEISTIC THEORIES OF THE WORLD.

Two antagonistic extremes in the province of religion, are to be met with in all ages of the world; and especially in our own times — Deism and Pantheism. The grand point of conflict between these two religious tendencies is, as is well known, the relation of God to the world. The former acknowledges only a far-distant God, whose infinite greatness prevents him from stooping to regard every little circumstance in the world, and renders it necessary that he should commit the preservation and government of the world to the so-called laws of nature. The latter recognizes only a God near at hand and in the world, who lives in and shares his life with all things, who unfolds himself in a blade of grass, who finds his highest development in the mind of man, and whose life is the life of nature, apart from which it has no existence. Both these tendencies or views are, when opposed to each other, in the right; for each one has at its foundation a deep religious necessity, of which its antagonist is wholly unconscious. But both are, when opposed to Christianity, in error; as the latter unites in a most comprehensive system the elements of truth contained in these antagonistic views, and yet shuns their one-sided features. We shall employ the terms

Deistic and Pantheistic to designate the false theories of the world to which these extreme and one-sided views have mutually given rise. To these we will then oppose the Biblical theory of the world, together with the results and views of modern astronomy, which serve to explain and establish that theory.

Let us first notice the arguments in support of the DEISTIC theory of the world. This theory, which claims to be "par excellence" the learned one, has in modern times borrowed the weapons with which it attacks the inspired word of God, chiefly from the armory of natural science. That its weapons thus acquired are not always those of the keenest edge, and that it has been forced to content itself, amid others, with some very blunt, inferior, and unserviceable ones, will be seen as we proceed.

It is maintained that ever since the earth has, by the reception of the Copernican system, been lowered from its proud imaginary height as the throne and centre of the universe, to the low and servile position of a mere satellite to one of the most insignificant suns, the Biblical theory of the world must be looked upon as antiquated and exploded. For the planets of our solar system are worlds like our earth, and are doubtless like it inhabited. All the millions of fixed stars in the Milky Way, are suns like ours, many of them vastly larger and more magnificent than it, and like it, encircled by moons, planets, and comets. And those nebulous spots, visible only through the telescope, are new systems of Milky Ways, whose resolution into millions of stars is prevented only by a distance which mocks the powers of our best instruments. Sir W. Herschel, even in his day, counted about 3000 such nebulæ; and how many thousands may still lie hidden in the depths of space, beyond the reach of the best instruments of the present day! As far as the eye can penetrate, we see planets revolving about suns, carrying their satellites with them. But the sun also, and the fixed stars, of a like nature with it, are moving in a direction probably common to them all. Perhaps all these bodies, with their accompanying planets, moons, and comets — bound together under the influence of the powerful laws of attraction and gravitation — are sweeping in wide and majestic orbits, round a mighty, all-controlling central sun, far withdrawn from the reach of the eye or the telescope. Such a harmony of movement must also be assumed as most probably existing in all the systems of Milky Ways. But, though the human mind survey systems so vast in scope, all is not yet exhausted, no end of worlds is yet to be found. How then shall this little spot, earth, dare, in heaven-defying boldness, to oppose itself to the whole universe, of which it appears the most paltry and insignificant speck? Who can persuade himself to believe that all these millions of worlds are so insignificant, that the Creator spent but a single day in their creation, while the superior value of our earth demanded five days of his time? And when we remember that light, with the astonishing velocity of 192,000 miles in a second, is from nine to twelve years in reaching us from the nearest fixed star, and that Herschel could distinguish by means of his colossal telescope, stars still one thousand times more distant, whose light must travel for 9000 years before striking our retina — when we reflect that, according to the computations of that great astronomer, many nebulæ are 300,000 times further distant from us than the stars by which we are immediately encompassed, and that not until it had swept through space for about 3,000,000 of years, could their light reach our eyes—how can the assertion, that all these bodies and systems were created with our earth, some 6000 years ago, maintain its ground? Who can be so devoid of reason as to imagine that all the countless and immeasurable worlds of the firmament, have been placed there for the service of this atom, earth; to adorn its nights, or in addition, perhaps, to exercise the astrological faculties of its inhabitants? Yea, who can believe what is incredible, that the Creator of all these systems of worlds should have condescended as a child, to speak with the children of men? and more, not only to take *upon* himself human nature, of earthly origin, but also take that nature *with* him into eternity, into blessedness. Thus, then, as a writer who has set himself up as the representative of the Deistic tendency, has assured us again and again,[1] since the Ptolemaic system has been overthrown, all the fundamental doctrines of Christianity, such as those touching the incarnation of the Son of God, his atonement, his ascension to heaven and his future advent,

---

[1] Compare *Bretschneider*, Sendschreiben an einen Staatsmann, p. 70.

the doctrines of a resurrection and a judgment, of a heaven and a hell — all these fall to the ground as the playhouses of children are overthrown by a storm.

It is not to be so confidently affirmed, however, that the triumph of the heliocentric doctrine, as such, furnished the occasion and served as a signal for such "en bas" clamor; indeed, it is more than doubtful that such was the case. It is certain at least, that the three greatest and most zealous champions of this doctrine, and those to whom it owes its common and intelligent reception among men, most strenuously protested against the honor attempted to be heaped upon themselves, of having by their teachings detracted from the validity and authority of those doctrines of the Scriptures; for *Copernicus*, *Kepler*, and *Newton*, were sincere, believing Christian men, who reposed their only hope in time and for eternity upon the truth of the doctrines of the Bible.[1] But if, on the other hand, we seek to trace

[1] The Christian sentiments of the latter particularly, are well known to the world, through his posthumous theological works. The cause of his enemies was, indeed, sought to be furthered by the base report, that he (*Newton*) had, in the closing years of his life, fallen into a childish and melancholic state of mind. Some even went so far as to commiserate the great man, that such a misfortune should have befallen him, and to show quite a fair amount of indignation, that any one should have been so unfeeling as to expose the decadence of so great a mind, before all the learned of Europe, by the posthumous publication of the productions of his childish old age. But let us hear what is said in this connection by the "*Conversations-lexikon*," ed. 8, vol. 8, p. 321 (comp. 9th ed., vol. 10, p. 735) which would be accused of religious bias by no one: "The remarks of the philosophers of the 18th

the true pedigree of the presumptive antagonist of our Christian faith; *Copernicus*, *Kepler* and *Newton*, deny all relationship with it; perhaps *Shaftesbury*, *Toland*, and *Tindal*, would not be so modest.

We are certainly not in error when we assert that just here lies a self-deception, which we have met too often to be ignorant of it. Two plants may draw their nourishment from the same soil, and yet their fruits be of entirely different kinds. The analogy of this is to be found in the moral and intellectual world. The Hegelian school of philosophy produced a *Strauss* and a *Feuerbach;* but also such men as *Göschel*. Both parties proceeded from the philosophy of that great thinker, the former using it as an instrument to overthrow all Scriptural belief, the latter drawing from it the strongest sup-

---

century, touching the disorder of Pascal's mind, are founded upon the same false basis as the report of Newton's mental calamity. The Christian sentiments of the one as well as the other, being of a decided and open character, when denial of the fact would no longer answer, it was attempted to account for it through mental unsoundness. Chronology furnishes a complete refutation to all such ungracious charges." Newton's theological writings owe their origin to the bloom of his manhood. As to *Kepler*, comp. his life, by *Breitschwert*, 1831, and the notice of the same by Tholuck, in his *vermischten Schriften*, II, p. 384–402. As to the religious sentiments of *Copernicus*, it is sufficient here to quote the epitaph, which may yet be read upon his monument, in the "*Johanneskirche*," in Thorn, and which was conceived by himself for the purpose it now serves: —

"Non parem Pauli gratiam requiro,
Veniam Petri non posco, sed quam
In crucis ligno dederas latroni
Sedulus oro."

port for the Christian faith. But how are these differences to be explained?

In the one case as well as in the other, that which appears as the last was the first: the cause appears as the consequence. The same will apply to the theory of the world. The Deistic theory, which would fain be regarded as the consequence of the Copernican system, was first in point of time. Not until man had succeeded in barring out from his creation, the living God, whose almighty word upholds as well the tiny atom as the huge globe, and had relieved him from the cares of the world in a "dolce far niente"—not until man was willing to acknowledge the infinite only, while he denied that He took finite nature upon himself, was he able to regard the universe as a machine; and to praise as highly sublime and alone worthy of the Infinite Being, that monstrous, or as *Fr. Baader* calls it, "tedious" idea of the heavens, as an endless, monotonous repetition of suns, planets, and moons, with their inhabitants, "tout comme chez nous." Thus in a convenient yet noble manner would man free himself from all belief in an incarnate God, and from all the uncomfortable attendants of such a belief. Starting out from the heliocentric theory, which well deserves the reception it has received, but forgetting that in the sphere of spirit other laws obtain than those of mere magnitude and distance; and that, notwithstanding the correctness of that theory, the earth in another—in a religious—respect, may be a central and important point in the universe, man fell to heaping worlds upon worlds, and solar systems upon solar systems, taking away in geometrical progression from the religious significance of the earth, just as he swelled the number and magnitude of the systems scattered through space. From Sirius, the earth appeared as scarcely worth naming, and by the time the nebulæ were reached, it had entirely vanished. When the mind became giddy amid those infinite heights, and the heart felt overwhelmed, desolate and forsaken, that was called devotion! The prodigious discoveries of the celebrated *W. Herschel*, completed the vast and dreary prospect; but he himself, though retaining as much as was possible the prevailing notions of his age, found it necessary, gradually, to give them up, as he made an increase in discoveries not in harmony with those fundamental notions. But to *G. H. von Schubert* especially, belongs the praise of having (in his ingenious work, *Die Urwelt und die Fixsterne*) followed out in an independent manner, the path trodden by Herschel, and "given the old song new words;" while he pointed out the way to more profound and comprehensive views, according to which our little earth, together with the breathing dust which inhabits it, may boldly and significantly oppose itself to the huge masses of all the worlds."

Directly opposed to the theory of the world we have just been considering, is another, which recently has been seeking to gain prominence in the world, and to which from the Biblical stand-point, we are compelled to offer no less decided objections. We shall call it, both from its basis and its nature, the PANTHEISTIC theory.

The former theory over-values the results of astronomy, and claims too much for that science; the latter contemns or ignores both the science and its results. The former is governed by a religious, or rather, an irreligious motive; the latter equally so. There the heavens are over-estimated; here the earth. In the one case the earth, when opposed to the universe, vanishes entirely; in the other, the universe retreats into the back-ground before the significance of the earth. The one loses itself in an imaginary infinity of worlds; the other feels at home and comfortable, only when upon its own earth, and devours a whole heaven of heavens at one speculative meal. The one delights to represent the earth and its inhabitants as small and insignificant as may be, so that the Lord of heaven and earth need have but little care for his creatures; and man have the consciousness of being but little regarded by Him whose eyes are as a flaming fire, and who trieth the hearts and the reins. The other delights to celebrate the deification of the human mind as the only and highest soul of the world, and thus add a significant page to the glory of man. The approach of the venerable Copernicus, is, under such circumstances, very inopportune; and the sublime views of a Herschel, a Bessel, a Strvne, and a Mädler, are exceedingly annoying. Hence the results of sublime and laborious research are persistently ignored, and man chooses rather to interpret the world "à priori."

The Pantheistic theory of the world stands in just as glaring contradiction to the peculiar doctrines of Christianity, as does its antagonistic theory, the

Deistic one. The one so peoples the universe, that it is impossible that the blessed God should become man; and thus wholly avoids Christianity. The other so depopulates it that man alone is God; that is, the highest unfolding or manifestation of Deity. The teachings of the Bible in regard to an incarnate God, are well enough, so long as they heap honor upon man; so long as they are regarded as involuntary prophetic intimations, in which the Divine self-consciousness of humanity begins to dawn; so long as they constitute a bold fiction, in which the soul of the world unconsciously shadows forth the history of its own development. Thus are sin and redemption, personality and immortality, the resurrection and judgment, heaven and hell, all at once, and without effort, done away with. The earth alone is the place where God reveals himself, the favored spot of his highest manifestation; beyond it no trace of spirit is to be found; and he must be a child or an idiot who would seek for reasonable beings, spirits and angels, beyond this world.

Deism, however, is still content to believe in the existence of angels—its vision is unclouded, so long as we keep at a distance with those abortions of a gloomy superstition: such as fallen angels, spirits of the abyss, and princes of darkness. Yea, it even flatters its disciples, with the confident expectation, that they who are here, "half beast, half angel," shall, so soon as they have "shuffled off this mortal coil," their animal nature, become complete angels, and be permitted, with powers of infinite perfectibility, to rove at will through the immensity of creation, gathering fresh acquisitions and new delights in each world. Pantheism, indeed, does not beguile itself with any such childish hopes and fantastic dreams; but it also looks upon all the Bible has to say about celestial worlds and inhabitants of light, angels and champions of heaven, dominions, principalities, and powers, as mere silly childish legends.

Therefore, in order that the stars of the firmament be not wholly useless, they are made to serve as gaslights at large, whose only purpose is to light up the birth-place of an ever-unfolding God; to deck a brilliant saloon in which the human mind may disport itself, and from whose countless glittering mirrors man may see the reflection of his own glory; and lest he lose himself therein, this showy saloon must be of only a moderate size and capacity. With all this, however, the Pantheistic theory arrives at such an impious denial of the true God, or indeed a God at all, that Deism looks fair and religious, compared with it.[1]

---

[1] It is Michelet who (in his *Vorlesungen über die Persönlichkeit Gottes und die Unsterblichkeit der Seele oder die ewige Persönlichkeit des Geistes*, Berlin, 1841) has most perspicuously and unreservedly advanced this theory of the world. The stars are to him, "*nothing further than bare rocks of light scattered throughout the seas of the heavens* (p. 227), and the whole idea of the starry heavens represents merely "that of abstract, unchangeable duration, as the vague, lifeless manifestation or image of eternity" (p. 228). "The earth surpasses the sun in dignity," and "it is evident beyond contradiction, that the highest and most complete manifestations in sidereal nature, are not to be sought beyond the sphere of our own planet, and no less, that no trace of spiritual

Thus, then, is there a complete antagonism between Deism and Pantheism, in all their fundamental views; and that portion of Divine truth which one receives—in the one case, of a God without and beyond the world; in the other, of a God within and about the world — that portion of truth is an exceedingly bitter draught for the other. Such fundamentally different views have not failed to give rise to fierce collisions and conflicts between these parties, in which vehemence, passion, and

life is to be found apart from this same terrestrial world" (p. 230) and like instances. It will be no matter of surprise, after what we have already said of Deism and Pantheism, that such a thoroughly Pantheistic theory of the world should sympathize in some points with the true Christian theory nevertheless. Thus, we find expressions of Michelet himself, to which, from the Biblical stand-point, we can subscribe, and which we can also defend, as, for example, when he asserts "that the earth, if not the sensible, is at least the spiritual centre of the system;" or when he says: "The *extent* of the space is a matter of no moment for the revelations or workings of Spirit, which often delights in crowding the greatest wonders in the least amount of space." But also modern philosophy in general, even where it has not yet strayed into open or virtual Pantheism, has all along shown an evident desire to detract as much as possible from the significance of the heavens in opposition to that of the earth. Comp. Schelling's *Sendschr. an Eschenmayer*, in *der Zeitschr. von Deutschen und für Deutsche*, 1812, and Hegel's *Schlussbemerkung zu der Kosmologie des Anaximander*, in his *Gesch. der Philos.*, p. 207. The latter aims directly, among other things, at showing the modern view of the vastness and infinity of the stellar worlds, to be a laughable and absurd fancy of narrow minds, as for example, when he would designate them to be "merely a *luminous eruption, no more to be wondered at than an eruption upon the human skin, or than swarms of flies*" (*Vorlesung über Naturphilos.* I, p. 92, comp. p. 461).

mutual contempt, have been very often exhibited. Especially has Pantheism — as was natural for it to do, since it is the philosophy of pride and self-exaltation—cast the most biting sarcasm and bitter contempt upon the "antediluvian theologians," as it characteristically designates its antagonists, holding itself to be the flood which has swept away all rubbish from the province of knowledge, and renewed the world of reason.

But it would now appear almost as if a time of peace was to follow the fierce conflicts between these antagonists; and that henceforth they are to be allied in common hostility to Christianity. *David Strauss*, even in his thoroughly Pantheistic "*Glaubenslehre*," has refrained from the customary haughty and contemptuous expressions of opposition to the results of natural science; and, taking his stand beside *Ballenstedt* and *Bretschneider*, has not blushed in the attempt to renew the old "hue and cry"[1] of the champions of Deism — the offspring of lamentable ignorance, and want of science on the part of the latter—in regard to an evident contradiction between the sublime results of natural science and the absurd Biblical history of the creation. And as in theory, so also in practice have the Pantheists of our day, who have heretofore kept aloof and by themselves, become enlisted in the thickly crowded battalions of the friends of progress — formerly vulgarly denominated Deists and Rationalists—in order to wage with them a common warfare against the Christian Church.

[1] Comp. *K. v. Raumer*, Kreuzzüge, vol. I.

The fact of such an alliance, which at the first view seems so strange, may easily be explained, however, on a more careful consideration of the circumstances of the case. Looking at the matter in a merely practical point of view, the case is not a new one. Pilate and Herod became friends; the Pharisees and Sadducees joined themselves together. Theoretically considered, Astronomy in particular seems to be bringing about such an alliance.

Pantheism is wholly unable to retain the pretended strong-hold of its ideal construction of the universe, against the daily accumulating results of astronomy which so irresistibly besiege it. Indeed, a still larger share of the most approved self-complacency than has yet been exhibited by Pantheism, which, it must be confessed, has not come behind in this respect, is demanded, in order that at the present day, and in the face of the wondrous revelations of astronomy, the infinity of glittering worlds on high, should be persistently accounted as "bare rocks of light scattered throughout the seas of the heavens," or even, as "the eruptions" produced by a transient scarlet fever of the skies. It will, accordingly, at length be forced to abandon the illusion also, that man is the only manifestation of spiritual life in the universe. But this it can well afford to do, as man will still have the consolation, that he, the philosopher, stands alone as the highest embodiment of the soul of the world or universe, even though the same soul, no matter, be embodied in a similar and befitting manner in the worlds on high.

But Deism, on the other hand, overthrows itself

in its attempts to swell to infinity the greatness and number of the celestial worlds, and tumbles, unconsciously, headlong into Pantheism. Its grand and only aim is to get rid of prophecy, revelation, miracles, and the fact of the incarnation upon the earth. Hence it delights in piling suns upon suns, and systems upon systems of Milky Ways; vainly imagining that with each step of its progress toward infinity, the absurdity of giving credit to miracles and revelations of the Bible, becomes more clearly manifest. But so soon as it has wrought itself up to the huge idea of the absolute infinity of creation, logical sequence of thought must carry it, unless bounds be set to logic just where it begins to overturn previous and cherished views, fully within the limits of Pantheism. For, connected with the idea of the infinity of space, and of the worlds which fill it, comes the correlative idea, hardly to be evaded, of the eternity of time. But when once we admit the infinity of space and the eternity of time, the idea of the creation, and along with it, the idea of a personal Creator, superior to both time and space, glides away from our minds and leaves no trace behind.

It is clear that a mutual approach and alliance on the part of two such antagonists, differing at the first, "toto cœlo," cannot take place upon the basis of the Scriptural truth they still respectively retain (in the one case, that God exists beyond the world; in the other, that he is present in the world). Nay rather, it is to be effected only by their mutually giving up what is peculiarly in accordance with the

Bible (in the one case, the Scriptural idea of the creation; and in the other, the Biblical view of the high and special significance of man and his history).

Such, then, has been the position taken by astronomy, or rather, the parasite speculation which has attached itself thereto, to feed upon it, and convert all its wholesome lessons into hostile attacks against the Christian faith; and that noble science which above all others should be an unceasing song of praise to the glory of the Creator, has been degraded to the purpose of casting into the dust, not only the precious jewel of Divine love and condescension, his incarnation in the person of Christ, but also, the majestic crown of his greatness and glory, his creative dignity.

Let us now inquire how theology, which has been set up as the guardian of the dishonored sanctuary, has regarded the results of astronomy, and also the perversion and misuse of them on the part of Deism and Pantheism.

It is well known that Rome anathematized the system of Copernicus, and persecuted it in the person of Galileo; nor is it so very long since that system was first allowed to be taught from the professor's chair. Protestant theology, also, has barely, with severe struggles, overcome the religious difficulties which presented themselves with that system.[1]

[1] One of the last reactionary movements of pretended theological origin, against the triumphing Copernican theory, is to be noticed in a book published in the year 1740, by *Hensel*, Rector of the Hirschberg Gymnasium. The significant title of the work runs thus: " *Cosmotheoria biblica restaurata;* or, The new Mosaic

A secret reluctance to absorb into its theory of the world, and adapt to itself, the heliocentric doctine, and the results of later astronomical investigations generally, has propagated itself in the Christian mind, even down to the present day.[1] This disposition, which is founded upon supposed irreconcilable differences between the facts of science and revelation, is now very rarely to be met with, and is to be found prevailing almost exclusively among certain Christian Gnostics, who think that they can or should, in the tranquillity of the Christian spirit, withdraw themselves from the tumult and strife of unsettled opinions.

---

system of the World; in which it is to be satisfactorily proven, upon sacred and natural grounds: 1st, that the earth stands still; 2d, that the sun moves; 3d, that the rapid movement of all the stars is not impossible or contrary to reason, but in harmony with the "principles" of the latest physical science; 4th, that the celestial bodies are indeed of great size, but not of such immense magnitude as they are generally represented to be in modern times; 5th, that the fine small planets have a periodical revolution peculiar to themselves, from which the "retrogradatio" arises and may easily be understood; together with plates to the praise of the great Creator, and the defence of the truth, designed for the profit of all, but especially that of youth engaged in study, written by . . . etc." But subsequently, in the year 1806, there was published in Paris: *Mercier sur l'impossibilite des systèmes de Copernic et de Newton*, (Comp. Mädler, astron. Briefe, p. 40).

[1] In addition to the chief passage, Joshua 10 : 12–14, it has been attempted to show that the following passages also, are not reconcilable with the Copernican system: Ps. 93 : 1; 96 : 10; 104 : 5; Eccles. 1 : 5; Is. 34 : 4; Judg. 5 : 20. It is hoped the kind reader will forgive us for not consuming time to show, that these passages of Scripture prove no more in opposition to, than in favor of, the heliocentric doctrine.

But theologians as a whole, on the other hand, have ceased to agitate the question of a contradiction between theology and astronomy. Whether they have completely solved the connection and relation of these two sciences, *i.e.*, brought their respective theories of the world into full accord, into one harmonious whole, without constraint or diminution of right on the part of either science; whether they have removed all difficulties which honest doubt or careless unbelief may have taken occasion, with or without reason, to draw from astronomy, is a matter not to be so confidently affirmed, however well the subject has been treated by other pens,[1] that remarks

[1] Comp. the writings of Fried. von Meyer, pertaining to this subject, in his *Blättern für höhere Wahrheit*, II, 4 seq.; IV, 354 seq. VIII, 342 seq.;the small work by J. P. Lange, *Das Land der Herrlichkeit*, written in such a warm and glowing style; further, the work of the brilliant writer and eloquent English divine, Thos. Chalmers, *Discourses on the Christian Revelation, viewed in connection with Modern Astronomy;* also, the treatise by T. Milner, another English divine, a work, however, which has not come into our possession: *Astronomy and Scripture*, or some illustrations of that science, and of the solar, lunar, stellar, and terrestrial phenomena of Holy Writ, London, 1843. Comp. also, the profound work of E. A. von Schaden, *Theodicee*, vol. I. (under the special title, *Orion, oder über den Bau des Himmels*), Carlsr. 1842, in which he attempts to represent the Christian and the Astronomical theories of the world in absolute harmony, by giving them both a more profound cast; a book, however, of which we are sorry to say, that we have not been able to adopt its profound speculativo-gnostic views: and finally, the in many respects excellent treatise by Aug. Ebrard, *Die Weltanschauung der Bibel und die Naturwissenschaften*, in the third year of a periodical by this scholar: *Die Zukunft der Kirche*, Frankf. u. *Zürch*. 1847. In the sphere of dogmatic theology, J. P. Lange (*positive Dog-*

which so peculiarly apply to the subject in hand, should be looked upon as useless, or as from the outset unwarranted.

---

matik, Heidelb. 1851) enters most largely into a comparison of the revelation of the Scriptures with the results of natural science, and in that of exegetical investigation, Fr. Delitzsch (*Auslegung der Genesis*, Leipz. 1852). The talented work of J. Richters (*Natur und Geist*, Leipz. 1850, seq., thus far 3 vols.), which also promises to treat of this subject, has not yet progressed so far.

CHAPTER THIRD.

A UNIVERSAL HISTORY OF THE COSMOS.

Universal History, or the History of the World, as it is generally understood, contemplates and comprehends the total earthly development of the human race, and estimates single facts according to their general significance and influence. All that is accidental and without influence, every merely vegetative or instinctive manifestation of life, is therefore excluded from the sphere of its contemplation; and it forbids, on the other hand, any division, dismemberment and separation to take place in the total organism of the development. On the contrary, it demands the closest examination of all the facts and phenomena, which to any extent may have had influence upon the direction, progress and results of the total development, which have had any bearing upon the education and advancement of the human race. No *nation* that has had its peculiar task to perform, and has held an important, influential position; no *time* which has left behind traces of its footsteps; no *occurrence* which has hastened or retarded the stream of the development, or has given to it a new and different direction; no *person* who has in a measure ruled his age or has advanced its interests in any respect; no *effort* of the human mind which has cleared the way for a further development, or has called it forth, may be disregarded;

all means and resources, all furtherances and hindrances in the movement, science and art, commerce and industry, religion and politics, and whatever else has influenced the indefatigable mind of man striving ever after higher advancement, after more universal sway, are subjects for this science:—to inquire into *all these,* as to their beginnings and progress, their causes and effects, their means and ends, to compact them *all* into an unique, articulate organism, and to comprehend them in their unity and polarity, their reciprocal and antagonistic influences, is the irremissible duty and task of the historian. History is the *physiology,* and not the *anatomy* of the past.

Thus does this science indeed appear as the grandest and most comprehensive of all human sciences; and the inquiring mind of man finds in the study of it a subject beyond measure worthy of his highest powers. With good reason, therefore, may we call it *Universal History,* or the *History of the World.* All other sciences, be it God or man, the state or the church, nature or art, in some of their relations, which is chosen as the subject of special investigation, must begin with this science; must enrich themselves and widen their intellectual horizon by means of its labors and results. It desires to grasp and understand the whole earth, the whole human race, all times which have swept over the face of the earth, a complete *world* in respect to time and space, a *universe* of life and action.

And yet how small, with all its astonishing magnitude and fulness, is the province of knowledge which such a universal history would comprehend! how meagre and contracted does it appear, with all its immeasurable extent, when we apply a higher, a yet more comprehensive rule of measurement, the principle of which, also, lies concealed in the human mind!

Is then our earth the *universe* in a proper sense? is it *the world* which contains *all* life? are its boundaries the boundaries of all time and space? is the human mind which rules and inhabits the earth, the only manifestation of a free, personal spirit, a spirit creating history, amid all the creatures of God? and is there no space beyond the earth for the swelling tides of spiritual activity, for the movement and development of self-conscious life?

Or, if we be compelled to say that such cannot be the case, are the operations and movements of other worlds and domains of life, of no significance or interest to us? Does the history of what is beyond the earth, of supra-mundane beings, take place wholly without reference to our history, or the reverse? or does no superior or subordinate, no reciprocal and influential relation exist between them? do the different members of the universe hold their position beside each other wholly devoid of meaning, without a higher unity, without being united together as a complete organic whole? Is there not rather a higher end attained by the design and special functions of the individual members, somewhat as the special purpose and office of the single members of the body serve the design of the whole?

Indeed, were it so—had beings existing beyond

this earth *no* reference to us, were there *no* essential relation between them and us—then would the eager research of the inquiring mind, which carries the venturesome ship of thought beyond the boundaries of mortal time and space, be a mere ignoble and censurable *curiosity;* for the desire to investigate and comprehend what does not concern us, what has no reference at all to us, or to our position and duties in life, deserves not to be called a thirst for knowledge, but much rather a mere vain curiosity. *Then* would the *astronomer* who turns his telescope towards the worlds on high, and the profound *thinker,* far from content with searching into earthly things, and also the *divine,* ever inquiring of the Scriptures concerning the future, and seeking to quench at its pure fountain his thirst after knowledge, faith and hope—*then* would these all deserve, not esteem and approbation, but disapproval and censure; *then* would those nameless longings after a precious and blessed, though distant and yet unknown abode, which stir in our inmost bosom, as we, full of mysterious earnests of a great future, gaze on the wide-spread glory of the nocturnal heavens, be but the sentimental conceits of silly fools.

There is—such is the answer of a deep-felt necessity which we cannot ignore—there is a history other than that of our earth, and that history has essential connection with our history; both, at least in the grand points of their mutual development, touch and penetrate each other; the last, the final goal of their mutual strivings and movements, is a common one, a grand and all-comprehending one. There is a universal history, or history of the world, which in a higher and a truer sense deserves that name, than that history whose object is to grasp all the developments and tendencies of the human race inhabiting this atom earth, and arrange them, under one point of view, in scientific order and harmony.

Just in the degree that the human mind, by means of its innate, its noble tendency toward freedom, becomes independent of this clod, from which man must earn his bread in the sweat of his brow,—just in the same degree does it enlarge and expand its world, and seek intellectually to take possession of it all, to penetrate it with its powers of knowledge and make itself at home therein. A quiet home with its sorrows and joys, a workshop with its labor and toil, or the village of his birth, is to many an one, all his world. And the history of *this* home or *this* village is *his* history of the world, *his* universal history.

But under the hand of more favorable culture, the vision of the aspiring mind extends, and its horizon retreats; its world and its history of the world, both within and without, become ever more extensive and comprehensive; it would grasp, search into and fathom the origin and existence of everything significant offered by the east, the west, the north, or the south; by the heights or depths of the earth, by the past or the present of the human race;—and soon, very soon do the boundaries of its subjective world begin to coincide with the boundaries of the objective, the *sublunary* world; it eagerly craves the knowledge of that history of the world which is none other than the universal history of all human and earthly developments.

But is the inquiring mind now at ease and satisfied, when in its bold eagle flight it has reached the ends of this sublunary world, when, as it were, it has taken the wings of the morning and reached the uttermost parts of the sea? Does it bow willingly and unconditionally to the stern command: "Thus far—but no further?"

Never. And just in that does the mind prove its origin and its high destiny, its dominion over time and space; and although bound to time and space, and hindered at every step by the bonds of the flesh, never in the infinite sphere of time and space does it permit limits to be set to its inquiries and its conquests.

The *student of nature* beholds through his microscope, a whole world of life in every drop of the ocean, upon each leaf of the forest. But this does not satisfy him. He appropriates to himself something from each of these worlds, as a preliminary possession, with the bold assurance that he shall ever more and more be able thoroughly to explore them, even in their most hidden details, and bring them under his own sway. Putting aside the microscope, he grasps the telescope, in order to review, examine and explore all those countless worlds on high, which at an inconceivable distance spangle the vault of heaven, and each one of which mocks his petty earth in magnitude, in fulness and might of cosmical forces.

The *historian*, who deals not with nature directly,

but rather with the developments of mind *therein*, and the influence of mind *thereupon*, summons all nations and all times which have left behind them traces of their existence, to pass before his sagacious and discriminating mind; in order to construct from the ten thousand events, developments and changes which have occurred, a unique, life-like image of the history of his own race; in order to determine the work that race has to perform, its destination, and its present nearness to or distance from that destination.

But he also catches the glories of the worlds above, and has kindled within his bosom the hope, the anticipation, that they also may have a history; that this history is not without interest to the human mind, that perhaps many a problem in the developments of this sublunary world which he cannot solve, may there find its solution. A *John von Müller*, whose large and comprehensive mind grasped the total development of the human race, with a depth of insight and a clearness of understanding which few historians have attained to, felt *far* from being satisfied with what he had accomplished; his lofty and aspiring mind yearned for a more grand, a more comprehensive, a higher history; he eagerly longed for the time, when, as he said, he might study the universal history of the solar system.

John von Müller was a Christian man. That hope he cherished sprang from his assurance of that eternal life, in which our faith shall become sight, and our present meagre attainments be perfected and exalted to the all-comprehending completeness of knowledge; in which all problems of the microcosm and of the macrocosm shall be solved, and we see face to face, and know even as we are known.

But is that knowledge which he believed awaited him only in the clear vision of the future, wholly beyond the horizon of the present life? Is it true, then, that we cannot here see into those mysteries, though it be but as through a glass darkly?—at least, so long as we remain children in the knowledge of the real nature of all things, may we not still be wise, as children, and have childish thoughts, childish views, and childish expectations? Shall then, with respect to those mysteries, even the meagre boon of this short life—*to know in part*—be denied us?

"*All things are yours*," says the apostle; but he wisely adds, "*but all things are not expedient.*" Are we then profited by knowing *in part?* are such *childish* thoughts and views of any service? Is it not rather useless, foolish and wrong, to expend time with them, since we can arrive at no certainty of knowledge by such means, since we can thus scarcely avoid misconstructions, distorted conceptions, and even positive errors?

We reply with another question. Shall we deny the child, because it is unable to fathom in their inmost nature; the world and the things that are therein, and to comprehend them in their various relations and their manifold indications,—shall we on this account deny the child every thoughtful and inquiring contemplation of these things, every endeavor to understand them so far as it can, or so far as its faculties will permit, and every attempt to explain them after its own manner? Shall we not grant it the privilege of enriching the little world of its thoughts and opinions, of widening its sphere of vision, even at the risk of one-sided, distorted and ridiculous conceptions being taken up, over which it will itself laugh in later years? What would otherwise become of, or how else should we secure, the education of its mental faculties, and its fitness for the duties of maturer life?

If, then, some knowledge of the history of the universe is to be obtained, however fragmentary and inadequate that knowledge may be, and we inquire for the surest sources from which it is to be derived, these three alone present themselves; philosophical *speculation*, the *telescope* of the astronomer, and the *Bible.*

The human mind is possessed of *prophetic* powers; —these are grounded in its divine origin; for a divine breath of life dwells in man, a breath which was breathed into the form made from the dust of the earth, by the God of the spirits of all flesh. But since man, in the perversity of his will, has cut himself off from the eternal source of his being, since now the finite mind of man is isolated from the infinite, the eternal mind, those prophetic powers are deprived of the only soil in which they can thrive, in which they can unfold their potential fulness and energy. Only vague, helpless conjectures, which grope around in the dark, remain within the power of man; the offspring of his futile efforts to re-assert his lost destiny. And it was only when divine fulness again poured itself into the soul of man, when through faith and grace the broken band of union was renewed, and the dried-up stream was replenished from the fountain whence it originally took its rise—in the Prophets of the old and the new Covenant—only then did these prophetic powers reach a development at all worthy of their end—only then was the human mind able to rise, on the wings of prophetic inspiration, to those heights from whence it was permitted to survey the regions of knowledge beyond the reach of mortal vision.

But what can mere philosophical speculation discover here? what can it bring to light? The philosopher may indeed, by means of his innate prophetic powers—powers which have not, however, been developed so as to be successfully and efficiently employed—have, as it were, by inward necessity, a presentiment that there is *something* to be sought after, *something* to be learned; but not a clear knowledge of *what* it is, or *how* it exists. Speculation is merely an instrument—it cannot derive from itself the material with which it would form a system of knowledge, but must receive it from without; it can at best deduce only the general from the particular, the *wherefore* from the *what* and the *how*.

If speculation be, therefore, of any significance in general, as a means of gaining knowledge of the history of other worlds and the relations they sustain to us, it can be so only in the degree in which it abides true to the established views of experience, or the representations of revelation; therefore, only so far as it makes the results of physical research or the data of revelation, the substratum of its reasonings, and thus deduces the unknown from the known, the general from the particular. It is clear, therefore, that speculation is at best unsafe, not to be relied on, and very prone to err.

But the telescope also, however magnificent and astonishing its discoveries—which bid fair to be increased every day—can scarcely offer us anything considerable for our present purpose. The telescope points out only the material cosmical masses; only the massive works of the creative mind, in their most general forms, relations and movements. But the individual is hopelessly lost from its view in the general, the small in the great, and the free acts of created beings in and on the worlds they inhabit; in short, all that begets, conditions, and forms history.

But still, however, the results of astronomical research are not wholly without significance touching the investigation and knowledge of the history of the universe. They teach us the general bases on which this history comes to pass, the peculiar constitution of which may more or less affect the history itself. They give us the power to deduce upon the grounds of necessary and universally existing relations between mind and matter, from a knowledge of the cosmical connections and relations of the different worlds of the universe to each other, more or less reliable conclusions touching a corresponding condition and relation of the beings which inhabit them; on whose account, also, those worlds *exist*, and exist *as they do*. The earth exists for man's sake;

its destiny is conditioned by his destiny, its development depends upon his development. We must assume a like connection between individual nature and the individual mind, as respects development and destiny, to exist in the heavenly worlds also. And wherever we perceive a vital connection, a reciprocal relation to exist between certain worlds, there we must also assume a corresponding relation or connection to exist between the inhabitants of these worlds — a relation chiefly of destinies, which the creative mind has affixed to them, but which they themselves must choose of their own accord,[1] unfold in their own proper development—in history —and gradually work out a full and satisfactory manifestation.

---

[1] The created spirit can, indeed, determine itself to some other end than that for which it was appointed by the Creator in the beginning; since as a *free* creature, it must possess the power to will as it chooses. But when once its godless self-determination carries it in direct and inveterate opposition to God, so as to cut off all possibility of its returning to its original destiny, of its recommencing the missed development with the divine potency of the beginning — when thus the bond of communion between the created spirit and the Creative Mind is irreparably broken, then also is the bond hopelessly severed, which bound that creature in living union and co-operation with all the creatures of God which have been *true to their original destiny.* The unholy creature shall at the *end* of its godless development—*together with* its nature, so far as that has been plunged into the abyss of iniquity, through the power of the spirit itself over it, or *without* it, so far as by virtue of its divine constitution it is not capable of such a calamity — be separated from the organic and harmonious family of faithful creatures, who find their end and rest in God, and be doomed to hopeless perdition.

But, notwithstanding, experience or empirical science is of little more use to us here than self-abandoned speculation—the united strength of both avails us little. Both speculation and empirical science lead us to imagine and take for granted the existence of a connected and widely-related history of the universe. But neither of them discloses to us the nature and the theme of this history; neither of them has the power to bring before the discerning eye of the mind the concrete, individual forms, the living images of this history, and the changes which underlie it. And even could they do this, the relation and reference of the details of the history to the whole, could ever be but a matter of vague and unsatisfactory conjecture, until the whole were come to pass, when the relations and reciprocal influences of the details might, indeed, be more clearly and satisfactorily discerned.

But there still remains a *third* source of knowledge. That is *Divine revelation* in the Scriptures. Here it is that the human mind is raised to the sublime heights of divine contemplation, and favored with the profound and penetrating vision of divine wisdom; so that it, according to the measure of its actual need, on the one hand, and within the limits of the sphere into which the wisdom of God has raised it, on the other—may know and declare what were otherwise hopelessly withdrawn from its most eager research, but what notwithstanding is wholesome, useful, and necessary to be known.

Here at last may we hope to find disclosures concerning that history, the knowledge of which is so foreign to our minds, but which yet so *closely* concerns us, that the thirst and ardent desire after it, manifested by the aroused mind, can neither be ignored nor appeased. For if this thirst, this ardent desire,is not an unnatural, sickly, feverish appetite; if it is rather a thirst springing from the nature of mind itself; if it be true that there exist mutual relations and influences between our world and the worlds on high, between the human mind and created minds in other worlds; if it be true that the developments and changes of our history stand in important relations to the design and end of the universe: are those conditioned in a manner by these, and these by those, so that the peculiar position and mission of the earth and its inhabitants as assigned them by the Creator, cannot be learned or understood without a knowledge of these important connections—then may we reasonably expect that the Scriptures should have given us such disclosures as would meet the wants of this mortal life and the present state of our knowledge.

Revelation is affected by none of the failings and defects of philosophy and science, none of the limitations and clogs which everywhere and on all sides meet speculation and empirical science in the investigation of the history of the universe, and which mock all earnest endeavors of the ardent mind, and all calculations of inquiring reason. But it, too, has its limits—for prophecy is, as all human knowledge which is gained in connection with human means, still of a fragmentary nature; it has its *subjective* limits, which are conditioned by the then existing capabilities and culture of the mind through which it comes; it has also its *objective* limits, which are drawn by the illuminating, instructing, and fostering wisdom of the Spirit of God.

Not all, indeed, that curiosity might wish to know, is opened to the vision of the prophet; but only that which is useful and practical. Nor was prophecy intended to satisfy depraved appetites, seeking after hidden wisdom, but only a real hunger and thirst for that spiritual nourishment which is just as necessary to the life of the soul as material food for the life of the body. But within the boundaries set to prophecy by its nature and design, by the wants and capacities of the mind, by divine foreknowledge and wisdom, it moves free, and untrammelled by any of the clogs which time and space impose on thought and investigation, themselves confined to time and space.

In prophecy the human spirit, made in the image and likeness of God, is raised to the eternal source from whence it proceeded, and satisfied from the fulness of divine knowledge, a source superior to time and space, though still pervading both. Proceeding from the present, its attainments and its still existing wants, which are both the result of past essential developments, and also the germs and conditions of all future developments—proceeding from *the present*, prophecy casts its divine glances backward into the past, and forward into the future;—proceeding from what is *near at hand*, yet bound to the distant by the unity of design and destiny existing everywhere in the universe, it casts its glances into the most remote regions of space. The facts of *the past*, though they lie buried under the rubbish of thousands of years, and have glided from the memories and tongues of men, yea, even though no mortal eye was witness to their occurrence, are revealed to its divine far-seeing eye. From what is seen it traces the origin of its being; from the condition of the present it divines the developments of the past; for the latter lie veiled and hidden in the former, but the divine energy of prophetic vision makes them *stand forth* in clear light. In a similar manner does it disclose the essential developments of the *future*, so far as they are conditioned by the state of the present, and lie concealed in it, as so many germs yet undeveloped, as so many attempts and beginnings still unseen; it also urges its way into *the depths of space*, borne by the really existing, if indeed not clearly perceived relation between what is near at hand and what is remote.

CHAPTER FOURTH.

THE BIBLICAL THEORY OF THE ORIGIN, DEVELOPMENT, AND CONSUMMATION OF THE UNIVERSE.

In the preceding chapter we have spoken of a Universal History of the Cosmos, founded upon unity of plan in all worlds, and striving towards a grand (einheitlichen) goal. That such a history should exist, has appeared to us more than probable. We have also found that, should there be such a history, neither philosophical speculation, nor empirical science, is capable of furnishing us with a knowledge of it.

But we have named a third method of obtaining the desired knowledge — a method by which, if by any, we could be satisfied: study of the Scriptures as the archives of Divine revelation. We design to apply this method in the present chapter. Perhaps we may here be able to discover the elements and fundamental features of such a universal history.

§ 1. *Origin, Significance, and Character, of the Biblical History of the Creation, and the primeval Age.*

There stands on the very threshold of sacred writ, an account of the primeval history of earth and man, of such significance to theology and science in general, of such depth and breadth of meaning, of such fundamental importance and manifold reference, that, in these respects, few passages of Scripture can be compared with it. It also offers us many grounds upon which to rest as we proceed with our present inquiry. Our investigations shall therefore proceed from it as a basis, and often return to it again during the progress of the inquiry. Let us first seek to gain clear views of the character and significance, the origin, position, and design of this record.

We shall perceive at the first hasty glance, that this record naturally divides itself into two independent sections. The first, which includes Chap. 1–2 : 1–3, treats of the origin of the universe, or, in the words of Scripture itself, of the origin of "*the heavens and the earth and all the host of them.*" The second, extending from Chap. 2 : 4, to the end of Chap. 3, records the history of the fall of man, its causes and consequences, its preliminaries and its results. The latter part of the second section (the consequences and results of the fall) is given as a foundation for all future sacred history; the former (the occasion, the causes, and the preliminaries of the fall) conducts this section back into the sphere of creation, whereby it sustains several points of contact, with the first or preceding section. For the present, we shall leave the relation of the two sections unexplained; in order first of all to take into consideration some questions which demand a general answer. (Comp. § 10.)

The three first chapters of Genesis treat of what in part lies outside and beyond all human experience and recollection; and on the other hand, of those first, quickly flown, morning hours of the history of the human race, the nature and circumstances of which have ever since that time been something foreign to and beyond all experience, observation, and analogy.

Is the representation which is here presented us, of the nature and developments of those primeval times, poetry or philosophy, tradition or history?

When *poetry* appears in the form of history, as the relation of something that has occurred, it is either *pure* poetry, the material of which is wholly drawn from the poet's own mind, or *historical* poetry, in which facts or events are moulded or re-constructed to serve a poetical end. In either case, the creation or new-creation of the poet is merely clothed in the *drapery* of history. But he makes no demand that his representation, as a whole, should be looked upon as one of real and substantial facts.

Such a poem may proceed from a poet enlightened by the Spirit of God; such a poem may therefore, when such is the case, be included in the Sacred Scriptures. As an example, we mention the Book of Job, in which historical or traditional material is poetically wrought into a canvass upon which to represent the wisdom and knowledge flowing from the depths of a soul enlightened by the Spirit of God. But *our* record offers nothing at all analogous. Here history is not the drapery or the canvass, but the body and the substance. The record evidently designs what is here represented in an historical form, to be regarded as real truth. In respect to the first section, this appears with indubitable certainty from its close, Chap. 2 : 3, where the hallowing of the Sabbath day, and the resting of God on the seventh day, are grounded on the six days' work of creation. There is no sense in this connection, unless both be looked upon as facts historically true. But neither will the second section pass for a mere creation of the fancy, as an unrestricted poem. Its whole conception, arrangement, and representation, shows clearly that it was intended to communicate what is substantially and essentially true. Both sections are looked upon and applied as having this undoubted design, by all the subsequent books of the Bible which refer to them.

We can, indeed, conceive of a poem which has other than merely poetical ends; and which, to secure *these* ends, must be regarded as *history*, though in reality it be only *poetry*. May not (to keep as close to the account of the creation as possible) the fact that this record was to serve as the foundation of the law of the Sabbath, and that it arose in the giving of that law, lead us to believe that such was the case *here?* May not some Sage of Israel, with the noble design of recommending as divine that highly significant institution, and of establishing it among the people, have written the first chapter of Genesis as a poem, but have represented its contents as historical facts, in order thereby successfully to attain his object?

Such a question can arise only so long as the writings, the history and the institutions of the Old Testament, are looked upon as of merely human origin—or an attempt is made so to regard them. But when we are forced to the conviction, by inward and outward necessity, by the witness of the Holy Spirit, as well as the results of our own investigations, that another than the human mind—the mind of God himself, has presided over and wrought in these books and the history they contain; we cannot but at once repel such an inquiry with indignation and disdain. When we clearly perceive that the history, teachings and prophecies of the Old Testament, all point to the incarnation of Christ, that in Him they find their end of fulfilment; then it follows directly that their truth must be fully attested and proven by the facts of Christ's life and death. The Mosaic history of the creation is the corner-stone of *that* temple which has been perfected and finished by the apostles of Jesus Christ. The divine structure of Christianity cannot be founded upon a delusion: it would spurn a fraud as its basis.

Philosophy, just as poetry, is the peculiar creation of its author, but it differs in kind. Starting from the known, touching the origin, significance, and end of which, neither historical nor empirical research can satisfy us, it attempts to fill up the chinks and chasms in human knowledge—the voids in history and experience—by the agency of reflection and speculation; and is often so presuming as to ascribe unconditional certainty and credibility to the results of its erroneous reasonings.

It has been thought that we should rather ascribe such an origin to our record, since really the origin of the world and the origin of evil, of which it would give an account, have ever been the foremost and most weighty problems of all philosophy and speculation. But apart from many other considerations which oppose such a derivation of the record, the fundamental position it holds with respect to the whole history of Divine revelations, and the history of redemption, and also the attestation it receives from the writings of the New Testament, sufficiently assures us that we possess in it something else and vastly better than the mere abortions of a brain philosophizing on the problems of the world and of human life.

Tradition is an account of something which has occurred, orally transmitted from generation to generation. It has to do merely with pre-historical times, circumstances, and events. But so soon as eye-witnesses or cotemporaries begin to record in writing the events of the present for the use of future times, the historical age of a nation commences. Whatever in the accounts of former ages was not written by immediate eye-witnesses or cotemporaries, whatever has for a longer or shorter time been transmitted from lip to lip, is Tradition. But such tradition may have a two-fold *origin.* It may either lead back, by an unbroken chain of transmission, to the times when what it delivers came to pass, so as to be the vehicle of historical recollections, however much these may be transformed, enriched, and adorned by the poetical vein of the nation's mind; or the chain of transmission may be broken off, and the public mind in which dwells a no less universal "horror vacui" than fruitful poetical faculty, have supplied the wanting links, have attached to actual events, or something which now exists, a poetical history of its own creation, touching the manner of their origin, which the next generation would unsuspectingly transmit as proper tradition, reaching back to the time of the events themselves.

The articulate position of our record in connection with the history of all revelations, forbids us to regard it as tradition of this latter kind. We may, therefore, have no scruples in holding it to be tradition of the kind first mentioned. Indeed, we must regard it, though it be forced to maintain the position and significance it possesses from being included among the records of the kingdom of grace, in some other way, as *pure* tradition, as a true recollection of primeval times; we must deny that any such a transformation has happened to it as would destroy the truth of its special teachings, its essential contents. But it may be tradition, and still answer all requirements. For though we may nowhere among other nations be able to find tradition in such purity, in such close harmony with the original facts: though it be not possible for any other nation to trace back a tradition, transformed by the lips of the people, decked with arbitrary poetical fancies, and garnished with philosophical speculations, to the pure historical account first started—yet may we, in case we must derive our record from tradition, with great confidence maintain, with respect to it, both harmony with the original facts and integrity as a vehicle by which they have been transmitted to us. Let us not forget that we are here in a province where Divine Providence has presided in a special, in a striking manner; so that it cannot appear impossible that a tradition, which was designed in process of time to be included in the Divine records, should have been preserved unchanged, under the supervision of the Spirit, down to the age of him who was called to incorporate it in the Sacred Scriptures, and *thereby* attach to it the Divine Sanction. But we are in no need of even such a supposition; we can, without hesitation, admit that the original tradition may have experienced many arbitrary or undesigned poetical additions or transformations, among the nation of Israel, and still ascribe to the written record, as it stands in Genesis 1–3, unconditioned credibility and truthfulness. For we know that the men of God, to whom the composition of the Holy Scriptures was entrusted, were under the guidance of the Holy Spirit, and that they may thus have been made competent to separate the true from the false, the genuine from the counterfeit, in those traditions which they were to receive as sources of knowledge in relation to the Divine counsel and history of redemption, and (we repeat it) thereby attach to them the Divine *sanction.*

Our record may be derived from *tradition*, but then the tradition it contains must of necessity be *pure, unadulterated,* or at least *refined* and *purified* — *such* a tradition as is one in essence with the proper history of the Bible, and can be distinguished from it merely in this, that it comes from oral transmission of facts and not from the writings of cotemporaries. It *may be tradition.* But whether it really is *tradition*, whether the author of Genesis did in fact derive it from oral transmission, or whether he acquired its contents in some other way, from some other source, cannot as yet, indeed, be positively affirmed.

But further examination will lead us to answer the question before us affirmatively. Simple combinations of indubitable facts necessarily force us to this conclusion. The author of Genesis either found the substance of the account already in existence, or it was imparted to him by means of revelation. But the latter supposition is wholly untenable, since the traditions of all other nations, in the north and in the south, in the east and the west, whatever fundamental differences there may be in their religious views, agree in so striking a manner, in respect to the substantial facts, and often even the minutest details, with the representations of our record, that we cannot avoid referring all accounts to the same source. For it is wholly incredible that the other nations should have derived those features common to the traditions of all nations, from the Israelites; neither can the author of Genesis nor any single Israelite have been the sole recipient of this record. We must assume the existence of a common source, from which both the Israelites and the other nations derived their accounts; and this original source must pertain to a time when the human race yet retained its original unity, in which it was not yet divided by varieties of language or abode, by marked distinctions of race, or differences in civilization and religion. The nations now isolated, must have derived such accordant recollections and traditions from those primeval times. According to the different spiritual channels through which this heritage of the Father's house was conducted, did it assume, on the lips of priests or people, manifold forms, but ever so that the seal of the Father's house, the unity of its origin, remains indelibly enstamped upon it. In Israel only, the nation of revelations, did tradition retain its original form, or only here were means and abilities to be found capable of tracing it back to that form.

But if we are forced to recur to the time when the tribes and peoples of the human race were still united, nothing prevents us, nay rather, many things compel us to go back still a few steps further, to the time of Noah, and from thence to the time of Adam. We believe we express what may well be assumed, and a more than probable conjecture, when we say that the contents of this tradition were propagated by oral communication from the earliest times down to the age of the author of Genesis.

But our record contains, as we have already seen, two sections, each of which forms a distinct, well-rounded whole, with its own peculiar arrangement and representation of what is possessed by both in common. Does not this twofold character of the record lead us to infer that the tradition was duplex, and that hence it had a double origin?[1] By no

---

[1] Such a view would fain be confirmed, from pretended or apparent contradictions between the two sections. I have shown in other places (*Berträge zur Vertheidigung und Begründung der Einheit des Pentateuches*, Königsb., 1844, I, p. 50–73 ; and,*Einheit der Genesis*, Berlin 1846, p. 2–14) that these pretended contradictions are of little account. It would lead us too far aside from our main object, and interrupt too much the progress of our work, to examine this matter at length, at the present time. It is, doubtless, true at the outset, that the author of Genesis as it now stands, though it be true that he came into possession of two dif-

means. It can only, in the worst case, point to a double form of the original tradition, to a twofold circle of tradition at the time of the composition of Genesis; but never to a duplex original source. The Israelitish tradition, at all events, found a proper central representative in Noah, and later again in Abraham, Isaac, and Jacob. Even if those rays of light from primeval times had been in the meantime decomposed, by the prism of oral propagation, into circles of different color, yet must they have resolved themselves into unity again in Noah and in Abraham, though it were with the loss of a few shades, (which was indeed possible, but not necessary). Since that time different concentric, yea, even excentric circles, may again have been derived from the original tradition, but these may not, on this account, stand in irreconcilable contradiction with each other, or the genuine original tradition. But the unity of the original tradition may just as easily have been preserved intact. In the first case, the author of Genesis may have really drawn from two different circles of tradition, in order to fill out and complete the one from the other. The more deep and heartfelt his consciousness in this case, that he found only truth in both, or rather, that he took only truth from

---

ferent records, or drew his materials from two different circles of tradition, saw no irreconcilable contradiction between them; for, otherwise, he would have certainly composed their differences, or have confined himself alone to *one* of the sources of information. If the author of Genesis could adopt two accounts as mutually completing each other, it assuredly does not become us, even in the worst case, to doubt whether the two be reconcilable.

them, so much the less would he have any motive for obliterating the duplex source of his materials. But, in the other case, it may easily be supposed that he himself formed two different groups, mutually completing each other, out of the different parts of the genuine and undivided tradition.[1] We shall

[1] The case remains still the same, even though it be true, as Delitzsch latterly maintains (*Auslegung der Genesis*, Leipzig, 1852), that the two sections are not the offspring of the same pen, but the second the product of another author (the so-called completer of the work). It must be said in favor of the arguments used by Delitzsch, that they are altogether the fruit of legitimate criticism, with no intermixture of dogmatic improperties. They give the hypothesis touching a completion of the Mosaic Books by some other hand, such a shape, that there is no call for opposition to it, from the purely Biblical stand-point. They still accord to the Book of Genesis, and the Pentateuch in general, the fundamental position in respect to the history and doctrines of redemption, which is the indispensable condition of a correct apprehension of either — the position assigned them by the collector of the canon, which Christ and his Apostles confirmed, and which both the Jewish and Christian Church have ever acknowledged as their legitimate one. The Pentateuch still remains the basis of all the other Books of the Bible, and is preparatory to them, even though it did not *all*, without any exception, proceed from the pen of Moses, but rather, was completed by some one or two of Moses' cotemporaries, who remained subsequent to his death. Its whole *contents* are still Mosaic, since they are the offspring of the spirit and school of Moses. (According to Delitzsch, Moses himself wrote, as the Pentateuch itself shows, the portion of the law contained in Ex. 19—24, and several smaller sections of the books of the law, as indicated by Ex. 17 : 14; 34 : 27 and Num. 33 : 2. He also wrote, according to Deut. 31 : 9, the whole of Deuteronomy, except its close. A priestly cotemporary of the lawgiver, perhaps Eleazar, wrote the rest of the laws, and, adding them to the portions of the laws written by Moses himself, gave

learn further on, in § 10, what may have led to such a grouping.

§ 2. *Continuation.*

From what has preceded, we have learned that the Biblical account of the creation and the primeval history of man, in Genesis 1–3, is derived from tradition—from that tradition which was preserved by oral transmission, from the earliest times of the human race, down to the time of the author of Genesis, and which was by him taken up, under the direction of the Spirit, and placed in the Holy Scriptures, as the foundation of all sacred histories and teachings, and thereby divinely sanctioned and approved.

But just here a fresh, a weighty inquiry meets us, to which we are compelled to reply. In what way and by what means did the first framer of the tradition attain to a knowledge of those occurrences which our account describes? Part of it may, without

---

the whole an historical foundation, through the composition of the Book of Genesis. A second cotemporary of Moses, of marked prophetic tendency of mind, completed the work of his priestly predecessor, by the addition of several matters of special importance in his view. He also more fully developed some portions of the work, revising others, and subjoined to the whole the Book of Deuteronomy, from whence he derived much of the information he possessed). It must be confessed that this hypothesis has much to claim our attention. Still, however, there are a few doubts and difficulties in regard to its correctness, which cannot be solved by what Delitzsch has thus far brought forward to establish and defend it. We shall refrain for the present, therefore, from a decisive judgment, and await with anxiety his promised more extended treatment of the subject.

doubt, be referred to recollections of his life and conscious experience on the part of the first man. But another part, and that which is particularly important for the purpose we have in view, seems not to be referable to such a source. The whole *first* section, and part of the *second*, treats of times and conditions, of occurrences and developments, which were never witnessed by mortal eye, which lay outside and beyond all perceptions and recollections of man. Other means and powers were demanded in order to gain a knowledge of that *pre-adamite* history, than those which now lie within the power of man when investigating the past.

We have these words from a highly respectable source:[1] "We take the account of the creation as it offers itself, for a statement of the knowledge the first man had, of what preceded his existence. But he may have acquired such knowledge, without the necessary interposition of a special revelation, if the then existing condition of the world lay as clear and transparent before his view, as the Bible leads us to believe it did. Just as, in our day, the primeval history of the earth is disclosed to the man of science, from the present constitution of the globe, would the then existing state of the world, which was clear and legible in all its relations to the first man, have resolved itself into a history of the origin of that world." "The account of the creation is offered, neither as the result of musings or reveries touching the origin of the world, to say nothing of scientific research, nor yet as a revelation *compensating for* re-

[1] Comp. Hofmann, *Schriftbeweis*, Nördl 1852, I. p. 232–243.

flection and investigation, but as an account of the transmitted *contemplations* or *views* of the first man."

This view of the case assumes in the first place, that the conception of the history of the creation belonged wholly to a time *preceding* the fall of man: and next, that man possessed before the fall, what he has since lost, the ability to perceive, with a clear, penetrating and unerring glance, the essence of all created things, not only as to their *existence*, but also as to the history of their *origin*; without being compelled as our modern students of nature "to break with the hammer and cut with the scalpel, in order to get at the core of things." "They were," as one says,[1] "transparent and plain to him, without any effort on his part."

Let us first try the correctness of the second assumption. It appears to be legitimate and well-grounded from what the record (Chap. 2,) says of man and his original state. We there perceive that man was able, on the first view of the animal world as it passed before him, to give to each creature its proper name; we further perceive, that his first glance at woman, just then created, revealed to him her origin, her nature, and her mission, with unmistakable clearness and certainty. Is not this view of the case hereby sufficiently established and justified? Are we not warranted in the conclusion, that he who at the first glance perceived both the origin and mission of woman, as well as the nature and properties of the animals, must also have been capable of understanding, in the same manner, the his-

[1] Fr. Delitzsch, *Genesis*, p. 49.

tory of the origin of the heavens and the earth, of seas and mountains, of plants and animals? It would, indeed, appear so. But if we examine the account (Chap. 1–3,) more closely, if we take into consideration all its points with that care and scrutiny which their importance demands, not separating individual parts, but rather apprehending them in their organic relation to the whole, we shall immediately come to a different conclusion.

A number of explicit disclosures of the text militate against such a supposition.

God left the naming of woman and the animals to man; but *He* gave names to the heavens and the earth, to day and night, to the land and the sea. Wherefore such a distinction in the important act of giving names? If the act of giving names was a revelation on the part of man, that is, an exhibition of the knowledge he possessed of the nature of what he named, so likewise was the act of naming on the part of God, a revelation of God. And yet, "revelation is not to compensate for reflection and investigation on the part of man." Why then did not God leave the naming of those other objects to man, if he was capable of perceiving their nature and history by direct contemplation or intuition?

And did the qualification on the part of man to give names to animals, really and necessarily involve *such* a knowledge of them, as that whereby he could by direct contemplation apprehend not only their nature and manner of existence, but also their origin and previous development — not only their *present subsistence*, but also the *manner of their origin?* May not the former without the latter be regarded as a sufficient basis upon which to ground the naming of the animals? But even this limitation will not meet the demands of the case. Among the animals to which man gave names, was the serpent also, for according to Chap. 2: 19, 20, he gave names to "all the beasts of the field." Man named the serpent at least, without having perceived and *thoroughly* understood its *whole* being, its position, and its significance. He named it, indeed, but still there was *one* part of its constitution that he did not understand—that it "was more subtile than any beast of the field which the Lord God had made." Had he from the first clearly and distinctly seen the liar and deceiver in it, which it afterwards showed itself to be, he would not so readily have listened to its deceptive words.[1] But did man indeed at the first glance fully understand the nature of woman? and not only her present nature, but also her origin in the past, and even her position in the future?

The former must be allowed; whether or not the latter also, is still at least a question.[2] At all events, any positive conclusion that because man could understand the origin and nature of *woman*, therefore he could also understand the origin and nature of all other creatures, with equal thoroughness and cer-

[1] The author would ask that objections to the views here advanced might be withheld until after the perusal of § 26.

[2] When we observe that Christ (Matt. 19 : 5) cites the words of the 24th verse, as the words of God, we shall be inclined, with Delitzsch (*Genesis*, p. 114) ,to regard them not as the words of Adam, but as the words of the narrator, intended further to develop what is said by Adam in the 23d verse.

tainty, must be rejected as arbitrary and illegitimate. For the creation of woman was not, as that of all the other creatures, prior to the time of his own existence; and although the very moment of her origin was during his sleep, yet the origin itself was of such a nature that he could not fail to divine it with certainty, without such faculties of universal knowledge.

But we have express witness to the fact that the first man did not understand everything he was conversant with, in its nature and origin. The tree of knowledge stood in the midst of the garden; but man was not able himself to divine its nature and design. He knew not that he dare not eat of it, as he might of all the trees of the garden. God himself had to point out the qualities of that tree, by revelation. Adam knew not that eating of that tree would be followed by death. He must be told so expressly by God himself.

But, *even allowing* that man possessed before the fall, such a clear and penetrating glance, that his vision pierced to the *inmost essence* of all things—such a happy faculty of making combinations, that the knowledge of the world as it was then constituted, would immediately transport him to the knowledge of its origin—the text still offers much that cannot be explained; it contains *even then* points, to the knowledge of which many could attain only by means of special revelation.

We can, for example, on such a supposition, conceive even that man may have understood the series of the creations, and the number of the creative acts, from the character or arrangement of what was created; but we can scarcely imagine how he could learn from what appeared before his eyes, the number of the days of creation, and the actual distribution of those eight acts of creation among the six days. But it is absolutely inconceivable how he could have learned, without revelation, though the world lay clear and transparent before him, of the blessing of the seventh day and its being hallowed as a day of rest for man.

But if we proceed from an examination of the details, to the more general points which here come into consideration, we shall see far more clearly and distinctly the inadmissibility and erroneousness of such a view.

Although God declared at the close of the six days' work of creation, that all he had made was good, "very good," still we soon discover that evil was already in existence. Man was designed to learn to know good and evil, without himself becoming sinful. Consequently, evil must have existed externally to man, which it was necessary for him to know and overcome. And from the very fact that his whole spiritual development, his self-determination, his probation, his voluntary activity, in a word, his history should and must begin with the knowledge and mastery of this evil — from this very fact we perceive with what significance, power, and all-pervading influence it bore upon man and his history. The antagonism between good and evil, for the *solution* of which, man before all others must become *acquainted* with it, must be so universal and so real, as to pervade and affect the whole intricate web of his life; so that he might never move or act in the whole circle of his destiny, without coming in conflict with it; so that there might be absolutely no point where he could begin to realize his destiny, without being surrounded by it. Acquaintance with this antagonism was, therefore, the necessary beginning of all knowledge. A knowledge of the present might and did exist before such an acquaintance, (and from this must we derive the ability to name the animals properly); but a just, true, and deep *knowledge*, a diving into the depths of being, into the secrets of what exists and what is now coming into existence, into the relations between the present and the past, was wholly out of the question without a knowledge of the antagonism between good and evil—without such a knowledge as would enable man to master and solve the strange contradiction. No just knowledge of things could exist *before* the knowledge of good and evil. The condition of all knowledge, was the knowledge of good and evil.[1]

Had man, therefore, acquired the contents of our account, by being transported from an insight into

[1] Man did, indeed, attain to the knowledge of good and evil, by means of the fall, but not in the *proper way*, and hence not to the *proper knowledge*. The knowledge of good and evil he had now acquired, was the very reverse of *that* knowledge he could and should have obtained. As he did not rightly apprehend and understand good, so neither did he truly understand evil. Not until, by means of redemption, he attains to a complete knowledge of good, shall he possess a complete understanding of evil. Progress in these two kinds of knowledge goes hand in hand.

the essence of all created things which lay clear and legible before his eyes, to a knowledge of the origin of all these, he must first of all have discovered the origin, nature, and existence of evil. Passing by entirely the fact that the subsequent test of his knowledge would *thereby* have become superfluous, we would merely remark that the whole account, both in the first and second sections, contains not the least hint as to the origin of evil, then assumed to be in existence; yea, even then about to make its impress upon history. The origin and all-pervading influence of evil could not have escaped a glance which pierced to the inmost essence of things, which fathomed, by means of its native powers of vision, the whole history of all origins and beginnings. It is, hence, wholly impossible that the account of the pre-adamite history should have proceeded from the individual contemplations of the first man. This silence concerning evil, even then in existence, can be explained only by admitting that the facts of the account were *communicated* to man; that the sovereign judgment of a wise Teacher and Instructor, yet for a while wisely set limits to the knowledge of his pupil. In short, the contents of the account, so far as they reach beyond the experience of man, must be a *revelation from God to man;* a revelation which made known to him *only so much* concerning the mysterious events of the past, as was at the time necessary and useful for him; which reserved the filling up of its chasms and the development of its hints, for a more experienced and mature age of the pupil.

The same result is obtained from other sources.

We, too, indeed, are convinced that man in his original state was called, and therefore *endowed with the capacity* to understand all things in their inmost essence, according to their relations to each other, and as to their origin and design. We are forced to assume this, from the position accorded to man by the account of the creation, and the mission assigned to him to rule the whole earth, with all its creatures. For, were he to rule over them, he must understand them; he must know what they are, whence they are, and what they were designed for.

We are even further convinced that man, had not the catastrophe of the fall impaired his original faculties and transferred him to a wholly different stage of development, would have attained to that knowledge in the way heretofore specified—that of direct intuition, or immediate contemplation—and that the essence of created things would have been fully disclosed to his sovereign glance, without using the scalpel of the anatomist, the hammer of the geologist, the telescope of the astronomer, or the microscope of the naturalist; in a word, without any of the astonishing yet necessarily feeble helps modern science makes use of, in order to understand the mere surface of things.

But, on the contrary, we must most decidedly reject, as false and erroneous, and as in direct conflict with the Bible, the view that in the short time man retained his original position, his innate *capacities* for such a knowledge had been *sufficiently developed*, and his *vocation* thus to know had already been *realized.*

Man was created *perfect*, and "very good;" but

his innate perfection was a perfection *needing* development, and *possessing powers* of development; for he was created a free personal being, with the design that he should determine himself, by an act of *his own* free will, *to that end* whereto God had appointed him; that he should unfold the energies and faculties his Creator had lent him, and thus realize his vocation. His powers of knowledge, just as all faculties and natural talents, demanded a progressive development, in order to attain to complete, all-comprehending, and all-pervading *knowledge.* What is set forth as the end of the development must never be looked upon as its beginning.

This is precisely the view exhibited on the face of our record. The first section[1] mentions this as the destination or appointment of man, that he should rule the whole earth, and all that was thereon. But that this destination was the *end* of his development, and *not* its *beginning*, appears clearly from this fact, that the multiplication of the race and the *replenishing of the earth* was plainly made the condition and foundation of this wide exercise of *authority.*[2] This is further established by the second section,[3] which describes not the end or goal of the development of man, but its beginning. *There* it is said, not that he should rule the *whole earth*, but merely that he should dress and keep the *Garden of Eden.* The realization of that authority man was to have over the earth, which presupposed a knowledge of what was to be ruled, was thus to commence at some *one*

---

[1] Chap. 1. [2] Chap. 1: 28. [3] Chap. 2.

point upon the earth. From thence it was to extend gradually, by a steady progress, over the whole earth.

Finally, that the view we have been controverting is an erroneous one, appears also from the fact, that, when consistently applied and carried out, there is no room left for the necessity of Divine revelation in the history *previous* to the entrance of sin; while the history before the fall, as represented by Chap. 2, bears witness to the habitual employment of Divine revelation, as a real, historical fact, and leads us to infer that it was needful and necessary, since it was really employed.

Had man been able by means of his native powers and ever-abiding endowments, to discern from the first moment of his existence, with a clear, penetrating and unerring glance, the nature and essence of created things, and to understand them and all their mutual influences and relations, in their cause and origin, in their beginning, progress and end—we cannot see how he could still need special Divine instructions and revelations, in order to fulfil his mission.

The Bible takes a wholly different view of the case. According to it, the first man appears as a being designed for a high end, and therefore highly gifted, whose gifts or talents were not yet developed and actively employed, whose vocation was not yet realized and maintained, but was to be hereafter. And in order that it might be, that his talents might be properly developed, that his vocation might not be missed, he was surrounded on all sides, and in every step he took, by Divine instructions, teachings, admonitions, and warnings.

Revelation, most assuredly, was never designed to *compensate* for investigation and reflection on the part of man, either before or since the fall; but it was intended to direct these into the right paths, preserve them from errors, strengthen them, restore and purify them, where this was needed, supply their failings and fill up their chasms. Sufficient occasion was offered for this design, not only in the perverted and degenerate state of man *after* the fall, but also in his undeveloped state *before* the fall, surrounded as he was by dangers, yet ignorant of them.

We now proceed to the examination of the other supposition upon which *Hofmann* founds his view—the supposition that the process of the creation, as mentioned in Genesis 1 and 2, was known and comprehended by the first man, *before* the fall.

Even if we are forced to admit the correctness of this supposition, the result of our previous inquiry—that the account of the creation is not to be referred to contemplations of nature on the part of the first man, but rather to supernatural revelations from a Divine instructor—still remains.

But we cannot admit its correctness, since the history of the first man, as described in Genesis 2 and 3, allows no room for it, since the unique and closely connected progress of this history excludes it.

Chap. 2, describes the primeval history of the developments of man—how he progressed under the guidance and heavenly teachings of a watchful and Divine instructor. When God placed man in the garden, he was yet without knowledge. This he was first to gain in the Garden of Eden. But least of all could man *already, at that time*, have possessed that profound discernment and comprehensive knowledge exhibited in the first Chap. of Genesis. The ignorance which the instructions given by God to Adam pre-suppose, would strangely harmonize with such a state of the case. The consciousness of man was yet at this time "carte blanche." He must, therefore, have acquired such an understanding of the whole previous occurrence of the creation, *during his stay* in the garden! But this is wholly out of the question, since the progress of his development during this time was so systematic, so coherent, so decisively and exclusively directed towards the *one* fixed goal—that of preparing man for his decisive trial—that there was no time for it. Everything that did not directly serve this end, would not have furthered his development, but hindered and retarded it, and *all* new knowledge which did not bear upon the point in view, would, for the time, have been foreign, irrelevant and distracting. But this much is clear, that the whole account in Genesis 1, contains nothing at all to the purpose, in view of the preparatives to the trial of man. Consequently, the conception of its contents cannot pertain to the time before the fall.

God placed man in the garden, for there his great trial was to come to pass, and caused him to progress step by step, in the attainment of all such knowledge as was required for his decision, and allowed him to realize those developments which were the preliminary conditions of that decision. There was, while there, no time, no occasion, and no grounds for the conception of what was so foreign to *this* design, as the contents of the first Chapter of Genesis.

If, therefore, the first man was at all acquainted with the substance of the first Chapter of Genesis, he could only have attained to such knowledge *subsequently* to the fall. He carried with him from Paradise, recollections merely of what he had there experienced, and of what he had learned from Divine Revelation. But the history of the creation of the earth was not included in these.

The recollections of his individual experiences previous to the fall, constituted the first germ of that self-propagating tradition, which after the fall began to take form, and passed from mouth to mouth, down to the time of Noah, to that of Abraham, to that of Moses.

This tradition was enlarged by the absorption of post-paradisiacal histories; but it was also enlarged in another direction, by the reception of the history *antecedent* to man's abode in Paradise, that is, the history of the origin of all created things, which could be disclosed to the mind of man, *only by means of revelation.*

We might next inquire whether this history was made known to Adam by revelation, or whether it was imparted to the next generation by some man of God—some one who, like *Enoch,* "walked with God,"[1] and to whom, divinely illumined, the facts of *pre*-historical times were disclosed, in like manner as the scenes of the future judgment were, according

---

[1] Gen. 5 : 22.

to an old tradition sanctioned by the New Testament, laid open to the illumined vision of Enoch himself:—but the necessary data are wanting upon which to venture a satisfactory reply to such an inquiry.

Still, however, we cannot repress a conjecture which appears at least to possess much probability.

If we examine more closely the account of the creation in the first Chapter of Genesis, we shall see that it is the offspring of a fixed, evident, and express design, or at least that it points to and serves such a design, namely, that of giving a foundation to the hallowing of the Sabbath day, as of Divine institution, as a day adapted to the religious duties and wants of man. As God rested on the seventh day, after the work of six days, so also, in accordance with the Divine example and will, must man work six days, and rest from his earthly labors on the seventh. We therefore think, and not without reason, that here, in Gen. 2 : 1, 3, may be found a hint as to the occasion and aim of this revelation of the history of the creation. If we inquire further for historical grounds upon which to rest such an origin of the account, we shall find in Gen. 4: 26, that at the time when a son named Enos was born to Seth, the son of Adam, "men first began to call upon the name of the Lord." These words are not equivocal. They give an account of the first institution of a formal, solemn, and public worship of God or Jehovah. Here we find, instead of a merely private, arbitrary, and irregular worshipping of God, as for example, the sacrificial offerings of Cain and

Abel, the introduction of a common and general divine service. The first exigency in such a case would be, the fixing of a time of worship, and this time would be determined by the Sabbath, the archetype, the model and key-note of all times of worship.

Are we extreme[1] in conjecture when we suppose that the revelation of the history of the creation pertained to this time, so that it might serve as a basis to this sacred institution.

Whether Seth himself, or Adam, who was then still living, or some other cotemporary, was the medium of this revelation, we cannot, of course, determine.

§ 3. *Continuation.*

The revealed history of the creation was given as the communication of a knowledge of the past—of such occurrences as had taken place *before* the conscious existence of man. In what form and in what manner may we suppose this communication to the human mind to have taken place? And how, regarding it as such, must we apprehend and interpret it?

The source of all human history is autopsy, or personal observation on the part of man, whether it be that of the author of the history himself, or whether he have transmitted to him the observations and experiences of others. Nothing but what has been

---

[1] Let it not be objected here that this worship concerned *Jehovah,* whilst the account of the creation recognizes the name *Elohim* only. The mention of Jehovah-Elohim, in Gen. 2 : 4 seq., is sufficient to remove all difficulty.

seen or experienced by man himself can be a subject of *human* historical description. History, such as *man* is capable of writing, can *commence* only where he himself, or his race, has arrived at self-consciousness, or a knowledge of the world, where (whether active or passive) he himself is the beholder of what occurs; and it must ever end with the present moment. But there lies a succession of being beyond each of these limits; a development, therefore also a history,—behind, as the past, before, as the future. For when man begins to observe or construct history, himself and all the attendants and circumstances of his being already exist, or have come into being; nor does the stream of the development stop with the present; the thread is not cut off, but spun and drawn out further by the countless bands and secret influences of both the visible and invisible world. All take part in spinning this mysterious thread, but no one is able to divine what form the common, the final product of all these workmen, shall receive. Both these histories, therefore, lie without the sphere of human knowledge, for it is confined to time and space, and is able to rule and take possession of the present only. God alone, who dwells beyond and superior to time and space, glances backwards and forwards, beholding as clearly the developments previous to the *first moment* of man's existence, as those in advance of the present moment. However different these two histories may be, yet are they both on the same level with respect to the ground of man's *ignorance* of them, or *acquaintance* with them. The ground of his *ignorance* is his finite nature; the ground of his *knowledge* lies in the *knowledge God possesses*, and the medium between ignorance and knowledge is, objectively, *Divine revelation*, and, subjectively, *prophetic contemplation on the part of man*, who beholds therein, with the spiritual eye, what is excluded and hidden from the bodily eye. Since, therefore, the source of knowledge is the same with respect to both histories, and also the manner and means of its attainment—spiritual prophetic autopsy—so must also in both cases the historical representations founded upon such autopsy, stand on the same level as to truthfulness, and be interpreted and apprehended according to the same laws. Thus, therefore, we come into possession of the very important hermeneutical rule, that representations of *pre*-adamite developments, founded upon revelation, must be viewed from the same stand-point, and interpreted according to the same laws, as prophecies and sketches of *future times* and developments, founded also upon revelation. And this is indeed the only proper stand-point for the scientific exposition of the Mosaic history of the creation—so long as we acknowledge in it a record which has proceeded neither from the speculations of natural philosophers, nor from empirical science, nor yet from abstract reasonings of men, but from Divine revelation.

But we must remember that the conception of developments lying outside of all human observation, is totally different from that of the facts of experience. There we behold with the spiritual, here with the bodily eye. Here we are governed by the sober and unimpassioned spirit of every-day life, with its keen and unfailing eye for the outward relations of things, while their inner character and deep significance escape notice; but there the beholder finds himself in a state sublimely surpassing all ordinary experience, and his vision rendered clear for beholding the secret connection of things. But at the same time he loses his interest for their outward relations and connections, and along with it, his ability to apprehend them clearly. The bodily eye looks upon that which is material; it is confined to the mere form of what appears, to the outward connections of things, which are indeed often wholly accidental, and without any necessity. It seeks for points of repose in outward circumstances; but as these are often merely deceptive appearances, it thus frequently loses the secret connection, the intrinsic worth, the higher significance and true position of things. It is directly the reverse with spiritual vision. This is directed to the spiritual element of what appears in external manifestation; it regards with indifference all outward, accidental, and secondary relations, which might be totally different without changing the nature of the thing, and hence possess no interest or significance for it. It pierces to the core of the matter, and thus often overlooks the *external* features, the *outward* connections of things, the circumstances of the manifestation. Besides, we must remember that the objective contents of what is divinely revealed to man, accommodates itself to the subjective posture and capacities, and also to the existing wants of the mind to which it is communicated; so that we should by no means expect to find in the history of the creation, solutions of all possible questions, especially of such as can only be raised in an advanced state of scientific inquiry; but only of such as are of general religious importance, and of equally deep interest in all ages of the world.

From what has been said, it is clear that the historical representation of the prophetic contemplations, and the reality of an occurrence as it comes to pass and is seen in vision, must absolutely agree, in their essential features—those essential to the mind beholding, and those essential to the deep significance of the facts themselves—but by no means that all outward, accidental and secondary circumstances of the passing event, must, of necessity, be closely apprehended, and minutely depicted in the account given by the prophet. We must ever keep this in mind in our apprehensions of prophetic histories; and it must be conceded that arbitrary interpretation cannot conceal the fact that this fundamental principle arises from the very nature of the matter itself. We must necessarily wait, in the explanation of prophetic *histories of the future*, until they are realized, in order to learn their *outward* features and characteristics, their outward connections and *accidental* circumstances. We can, of course, entertain no such hopes with respect to the (past) *history of the creation*, but still, perhaps we possess a substitute for them, in the progress of modern empirical science, which, *so far* as it is able to draw, from a thoroughly explored "statu quo"—from the autopsy of what now exists—RELIABLE conclusions concerning the history of its origin, can also, in a certain degree, recur to the outward reality of the creation.

"The Mosaic record," says *Eichhorn*,[1] "is improperly called the history of the creation; it should be called a picture of the creation. Every feature of it appears to betray the pencil of the painter, not the pen of the historian." And *Ammon*[2] says: "According to the record, the author himself must have been an eye-witness to the creation."

Both remarks are (apart from the consequences attached to them by their authors) apposite enough. The record bears, with unmistakable clearness, the impress of a proper personal contemplation, and is, indeed, due to such, if it be what the Jewish and Christian Church have ever held it to be. If it be the offspring of Divine revelation, then also was it—according to the analogy of revealed histories of the future—conceived through the medium of prophetic contemplation, and its author (whoever he might be) *framed in words what he had beheld* with the *eye of the mind;* he described *what* he had seen, and described it *as* he had seen it. Hence the account assumes so panoramic, so plastic, and lifelike a character. It consists of prophetico-historical tableaux, which are represented before the eye of the mind, scenes from the creative activity of God, each one of which represents some grand division of the great drama, some prominent phase of the development.[3]

---

[1] Compare his *Repertorium*, IV, 131.

[2] Compare his *Bibl. Theologie*, I, p. 269.

[3] This view has been received and approved by Ebrard (*Abhandl. über Bibel und Naturwissenschaften*, 3, p. 167), and by

One scene unfolds itself after another before the vision of the *prophet*, until at length, with the seventh,[1] the historical progress of the creation is fully represented to *him*.

J. P. Lange (*Posit. Dogm.*, p. 243). On the other hand, it has been contested and rejected by Hofmann (*Schriftbeweis.* I, 231) and Delitzch (*Genesis*, page 42). The arguments of the two latter, so far as they touch the subject, do not affect the substance of my view; but merely some erroneous assertions connected therewith, in the second edition of this work, which now, however, are abandoned, Compare below, note 1, p. 115.

[1] "Taken strictly, therefore, the term *Hexæmeron* is incorrect, it should be *Heptæmeron.* The creation of the world was completed, according to the Biblical view, not in *six* but in *seven* days. For the seventh day also, the day of God's rest, was essentially connected with the days of creation, since it is expressly said, Gen. 2 : 2: 'And on the *seventh day* God *ended all his work* which he had made.' The rest of the seventh day was the keystone of the structure, the seal of its completion, and thereby the completion itself. It was a vain and shallow appliance the Samaritans and Syrians had recourse to, in altering the text of the above verse, and reading the *sixth* instead of the *seventh* day."

The above, precisely as it stands, was contained in the second edition of this work. With astonishment, therefore, I read in J. P. Lange's *Posit. Dogm.*, on page 232: "KURTZ *objects to the term* HEXÆMERON. *He would speak of a* SEVEN DAYS' WORK. *What, therefore, was the* WORK *of the seventh day?*" Did not the repetition of my words in this place furnish the opportunity, and call me to justify myself against the insinuation of *Lange*, I should still, as heretofore, have remained silent about the matter. But now I may be allowed a few words in my own defence.

Prof. *Lange* knows just as well as I do, and as every youth should know, that the term "*Heptæmeron*" is not to be interpreted, "*seven days'* WORK," but that it denotes a *complex period of seven days.* Hence he is guilty of *falsifying* my words — not from ignorance, and still less from evil intent — but through *haste* and *thoughtlessness.* But that, in itself, might be over-

§ 4. *Limitation and Duration of the Days of Creation.*

The first Chap. of Genesis mentions *eight acts* of creation, each one of which is introduced with the words, "God said: Let there be!" and on the other hand, speaks of but *six days* of creation, upon which these acts take place. Each of these days of creation begins with a *morning* of creation, which is marked by the Divine, "Let there be!" The day progresses, the wonderful commands of the Creator are effectually carried out, and at length, after the occurrence of evening and morning, a new day of creation is introduced.[1]

---

looked, for: "Quandoque bonus dormitat Homerus." But if, on the other hand, Prof. *Lange* cites the half-read and wholly misapprehended passage in wrong relation, merely for the purpose of bringing in the taunting question: "What therefore was the *work* of the seventh day?"—*i. e.*, to convict me of *absurdity* and *want of thought*, such as I should well be ashamed of, were the imputation true — then he most unreasonably casts a stain upon my literary character, in the eyes of all his readers, who have neither time, opportunity, nor disposition to examine for themselves the passage in my work so contemptuously treated, and thus converts a *pardonable haste and carelessness* into the *most unpardonable want of consideration.*

I deeply regret being called upon to complain so severely of Prof. *Lange*, to whose writings I owe so much pleasure and advantage. But there is a literary honor which must be *preserved intact*, and which, in the spirit of the *eighth commandment*, cannot be invaded at will.

[1] I cannot see that the common understanding of the words: "*And it became evening, and it became morning*," [according to Luther's rendering of the Hebrew, in Gen. 1 : 5] as containing within their compass a whole day, is the correct one. Such an interpretation is ungrammatical and contrary to the sense. The

Here we are met by *two* fresh inquiries. It may be asked, first, whether the number seven of the prophetic visions, in which through the medium of Divine revelation, the history of the creation was

---

"vav consecutivum" in this section, where order of time is so strongly marked, and in the *connection* in which it is used, must certainly denote *order of time*, so that what precedes in the order of the narration, must be regarded as also preceding in order of time. "God said: Let there be light!—and there was light.—God divided the light from the darkness. And it became evening, and it became morning." All here moves forward in an order of time the most strongly marked. Hence I cannot but regard the remark of *Delitzsch* (*Genesis*, p. 60) as founded in error, when he says: "The darkness preceded the light, hence the whole day began with the *evening*." For, on the one hand, the darkness is not called *evening*, but *night*, and on the other, when it is said: It became *evening*," it must be meant that a day had gone before, the place of which was now taken by the evening. The day of creation cannot, therefore, have begun with the evening; it *must* have begun with the morning. It were inconceivable how such a misapprehension should have been retained for so long a time, and with such general consent, were not its origin capable of explanation. It is not to be referred to the 1st Chapter of Genesis, but rather to the notorious fact, that not only the Hebrews, but almost all the nations of antiquity, began their ordinary day with the evening. It was hence thought that the same view must underlie the ancient sacred record, or must be attached to it. I myself believe, indeed, that this view has its foundation in that record, but in a wholly different manner. It has its basis, not in the days from the first to the sixth, but in the seventh day. The working day began with the morning, in conformity to its nature; but the day of rest, no less in harmony with its nature, with the evening. As now, the sabbath was the rule and measure of all civil and religious divisions of time, and was naturally begun in the evening, it was demanded for the sake of regularity, and by the typical character of the sabbath itself, that reckonings of time

represented, was essential; or whether it was merely accidental, that is, whether the creation might not have been represented in a greater number or fewer phases of development; whether this division is founded upon the *objective fact* of its really having *so occurred* as represented, or merely upon the *subjective views* of the prophet touching the manner of its occurrence? Even were we compelled to take the latter view, the record would no more forego its Divine character and authority, than prophecies concerning future history suffer abatement of their value and significance, from similar circumstances. Were not strong grounds against such an apprehension of the account, to be found in the record itself, or in subsequent revelations, we must at once admit it as legitimate.

But such is not the case. The record itself contains one explicit datum which compels us to regard the number seven of the visions as essential, as answering closely to the reality of the occurrence and the division of the work of creation. It is the foundation accorded to the division into weeks, and to the blessing of the seventh day, in Chap. 2: 3, an argument which derives much strength from those passages of the law (Exodus 20: 9–11, and 31: 12–17) which were to enforce the observance of the

---

in general should be made according to this rule. But the day of labor, as such, naturally began afterwards as before with the morning. We have in this view, which I am convinced is the only correct one, also a new proof that the "myth" of the creation is not derived from the division of time into weeks, but that the latter derives its origin from the "history" of the creation.

Sabbath upon the Israelites. No purely subjective, unessential, and therefore arbitrary limitation of the various phases of the process of creation, could possibly have been the occasion, the archetype, and the pattern of a Divine law or provision, particularly one of such significance and importance. The weight of this argument suffers not in the least from an appeal to the significance and sacredness of the number seven, founded both in nature and in the laws of the human mind.[1]

---

[1] This is the only argument brought forward by *Hofmann* and *Delitzsch* against the opinions expressed in the second edition of this work, to which I can concede the force of demonstration. But my view that the historical contents of the first chapter of Genesis were *originally* conceived in prophetic contemplation, still remains fundamentally and essentially the same, even after yielding an erroneous assertion or two connected therewith. I may here be allowed to notice at some length the remarks of *Delitzsch* in opposition to my view, since his (Genesis, p. 40–42) are the most extended.

Delitzsch says, 1st: "*Prophecy which so reproduces the facts of the past, that they again seem to occur and pass before the eye of the mind, is without a parallel in the Old Testament.* I reply: Without a parallel, true enough; since the history here to be gained stood alone, and hence there was no possibility of an analogous case. Every subsequent reproduction, under the influence of the Spirit, of *what had occurred* in past time, might rest upon human recollection and communication. But as for the history of the creation, it could not be reproduced by the intervention of man's natural faculties merely, since it lay beyond the reach of these faculties—anterior to the time of man's creation. 2d. "*The Spirit of God endowed the prophet, not with a knowledge of what had come to pass, but with a spiritual understanding of facts historically communicated.*" But still the *knowledge* of *future* events was imparted to the prophets of old by means of the Spirit. The know-

We now pass to the *second* inquiry, how we are to understand the *limitation of time* as represented by the prophet in the several phases of the creation: whether the days of creation there mentioned are to

---

ledge of the events themselves was, indeed, not the important point in these cases, but rather a spiritual apprehension of them, so that they might influence the life and promote the welfare of those who possessed them. But a knowledge of the facts themselves was preliminary to such a spiritual apprehension of them, and was the necessary condition of it. If such knowledge was not to be gained by the natural course of actual observation (on the part of the prophet himself, or from the observations of others communicated to him), then must it be imparted to him by internal vision, under the influence of the Spirit of God. But the same, precisely, is the case in regard to past events or facts, which were witnessed by no mortal eye. 3d. "*The section contains no evidence of its being of a prophetic character. The author says nothing of his having seen what he relates, in prophetic vision; his ideas are not framed in prophetic language.*" In reply, I shall turn this argument, in the first place, against the author of it himself. *Delitzsch* assumes that the first man derived the matter of the record from God himself, by an immediate revelation; or, more distinctly, that it was orally communicated to him by his Creator. *Where* is there any indication of *this?* Where does the author drop a word to indicate that he was thns taught of God? Where do we find that historical structure in which, ever since, permanent indications of such express oral teachings on the part of God are retained? But, apart from such considerations, is it absolutely necessary that what is seen in prophetic vision must always be expressly indicated as having been so beheld, particularly when it is perfectly clear that it cannot have been seen with the bodily eye? Could the application of such a canon, to accounts of future history conceived in vision, be shown as perfectly legitimate? But, granting even that all this be so, have we indeed the account in the very same form in which it was delivered, by its first author, to his descendants? May it not originally have had such a prophetic structure, which, being unes-

be regarded as true, natural, common days of twenty-four hours, so that precisely six times twenty-four hours must have been spent in the creation and more complete formation of the earth and its whole

---

sential at best, could easily be lost or removed in the course of twenty centuries, during which time it was dependent alone upon oral transmission? *Delitzsch* himself maintains, indeed, that the account may, during this long time, have suffered many injuries, and have lost essential facts (facts touching the substance of the account). 4th. "*The account belongs, if we recognize two channels of historical description in the Pentateuch (a prophetic and a priestly), not to the prophetic channel, but to the priestly.*" This remark is founded upon the above-mentioned hypothesis (page 88) of the origin of the Pentateuch, the correctness of which we have no need to call in question, so far as it influences the matter to which it is here applied. For, even if the account not only was originally the offspring of prophetic contemplation, but still had borne upon its face the unmistakable impress of such an origin, the assumed priestly author would doubtless have found in Gen. 2 : 3, ground enough to warrant him in incorporating it in his sacred history, in preference to many other traditional descriptions of the creation. 5th. "*If Gen. 1st be an account of what was seen by a prophet of Israel, whence then the surprising accordance with it to be found in the traditions of the heathen?*" This inquiry does not affect my view, for I have never maintained that the author of Genesis was the first recipient of the facts divinely imparted.

Delitzsch agrees with me in the fact that the account of the creation, as recorded in the first chapter of Genesis, is to be referred to divine revelation. We merely differ in opinion as to the form in which this revealed knowledge was communicated to man. Delitzsch believes (if I rightly apprehend him) that God imparted it to man orally; I believe that God communicated it to him through the medium of prophetic vision. I am compelled to take the view I do from the panoramic, life-like character of the account, which must be referred, as it seems to me, to circumstances of *personal observation* with which it arose; and also from the fact that information in regard to *historical* matters of the future, was

organism — or whether this limitation existed only in the mind of the prophet, having no foundation in reality, so that the days are to be regarded merely as *prophetic* days, spaces or periods of time of indefinite length.

That such periods of time *might* be styled days in the concrete representations of prophecy, no one will dispute. But we dare not maintain as a foregone conclusion, that since the record was conceived in the spirit of prophecy, therefore the six days *must*, of course, represent so many periods of indeterminate length. As in the predictions of the prophet Jeremiah, the seventy years were proper, natural years; so also the six days mentioned in the account of the creation, may very easily have been *natural days* of twenty-four hours. There are but two modes of deciding how to understand the term *day* in this as in all similar cases. Either the record itself must contain other points which decide the matter, (as in the case, for example, of the predictions of the pro-

---

never given, according to any representations of the Scriptures, by oral communication from God, but always through a prophetic medium under the influence of the Spirit. For I abide by the affirmation that the conception of a history lying antecedent to all human experience and recollection, is subjected to precisely the same conditions and laws as that of a history yet future. The oral teachings of God to the first man, chap. 2, and similar communications to the patriarchs (Noah, Abraham), are of an entirely different character, of a different form, import and design, from the revelations contained in the first chapter of Genesis. The latter would be, apprehended as outward, oral communications from God, without any analogy either in the Old or the New Testament.

phet Jeremiah, Chap. 29, where it is clearly enough indicated that the seventy years were to be understood as natural, historical years), or we must found our decision on the facts of experience—in the case of predictions concerning the future, upon the fulfilment of what is foretold; and in regard to the primeval history of the world, upon the results of scientific research.

Not unfrequently do we hear the over-hasty remark, that the results of scientific investigation speak decidedly in favor of the view that the days of creation were long periods of time. For *Astronomy*, it is argued, will not permit us to limit the time spent in the creation of the heavens above, with all their stars, or even that spent in the creation of our own planetary heavens, to twenty-four natural hours: neither can we, in the face of the results of *geological* research, believe that the production of the primary and secondary formations, and the origin, course of life and death of the organic beings they enclose, took place in a single day of twenty-four hours length, or even in six of them.

Delitzsch affirms that he has heard such positive assertions as the following, from the lips of distinguished and prudent scientific men, and those who have the deepest attachment to Christianity; that "millions of years" (?!) must have preceded the present condition of the earth, and the animal and vegetable kingdoms it contains.

But we must not allow our minds to be unsettled or turned from an impartial examination of the record before us, by any such assertions. The first and most significant inquiry should ever be, how does the record itself regard the days of which it speaks? If it contain reliable data, from which we cannot but infer that the days are to be understood as natural days, neither Astronomy nor Geology has the right to a single word in the whole matter. We believe most firmly, that were this record explained merely on its own merits and with the aid of other Scripture, and were there no outside, no foreign influences at work, the days could only be regarded as natural days. But we also believe that natural science can be harmonized with the Bible, in spite of such an exegetical result; even though it abide by its exorbitant assertion, that millions of years must have preceded the present form of the earth.

Delitzsch, indeed, believes that the position can also be maintained, "that it cannot possibly have been the intention of the account of the creation, to crowd the six days' work, together with the Sabbath which followed it, within the space of an ordinary week. The days of creation must have been periods of creation. Probably the author of the account himself did not intend to state their length. He may have meant days according to a Divine standard of measurement."[1]

The record itself shows what it would be under-

[1] I myself, also, in the previous editions of this work, interpreted the days as prophetic days of indefinite length; but merely in this view, that I thought there was nothing in the record itself, rendering necessary a decision either the one way or the other.

stood to mean by the term *day*, where it begins to note the number of the days of creation, in verse fifth: God divided the light from the darkness, and called the light *day*, but the darkness he called night. Then came evening and morning. Thus the *first day* came to a close, and the second was introduced.

The word "day" is certainly here used (not in different senses, but just as the term is now used among all nations) with more or less limitation of meaning. It first designates the period of time we call day proper, the time between the dawn of morning and the darkness of evening; and then the whole day, including the night and the periods of transition between day and night. It is clearly manifest, therefore, that the whole day, which was called the first day, thus included four divisions of time (day and night, evening and morning) which, within that period, succeeded each other. Now, there is no question but that the division of time which is here called *day*, was conditioned and limited by the *presence* of *natural light;* consequently, the evening which followed such a day, and the morning which preceded the next day, must in like manner be understood as parts of an *ordinary*, *natural* whole day; and the latter can only be measured according to the natural, every-day standard still in use—the occurrence of one regular, natural change of light and darkness (of day and night).

The days of creation were thus measured by the natural advent and departure of the light of day, by the occurrence of evening and morning. This standard of measurement is given by the record itself, and must be applied alike to *each* of the *six days* of creation. But whether each of these days was a natural day of twenty-four hours length, we cannot, of course, determine. Most probably it was, from the fourth day onward; since from that time the sun began to rule the day, and the moon the night, introducing in all probability the same order which abides undisturbed until the present hour. But the length of the three first days, when the present order of things did not exist, when the duration of the light of day and the darkness of night was determined by wholly different laws, cannot, so long as these laws are unknown, be divined. The days of our record were measured not by the hours of the clock, but by the four divisions of the day.

In opposition to this, *Delitzsch* appeals to Gen. 2: 2, 3. He says (page 61): "The Divine Sabbath does not favor, but bears witness against the correctness of the apprehension, that the days of creation were of but 24 hours' length. For, if the Divine Sabbath of rest was of much greater length than a common Sabbath, and yet was the archetype and pattern of the latter, so also may the six days upon which God wrought in the work of creation, have been vastly longer than common secular days, without, in the least, losing their significant typical character." Plausible enough, indeed, but still untenable! Where are we told that the seventh day, on which God rested from all his work, was much longer than a common civil day? Was it not, too, called a *day*, just as all those which preceded it, and numbered in a regular series as the *seventh?* The record itself, in the description of the first day, points out unequivocally the proper interpretation of the word day. It is not, indeed, said, as it is with respect to the preceding days, that an evening and morning followed the seventh day also. But may we hence conclude that this was certainly not the case? If so, then it was not a day like those which preceded it, and could not properly be called a *day* in the same sense and connection. The error of the argument lies just here, first, the seventh day is arbitrarily interpreted —the record says nothing in regard to its duration or its limits—and then, on the basis of this interpretation, is built a conclusion as to the duration of the other days, while the record itself marks the duration of the previous days, but says nothing at all in regard to that of the seventh.[1]

---

[1] The reason that the closing words: "And the evening and the morning were, etc." are not appended to the description of the seventh day, is simply this, that no new day of creation followed this seventh day. Those words form in every instance, the transition, the connecting link as it were, between one creative day and that which followed. Hence, as no such day followed the seventh, those words applied to that day would have been wholly out of place. *Delitzsch,* indeed, explains their absence in a wholly different way: "The divine sabbath had no close; it extends forward over all history, and is to absorb it into itself, and thus, having become the sabbath of God and of his creatures at the same time, is to endure for ever and ever." A beautiful remark and true enough, but not in place here. "The divine sabbath had no close." But had not the *seventh day?* Undoubtedly it had, just as well as the six days, to which it belonged as the seventh of a regular series.—But we cannot, however, concede that the divine *rest, in the sense of the record,* never had a close. The record looks upon the divine resting as the consummation of the creation. For

Delitzsch continues: "It is highly proper that the copy should answer to the incommensurable greatness of the original, the archetype, only in a very limited degree." Both the Divine working and resting are indeed incommensurable, and signally so, for this simple reason, that they took place within the *very same limits* of time which are accorded to the working and resting of man. *Delitzsch* again remarks: "It is enough that the characteristic features—in this case: And it became evening, and it became morning [following the German]—should pass from the original to the copy." This we freely admit, for it establishes our view, and involves the opposite one in a contradiction, since it virtually denies the real objective character of the record, which even *Delitzsch* would not under any circumstances renounce. For his arguments would only gain their end, by assigning a different cause for the occurrence of evening and morning in the days of creation, than the natural change of terrestrial light

---

it is said: "On the *seventh day* (not the sixth) God ended his work, and rested on the seventh day." So far, therefore, as God's resting finished the work of creation, *it* also had a close — it belongs to the past. Finally, as to the mention of the eternal Sabbath, and the entrance of God's creatures into it, I regard that as eisegesis and not exegesis; since the idea, legitimate as it is in itself, belongs entirely to the New Testament. The law of the Sabbath, as found in the Old Testament, could not be founded upon something there still unknown. Believers, in Old Testament times, knew nothing of man entering into the *rest of God*, at least for many centuries; but only of his entering into the rest of Scheol (Hades), since they were yet in great measure ignorant of the work of Him who was to burst the gates of Hades, in order to conduct his people *from* the rest of Hades into the rest of *life eternal.*

and darkness, (which at least since the fourth day, has ever been brought about according to the same laws that still bear sway). But with this we should have renounced the fundamental belief of the objectivity of the account.

"Again, let us remember," continues our worthy friend, "that the six days of creation are called immediately after the history of the creation, in Chap. 2 : 4, *one day*, and thus a constrained literal interpretation is forbidden by the Scriptures themselves:—further, that Psalm 90, composed by Moses, gives expression to the great truth, that a thousand years in God's sight are but as yesterday when it is past; and lastly, that as prophecy has its own methods of computing time, we may not forbid the application of a similar principle to cosmological questions."

But surely it is not a too literal interpretation, to understand as proper, natural days, those that are described as such. We might with reason make this objection to such an interpretation as would force us to read the passage involved, Chap. 2 : 4, thus: "*In the day that* the Lord God made the earth and the heavens."[1] But what do we gain by *this* literal interpretation? A wholly irreconcilable contradiction between the first and second sections of the record—the heavens and the earth being created in *six* days according to the one, and in a *single* day according to the other—a contradiction which can

[1] The word בְּיוֹם (with the following infinitive) literally translated, is—"*in the day that*"—but it has throughout, in usage, the force of a mere conjunction or adverb—*then, as* (at the time when).

never be solved by passing from a close, literal interpretation, into the province of arbitrary spiritual interpretation, with the remark, that the measurements of time here spoken of are Divine measurements, six of which may be equal to one.

The argument drawn from the great and immovable truth of Psalm 90 : 4, is altogether inapplicable here, for the days of creation here spoken of are described not as they appear to the mind of *God*, but as they are to be understood by *man:*—and the inference that, since prophecy has its own methods of computing time, nothing may forbid the application of a similar principle to cosmological questions, springs from the confounding of two wholly different things. For the unusual measurements of time that occur in prophecy, are conditioned by the subjective posture of the prophet's mind; but measurements of time as they occur in cosmological descriptions, are founded upon the objective nature of the real occurrence.

This then is the final result of our inquiry: the days of creation are, according to the record itself, to be understood as spaces of time, each one of which included a single change of light and darkness, of terrestrial day and night. They had precisely the same limits as a modern natural day. But whether the space of time included between these boundaries was, even then, of just twenty-four hours in length, we are wholly at a loss to determine. We are not to be swerved from this our final conclusion, by the self-confident remark of *Ebrard*, (p. 171), that it would bespeak "LAMENTABLE NARROWNESS OF MIND," to understand the days of creation as natural, physical days, instead of interpreting them symbolically.

We have undertaken the task of demonstrating that the Biblical account of the creation may be harmonized with the results of Astronomy. From what we have already said, the task has evidently increased in difficulty, and the basis of our operations become narrowed. But it has also become evident that the harmony we are endeavoring to establish, is to take place, not on the basis of mere fancy or dogmatic assumption, but on that of impregnable truth.

§ 5. *Creation of the Heavens and the Earth.*

The history of the creation begins with these words: "In the beginning God created the heaven and the earth." If we examine these words wholly on their own merits, without any reference to their connection with what follows, we cannot stand in doubt for a single moment as to their design and import. Nothing so pervades the whole body and spirit of the Old Testament as this great fact, that the universe did not exist from eternity—(either as crude, shapeless matter, or as a concourse of matured heavenly bodies)—but that God, who alone is from eternity, and the first great cause, created it in time, or rather along with time. This grand, this fundamental principle of Old Testament knowledge of God, stands on the very threshold of the Divine records, on the very threshold of *that Book* which was to furnish the Israelitish people with their own early history, together with a sketch of the wondrous events which had preceded it. This great principle was the distinguishing mark of that chosen people, the stand-point from whence they viewed the whole subject of religion, and also the basis and preliminary of their history. By means of its possession and application, they were distinguished from all other nations of antiquity, who one and all were ensnared in such a worship of nature as deified the world itself, who believed the world to be self-existent and eternal, who knew not, nor desired to know anything concerning a personal God, distinct from the world and infinitely exalted above it. Against these monstrous and destructive errors of heathendom, the first word of the sacred records of Israel uttered a most strenuous protest.

But greater difficulties are met with in the explanation of these introductory words, so soon as we attempt to apprehend them in their connection with the description of the six days' work which immediately follows.

They are regarded by many as a species of superscription to note the contents of the whole chapter, as a summary statement of the six days' work described in detail by the remainder of this chapter. The fact that the creation of the heavens is specifically mentioned, as the narration proceeds, in the 8th verse, and that of the earth in the 10th, would seem to harmonize well with this assumption, and even demand that it be admitted as legitimate. Yet still the passage cannot be so apprehended, taken in connection with what immediately follows. For the word "*and*," with which the following sentence ("*and the earth was without form, and void: and darkness was upon the face of the deep*") begins, proves the

2d verse, and indeed the whole chapter which follows, to be a continuation of the account *commenced in the 1st verse*, and places beyond doubt the fact that the creation of the heavens and the earth, mentioned in the 1st verse, took place previous to the six days' work. If the 1st verse be indeed a summary of the account of the whole chapter, then the narrative proper commences with the word "and," of the 2d verse. But it is clear that the word "and" cannot introduce a proper and absolute beginning. And besides, such an apprehension of the passage would border closely on the false view, that the words "without form and void," of the 2d verse, refer to an eternal chaos; so that, in such light, our account of the creation can refer to no other creative act than the mere transformation, arranging, and quickening of a chaotic mass of matter already in existence. A creation *out of nothing*, which is so imperatively demanded by the whole spirit and substance of the Old Testament, and assumed by it as the best settled of all facts, would thus not be plainly taught, but left in anxious doubt, by silence in the very place where there was every motive and consideration for expressing the fact in the most unmistakable language.

Though on the one hand we are forced by the word "and" of the 2d verse, to regard verse 1st as truly a part of the account of the creation, as a sentence of a connected whole, and not a mere heading to the chapter, we cannot, on the other, deny that there exists an obvious difference in tone, manner, and mode of representation, between its formal statement and the narrative which follows. The first verse evidently contains none of that life-like spirit and descriptive character, which is so decidedly impressed upon the whole remainder of the narrative. We there discover none of that concrete delineation of details which so characterizes all the rest of the chapter; and hence infer that we are warranted in not including the first verse in that history which was conceived in prophetic vision, and which constitutes the ground-work of the Mosaic record of the creation.

Nay rather, the *first* glance of the prophet, as we conceive, revealed the earth to him as already in existence, though in a waste and desolate condition. But as the vision progressed, he beheld how through the mighty energy of the Divine command, our present earth, with all its fulness of light and life, arose from the dark and lifeless earth first disclosed to his view. This is what the prophet saw, and this is what he has communicated.

But whence came that "waste and desolate" earth, from which God formed one so beauteous, and teeming with all manner of life? Heathendom of later ages, whose original consciousness of God had become so obscure, that the very idea of a living and personal God had been wholly lost, regarded it as a crude, unwrought chaos, existing from eternity.

In order to completely overthrow this sad and dangerous error, the prophet who first conceived the history of the creation, or some one who subsequently revised the tradition, perhaps the author of Genesis himself, prefixed to the sacred account that weighty and important introductory verse, testifying both of a personal God and of a creation in time.

It mentions what was preliminary to the six days' work, affording a foundation for the description of what took place on these days, and also guards us against a misinterpretation of the six days' work.

The first verse, therefore, as we regard it, is not a heading to the chapter, but an introduction to the six days' work; not an account of something first coming to pass within the six days' work, but of a fact preceding it both logically and chronologically.

In thus making a distinction as to origin and character, between the 1st verse and the narrative which follows—looking upon the one as the offspring of prophetic vision, and upon the other as a necessity or result of reflection on the part of spiritual and religious mind—we are by no means to be understood as making a distinction between the two in regard to their value in a religious point of view, as though one were Divine revelation and the other mere human opinion. Nay, we regard both alike as revelations from God, and differing only in the mode of their conception—the former as the offspring of *divinely enlightened reason*, the latter as the fruit of *prophetic vision.*

§6. *Condition of the Earth prior to the Six Days' Work.*

There are two conceivable modes of explaining the first verse in connection with the account of the six days' creation which follows. It may either be regarded as an account of the creation of the *elementary and primeval matter* of the universe, from which were formed in six days the systems of worlds as they now exist in their matured and completed forms —so that the waste and desolate condition mentioned in the second verse must be understood to involve merely the absence thus far of light and life, a stage of development not yet advanced to this high prerogative: or it may be looked upon as an account of an *original creation complete in itself*, into which, through the medium of a catastrophe hereafter to be considered, came that devastating process which gave rise to the darkness and the waste condition mentioned in the second verse — so that the six days' work can only involve the restitution or new-creation of the earth which had been laid waste.

Our record does not decide between these two modes of intrepretation. As its author, acting in the capacity of a medium of revelation, could only report *what he had seen in vision*, he does not nor could he say, whether the earth was created in this waste and desolate condition, nor by what means, if its desolation belonged to a subsequent period, such devastation was effected.

To the first mode of interpretation, it has been attempted to oppose the expression "heaven and earth" in the first verse; for these words, it is said, cannot refer to the universe in its elementary condition, because the heavens and the earth could not exist before those original elements had been subjected to a process of individualization and more perfect arrangement. But this objection goes but half way, and fails to prove just what needs proof. The expression "the heavens and the earth" involves the fact at least of such an individualization, but by no means the necessary maturation and completion of the individual worlds. Verse 2nd places the truth of this assertion, as far as the earth is concerned, at least, beyond all doubt; for there the waste and desolate mass of the earth even, from which the present earth was to be formed by the work of six days, is already called *the earth.* And it may justly be so called, for it was individualized even then—it existed as a world by itself in distinction to other worlds. The same must be assumed with regard to the heavens and the celestial bodies in general—even though the account, which specifies the case of the earth only, and what intimately concerns it, does not speak expressly on this point.

A further argument *against* the first, and *in favor of* the second mode of interpretation, has been sought in the words, "without form and void," ["waste and desolate"] (tohu vabohu), of verse second. This expression, upon which etymology can throw no satisfactory light, designates beyond all doubt, wherever else it occurs (Is. 34 : 11; Jer. 4 : 23), a *positive* DEVASTATION and DESOLATION, which has *succeeded to* previous life and fruitfulness; but never a mere negative want of life, a low stage of development which *cannot yet* claim the prerogative of life. Consequently, it has been said, the expression *must* be *similarly* interpreted here. But it is clear that this conclusion is quite too precipitate, when we reflect that the expression, just like our term "waste," may be so comprehensive as to be applicable alike, both to a state of positive devastation, and a state of mere negative want of life. The very fact of the author's silence in regard to the origin and nature of this waste and desolate condition of the earth, proves that he used the expression in this loose, undefined sense.[1]

Further, it has been said: "God is a God of light and of life; he would not create a dark, dismal, and lifeless chaos, but only worlds of light, life, and order, in which he might behold the reflection of his own glory and blessedness. And that were we to imagine a work of *God* not yet fully completed, it could never be one in such a condition as verse second describes the earth to have been; for a work or creation coming from the *Divine* hand, though it be not yet perfected, must, according to the measure of its development and capacity, reflect a Divine harmony and order, light and life."—We might, indeed, allow that the author used the expression in a loose, comprehensive sense, and still deduce from his words the conclusion, that they can only be *properly* understood, by assuming that the primeval earth had been subjected to a devastating process,

---

[1] *Delitzsch* (p. 55–63) accedes to my view in this connection. He says: "The sound and significance of these two words are portentous." . . . "Still it is true that the etymological significance of the phrase 'tohu va bohu' is not satisfactorily met by the purely privative idea of want of form and order." He attempts to do justice to the import of the phrase by a speculative deduction in which I cannot acquiesce. He would see in the "tohu va bohu" the pure original matter of the world, which, being found in a non-divine state, but not in a positively anti-divine, nor yet in a merely negative non-divine, but in a positive non-divine state, was to be reduced and brought under, etc.

anterior to the six days' creation. This view of the case still retains some weight in the author's mind; but he cannot any longer attach to it the importance (as in the first edition of this work) of a satisfactory proof. It can at best but add to the weight of arguments drawn from another quarter.[1]

The view that the earth was subjected to a devastating process, between the time when the heavens and the earth were originally created, and the time of the six days' work, and that this process gave rise to the necessity of a restitution, a new-creation, *cannot, therefore,* be established from this first verse of the Bible—but neither does the whole first chap-

[1] The assertion so often made, that the second verse can or should be translated thus: "And the earth *became* waste and desolate," is directly in the face of the true grammar of the clause. Had such been the meaning of the author, he should certainly have said וַתְּהִי הָאָרֶץ, instead of וְהָאָרֶץ הָיְתָה, and have supplied, to avoid all ambiguity, the preposition לְ with the verb היה. Drechsler (*Einheit und Echtheit der Genesis*, pp. 66, 67) attempts to show that the second verse cannot, from its structure, be intended to describe the condition of the earth as it came from the hands of its Creator, according to the first verse. The second verse consists, he says, of three clauses: The earth *was* without form and void, and darkness upon the face of the deep, and the Spirit of God moving upon the face of the waters. The effect of the copulative continues from the first clause forward to the others also, so that if we translate: God created the earth without form and void, and covered with darkness, we must also say he created it with the Spirit of God moving upon the face of the waters. But this argument avails nothing against the view it would combat. It may be replied, that the second verse says not in what condition the earth was, *as* God created it, but that it states the condition it was in when he had created it.

ter *contain a single word* that *militates against such a view.* Both opinions may still, from this stand-point, be held as equally legitimate. But we must leave the matter undetermined for the present, and patiently wait to see whether subsequent revelations do not offer us something more tangible and decisive on this point (Compare § 25).

§ 7. *The First, Second, and Third Days' Work.*

The earth was without form and void, and darkness was upon the face of the deep. The raging elements toss upon each other in wild disorder: nowhere does the eye of the Prophet behold the first beginnings of order and harmony, of light and of life. But it is not ever to be so. Already is the Spirit of God seen moving upon the face of the wild and restless waters, brooding over the dreary and lifeless waste, awaking it to life and fruitfulness. The barren waste must vanish *before his* presence, and the desolation before the breath of *his* mouth. The fettered germs of life, made fruitful by *his* breath, await the hour of their freedom and development. Then is heard over the dark and raging waters, the potent command of the Almighty: "*Let there be light!*"—"and there was light." Suddenly, and from the midst of the thickest darkness, light breaks forth in unshackled freedom, the first expression of life, and the condition of all further developments of life in the yet waste and barren earthy mass. Light, the first created of God upon earth, the resplendence of the Divine in the sphere of the cosmical, bears upon its front the seal of the Divine good-will; it ever greets the favored beholder as a messenger of Divine love and grace. "*And God saw the light, that it was good.*" Darkness had enshrouded the light over the face of the waters; but, "*God divided the light from the darkness.*" Thus light was set at liberty, and called forth to an independent existence. No longer does it lie enshrouded in the darkness; but, sweeping over it and through it, gives to it bounds and fills it with life. The light is called *day*, the darkness *night.* The first day's work is brought to a close. Evening has come, the morning dawns. Thus the first day is completed, and the second is introduced.

A new day breaks forth. The laboring earth begins to move amid the heavy waters: a new command from above has excited her, and *she* also must soon bring forth the hidden treasures of her fruitful womb. God said: *Let there be a firmament* (Ausdehnung. Heb. Rakiah,) *between the waters; and he called the firmament Heaven.* This was the ethereal heavens, the pure, clear, transparent expanse of air over-head; the atmosphere with all its unfailing springs of life and bliss, as inexhaustible as they are indispensable to all kinds of animated beings which were to appear upon the earth. It rests upon the waters of the earth, and, like a firm arch, supports the oceans of the heavens. Thus it divides the waters below from the waters above; the seas from the clouds which were to carry their waters, laden with the richest blessings, to the dry land, so soon as the latter had freed itself from the dominion of the all-engulphing floods.

The *third* day includes two acts of creation, but they are closely connected, both from inner significance and also the fact of their having followed immediately one upon the other: the separation of the seas from the dry land, and the clothing of the latter with the vegetable world. As the task of the first day was to liberate the light from the bonds of darkness, and that of the second to separate the heavens, so laden with blessings, with rains and fruitful seasons, from the chaotic floods of the primeval earth, so also the creative word of the third, freed the firm land from the constraint of the seas, which heretofore overflowed and engulphed all things. For as the polar opposition and well-established reciprocal relation between light and darkness, between day and night, as, also, between earth and air, seas and clouds, lie at the foundation of all life and prosperity upon our globe; so also is a complete and permanent division of land and water, the foundation of all further developments of life upon the earth, and also the guaranty of the prosperous and undisturbed life of the inhabitants of both land and sea. The land, indeed, is the favored abode of the noblest work of God; therefore must it be freed from the dominion of the sea by the creative and all-disposing word of God, and oppose to the latter firm, immovable barriers. The tumult caused by this sudden separation is depicted by the Psalmist in the following words (Ps. 104, 5–9):

"He laid the foundations of the earth,
That it should not be removed for ever.

Thou coveredst it with the deep as with a
garment:
The waters stood above the mountains.
At thy rebuke they fled;
At the voice of thy thunder they hasted away.
They go up by the mountains; they go down by
the valleys,
Unto the place which thou hast founded for
them.
Thou hast set a bound that they may not pass
over,
That they turn not again to cover the earth."

As the waters retire and the dry land appears, the pregnant earth immediately brings forth, through the energy of a fresh command from God, the wonders of the vegetable world, with all their beauty, brilliancy of color, and abounding fruitfulness; whose seeds and germs had been implanted by the vivifying breath of that mysterious Spirit which moved upon the face of the primeval waste and desolate earth. The vegetable world which eagerly clung to the parent earth, covering her nakedness with its magnificent robes, had no separate life, no independent existence. Therefore it originated on the same day which gave to the land from which it drew its sustenance, a separate existence.

§ 8. *The Fourth Day's Work.*

Thus, as we have seen, was the formation of the earth as a *globe*, separate and distinct *in itself*, completed. On this, the *fourth* day, its relation to the rest of the heavenly bodies is to be determined and permanently established.[1]

In the "rakiah" or expanse of the heavens, which was the result of the second days' work, the *sun*, *moon* and *stars* were placed, by the word of the Almighty. *They were to divide the day from the night, and be for signs, and for seasons, for days, and for years; and also, to be for lights to give light upon the earth. The greater light was to rule the day, and the lesser light to rule the night.*

The question, whether we are to understand by

---

[1] Hofmann (*Schriftbew.* p. 243) and Delitzsch (*Genesis*, p. 71) regard the remarkable progress in the work of creation in an entirely different light. With them the fourth day rises in the scale of creation, "in that the celestial bodies, separated from the general mass, and hasting in their wide and unwearied orbits through the heavens, offer to our view a higher stage of individualization than plants helplessly confined to the surface of the earth:" whilst, again, the creations of the fifth and sixth days, animals and man, from their capacity of voluntary movement, exhibit a still higher degree of individualization than the stars. I very much fear that the record, in spite of all that is said about its objective character, is thus to be degraded into a poor, miserable treatise on science. For, if the record is to be interpreted as an account of objective facts, and in such manner that the mind of the author is ever supposed to be guided by the Spirit, and so that the plants being created *before* and man and the animals *after* the stars, should show that, in the mind of God, the one hold a position *below* and the other *above* the stars — then will natural science, as I apprehend, be so fully able to show that such a scale of being is altogether false and contrary to nature, that all attempts to defend it will be utterly futile. Such a view must either be given up, or we must at once concede that the account is not objective truth, but the production of some philosopher or other, (and rather an unskilful one too).

the expression "stars," as used in the description of the fourth day's work, the whole starry heavens, with their countless millions of fixed stars, their milky-ways and vast groups of stars, or merely the stars of our solar system, has given rise to no little controversy. In the first edition of this work, the author, in accordance with the opinions of many predecessors, expressed himself in favor of the latter view; but now the most weighty arguments force him to regard the former as alone admissable.

His apprehension of the waters above the firmament, as mentioned in the second day's work, in connection with another view since acknowledged to be erroneous, was as follows:

"The massive waters of the beginning (verse 2d) being polarized and separated by the energy of the Divine command, those above the firmament furnished the substratum for the formation of the celestial bodies, those below the substratum for the formation of the earth. Immediately after the second day's work, began the individual development of each sphere—that of the earth under the firmament, that of the stars above it; — and the formation and completion of each progress with equal pace, as we might naturally suppose, and as the account itself intimates. The waters above the firmament soon withdrew from the eyes of the prophet, his whole attention being attracted by the *earth* to which *he belonged*, and in which he had such a special interest; hence the prominent place its formation and completion held in his mind. Having finished the description of the earth's formation in particular, he now proceeds in the next, the fourth day's work, to mention the completion of the celestial bodies. These were already so far developed as to be fitted to assume that relation to the earth which they were destined to hold, since they had progressed in formation equally with the earth. That we are warranted in this assertion, is shown by the fact that the earth on the third day brought forth the vegetable world, whose origin and subsistence certainly depended upon a real and settled, though perhaps at that time still feeble influence of the sun upon the earth. But the prophet, whose mind was fully engaged and filled with the wonders of the formation of the earth, could not mark the *simultaneous facts* going on in the formation of the stars—he could not regard the earth and stars at the same time and progressing in formation with equal pace—but one must be seen *after* the other, and of course be described in the same relation. The finishing of both earth and stars consisted in the fact, that the preliminary relation, introduced on the first day, between bodies giving and bodies receiving light, was now *permanently established*, and brought out in the contrast or opposition of the *solar* and *planetary principles*."

The assumption that we are to understand by the stars of the fourth day, merely the planets of our solar system, was attempted to be justified on the following grounds:

"The whole scope of the account of the creation is so evidently confined to the earth and what pertains to it, that we are irresistibly forced to believe that the fourth day's work also, had reference only to such heavenly bodies as are essentially connected with the earth; to such as stand in immediate or close relation to it, and form with it an articulate, unique system. Again, the sun and moon, the 'two great lights to give light upon the earth,' so monopolize the attention of the prophet, that the stars in the firmament are scarcely even observed. All he says touching the design and mission of the celestial bodies, and their relation and position with regard to the earth, refers so specially, indeed so exclusively, to the sun and moon, that the claims of the heavenly bodies in general to be 'for lights to the earth,' as expressed in verses 14, 15, and 18, are completely overshadowed by the high perogatives of the 'two great lights,' as stated in the 16th verse. The additional words, 'the stars also,' at the close of the 16th verse, which are subjoined in a manner so supplementary and unimportant, without any further explanation of the design, mission and position of these distant and unknown bodies, after the mission of the sun and moon had been stated with the utmost distinctness, were evidently regarded by the prophet himself as of mere subordinate importance. And they are so little dwelt upon, so left in the background of the picture, that their interpretation should not be attempted upon their own merits, but left to depend upon their connection with the whole account, their tendency in general, and particularly to the relation of the work of the fourth day to that of the rest. Perhaps we may thence draw some conclusion to aid us in determining the design and true import of words so vague and unsatisfactory in themselves."

The above reasoning is liable to weighty objections; and, as we are now convinced, has been overthrown.

In the first place, it places the whole stand-point of the author of the history of the creation in a false and unpropitious light, by attaching to it a scientific interest, which was wholly foreign to it. The account of the creation was given alone for religious purposes, and would scorn to be called a compendium of Astronomy or Geology. Its design is a threefold one. It would, first of all, show the relation of *God* to the world; then the relation of *man* to the rest of creation, his high, sovereign position in the scale of being, by which his mission in history is conditioned; and finally, the typical reference the work of creation bears to the disposition of his duties and labors in life.

The first design is clearly revealed in the first verse: *In the beginning God created the heavens and the earth.* Each word of this weighty verse bears with signal force upon the scales of religious knowledge, from the fundamental truths it *teaches*, and the dangerous errors it *excludes*. The author of Genesis might have satisfied himself that he had gained his end, by stating this general proposition, had not its general and abstract character involved in some degree—particularly in the case of the Orientals, whose minds grasp and retain only the concrete, the embodied — the danger that the infinitely important truths of the expression might be overlooked or feebly apprehended. Hence the necessity of dressing them in concrete forms, of developing them in detail, and attaching to them the character of tangible, living realities, that they might be indelibly imprinted upon the mind of the reader. But still more decisively would a further development of these truths be demanded by the two other designs of the account of the creation, which were, to bring to the consciousness of man, in living, tangible forms, his relation to nature and to his fellow-beings.

Such motives of a religious nature, and none other — least of all, motives drawn from the science of Astronomy or Geology — induced the prophet, or rather the Spirit, whose instrument he was, to give a more detailed account of the process of the creation, as it occurred in the six days. We cannot, from this point of view, believe that he kept in mind the distinction which Astronomy makes between the heavens of the planets and the heavens of the fixed stars, since there is not a word hinting at or expressing such fact. We may, indeed, easily conceive of such a distinction being of significance in a religious point of view; but had such been the case here, the distinction would have been clearly made, particularly as its legitimacy was in all probability matter of easy discovery and common belief, in the earliest times of the human race. But the very fact of its *not being mentioned*, indeed, not even hinted at, proves clearly that it was a matter of no significance or interest to the prophet, nor promotive of his design, as he viewed the facts of the creation; but this does not necessarily disprove the fact, that it might be of importance in a later or more advanced stage of revelation, or serve other and more comprehensive designs of the latter.

If it be true that the author, in the 16th verse, speaks of the *stars* in general, without restricting his meaning to any particular kind of stars, we are not allowed, however much the expression "the stars also," retreats into the back-ground, and appears of but secondary importance—indeed, on this very account the less—arbitrarily to restrict, limit or determine its meaning; the expression *must* retain the broad general character in which it appears.

Nor is, on the contrary, the objection of any avail, that the sun, moon, and stars, of the fourth day, were placed in the "rakiah" (firmament) made on the second day, which sprang from the earth; and that therefore they must be regarded as belonging to the earth, in a physical point of view. For the "rakiah" or ethereal heaven, is identical with the "terminus technicus" of the earth's atmosphere, merely in a scientific sense, whilst in *common* speech it comprehends, and ever has comprehended, much more than the latter, so that it includes also what in modern times is called the cosmical ether. So long as we keep distinctly in mind, that we here have to do with something wholly different from a text-book in Astronomy or Physics, we shall no more be stumbled by this scientifically incorrect, indeed, even positively erroneous manner of regarding and expressing physical facts, than we are by the common expressions concerning the rising and setting of the sun, which, scientifically speaking, are no less incorrect. The prophet depicted *what* he saw, *as* it was seen; he doubtless saw the fixed stars in the same heavens with the planets.

Finally, if upon the incontestable fact that the whole scope of the history of the creation is confined to the earth and what pertains to it; and that this history gives definite information only in regard to what has reference to the earth—if upon these facts, the limitation of the fourth day's work to the creation of the stars of our solar system is to be justified, by our being forced to believe that the 16th verse can only refer to such heavenly bodies as are essentially connected with the earth, and form with it a unique physical system—this reasoning also, is wholly inadequate. For there is not the least intimation given, that the sun and moon are here brought prominently into view, because they belong to our system in a physical and astronomical point of view. Were such the case, it would be in the face of the whole character and design of the sacred record. It regards the question of an existing *physical connection* as matter of no importance, and mentions only this one point, that the *sun* was to give light by day and the *moon* by night. But the same holds good with respect to the stars also (verse 17: "that they might give light upon the earth.") And it is plain that the fixed stars serve *this* end in higher measure even than the planets.

For the very reason that the description of the sun, moon, and "the stars also," of the fourth day, is exclusively confined to what these bodies are in their relation to the earth, and does not give the least intimation of what they are in themselves—for this very reason must we, keeping in mind the whole tendency of the record, and its origin from the contemplation of facts as they appeared to the senses, oppose the unwarranted assumption that the sun, moon and starry heavens were first created on the fourth day—then first called into being out of nothing, after the earth, as an independent globe, was completely finished. As the account says not a word in regard to what these heavenly bodies are *in themselves*, so neither does it say a word as to *when* and *how* they were created to be what they are *in themselves*.

The work of the fourth day was indeed introduced, like that of all the rest, with the creative "*Let there be;*" but this command was directed to what the stars should now become, and the end for which they should thus exist—that they should be lights to give light upon the earth. If they had never yet fulfilled *these* conditions, but were now about, for the *first* time, *so to do*, the words of the account are fully justified; for this relation of the starry heavens to the earth, just now being for the first time introduced, regulated and established, was as much an act and a result of creative power as the establishment of the relation between light and darkness, or of that between land and sea. Further, it is said: "*God placed them in the firmament of heaven*"—and naturally enough; for as the firmament meant the terrestrial heavens, which were formed on the second day, the stars, supposing they existed before the second day, could not be regarded as stationed in those heavens, but could assume their position there, only at the time they began to assume a significant relation to the earth.

Neither do the words "God made the sun, moon and stars," of the 16th verse, require a constrained or forced explanation; for he now for the *first time adapted them to the earth*, and they now first began to exist in relation to it. But this by no means destroys the correctness of the view that they may have been created long before, to exist in their *own capacities* and for their *own ends.*

We may sum up the results of our present inquiry as follows: The fourth day's work refers to the whole starry heavens, including the fixed stars; but we are not necessarily forced to assume that these were *first created* to exist in their *own capacities*, after the formation of the earth was completed. From this point of view, it still remains undetermined whether the sun, moon, and stars, were first created *after* the earth was finished; or whether they already existed in a perfect state *before* the creation of the earth, but now for the first time assumed their relations to the latter; or finally, whether their formation progressed simultaneously and in equal pace with that of the earth, so that by the fourth day both they and the earth were so far perfected, that, from that time forth, they might sustain the important and established relations which were designed to exist between them.

There now remain still three points to be explained:—the relation of the results we have obtained to the creation of the heavens, of Chap. 1:1,—

to the creation of light, of the 3d verse—and to the separation of the waters that were above the firmament, of verse 7th.

We shall commence with the last point. *Ebrard* and *Delitzsch* hold the upper waters to have been the substratum for the formation of the heavenly bodies of the fourth day, but with this difference, that *Ebrard* confines the fourth day's work to the bodies composing our solar system, while *Delitzsch* includes also the creation of all the fixed stars and systems of milky ways. We hold this view (although ourselves once attached to it) to be erroneous. The account of the creation nowhere even intimates that the heavenly bodies of the fourth day were formed out of the upper waters. This assumption contradicts, also, subsequent portions of Holy Writ, according to which the upper waters are *still* in existence.[1]

If we believe that the work of the fourth day must refer, not only to the formation of these heavenly bodies in their relation to the earth, but also to the same as they existed in their own capacities, and are in search of a substratum for such formation, according to the analogy of the formation of the earth, we must doubtless look to the first verse for it, but not to the seventh. The collected waters, which were subsequently divided into those above and those below the firmament, are called in the second verse, "*the earth*," but not "*the earth and the heavens;*" consequently, they cannot have been the substratum for the formation of the earth *and* the heavens, but *only* for that of the earth. If there ex-

---

[1] Ps. 148 : 4; 104 : 5; Job 26 : 8.

isted a corresponding substratum for the formation of the heavenly bodies, it could only have been the heavens mentioned in the first verse, which (since according to § 5, the first verse cannot be regarded as a mere heading) were in existence before the six days' work.

As to the relation of the lights in the firmament of heaven—particularly the relation of the sun—to light as created on the first day, the Bible leaves us in no doubt as to its meaning. *Light* ("OR") was called into existence on the first day, but not until the fourth day did the *lights* or bearers of light ("MAOROTH") appear. The power of giving light was not originally confined to the sun, but first became so when the earth was so far advanced toward its completion, that the solar and planetary polarity might be established. But as there had already been changes from light to darkness, from day to night, these must be referred to a telluric action and reaction, which ceased so soon as the contrast of solar and planetary functions was established. More the sacred record does not say. More it could not say, without compromising its character as a record of Divine revelations, and becoming a text-book in physical science.

§ 9. *The Fifth and Sixth Days' Work.*

So soon as the *cosmical* conditions and supports of *organic* life were provided, so soon as the chaotic confusion of elements and agencies was resolved into a harmonious and well-established relation and play of forces, the germs of life began to be developed in the bosom of the virgin earth, and she brought forth, at the nod of the Almighty, all the wondrous and varied revelations, grades, and potencies of life, which we now behold. Already was the vegetable world called into existence, by the command of the third day; and now the fifth and sixth days mount up in the scale of creation, from the fish in the depths of the sea to the eagle of the air, from the worm of the dust to man, "who walks majestic, with countenance erect"—to man, the crown and glory of this lower creation.

The account of the creation represents *man* as the last of created beings, and, since the series of the creations as they appear, seem ever to reveal a higher and still higher grade of life—as the crown and glory of creation. This progress is *physically* represented in the fact that each higher grade of life includes within it all previous and inferior grades, which have been realized and quitted for a now higher one, and is characterized by the addition of a new and higher development of life. Thus, the purely *cosmical* elements and potencies serve as the foundation for specific grades of life—such as belong to the vegetable world. The animal world or kingdom includes both; for besides voluntary life and action, which are its characteristics, it includes, also, an extensive and closely interwoven sphere of vegetable life—all the innumerable *involuntary* functions of life. Finally, there is added in man, to the three inferior, dependent grades of life, the *cosmical*, the *vegetable*, and the *animal*, a fourth, and a higher—

the sphere of personality and moral freedom, the *image of God* in the creature.

The Bible represents the creation of the universe in pyramidal form: heaven and earth constitute the broad base of this pyramid; man is its unique top-stone. He is the representative of all inferior grades of life, the unity in which the multitude and variety of all earthly creatures converge and find their end. Although it is not expressly so said, since the *turn* of the thought and *form* of the expression is foreign to the Bible, that man is the microcosm, the central point of this lower world, where all grades of life converge,[1] doubtless this *idea* underlies its whole import and tendency. The 26th verse of the 1st chap. of Genesis, expressly designates man as the king and lord of this lower creation, together with all its ma-

[1] The remark of Theodorus in this connection: *Theodoret, quaest. XX. in gen.*, is very appropriate: "God finally created σύνδεσμον ἁπάντων τὸν ἄνθρωπον;" and no less truthful and beautiful is that of Augustine: "Nullum est creaturæ genus, quod non in homine posset agnosci." Yea, even that Rabbinic saying, which appears so quaint and absurd, "that Adam was so large when he came from the hand of his Creator, as to reach from earth to heaven, and from one end of the world to the other; but that, when he sinned, God, by a touch of his finger, reduced him to his present insignificant stature,"—is designed to symbolize and express neither more nor less than the same by no means strange or absurd idea. The *name* also which the record gives to the first man, *Adam*, from "adamah"—earth, designates him, if the thought be carried over into our modes of expression, as the microcosm of the terrestrial world. *Umbreit* strikingly remarks in this connection (*Theologische Studien und Kritiken*, 1839, p. 201): "In the name of the man lay the significant idea that he was the representative of the whole earth, comprehending it as its lord and ruler, in his own form."

terial forces and all its creatures. His calling to and fitness for this princely dominion is no less unequivocally evinced. He is the last and most complete creation sprung from the bosom of the earth. He belongs to the earth; all grades of life are repeated in him—"nil terrestre a me alienum puto," it becomes him to say, since earth with all its creatures is closely related to him—therefore does he become their fit representative, and the mediator between them and all that is above or beyond the earth. But he is also the *offspring of God*, created in the image of God, and thus far exalted above earthly nature, and thus, also, becomes a representative of God towards *them*, lord and king, priest and mediator.

When creative power had thus attained its culminating point in the creation of man, then did God "*look upon everything that he had made, and behold, it was very good.*"

§ 10. *The Primeval History of Man.*

The drama of the six days' creation has attained its last and grandest scene in the resting of God on the seventh day, and his hallowing of the same as a day of rest for man. It has thus become a complete, symmetrical, well-rounded whole. Passing on, we meet with a new portion of Divine revelation, whose tendency is wholly different from that of the section we have just been considering, but which is no less weighty and full of meaning; in many respects, indeed, vastly more significant and important. It is a record which has employed the most profound and sagacious powers of interpretation for ages, over which superficial knowledge and skeptical indifference have fluttered and trifled for thousands of years; a history from which faith has drawn its strength, and the wisdom that is from above, ever-increasing light; over which infidelity has vexed and chafed itself in the most bitter and contemptuous spirit. It is the foundation upon which the whole structure of Divine revelation, closely bound together, has grown to a hallowed temple of the Spirit; it shows us the root whence sprang the salvation of God in Christ, with its buds of promise under the Old Covenant, and its mature fruits under the New.

The first section serves as a foundation for the history of the world in general; the second (Chapters 2 and 3,) as the foundation of the history of redemption in particular. The former shows us the sovereignty of God OVER the world, as the Creator of the heavens and the earth; it assigns to every creature, and particularly to man, his position, mission, and destiny, in the wide and general plan of the world; it points out to him his normal path of development even to its ultimate goal. But it designedly and according to its plan, says nothing of what shall be the development realized; for this would be adding what was foreign to the present object, and destruction to the unity and harmonious realization of the whole design. Of proper scientific teachings it contains nothing at all.

The aim and tendency of the second section is wholly different. It rests upon the first section, and presupposes it. It represents God as dwelling IN his own world, as a Father and Instructor, and from condescension and love, adapting himself to the state of his pupil, and advancing with him as the originator and the announcer of salvation.[1] The first section represents the work and idea of God in the creation, as also the Divine mission and destiny of man founded thereon; the second, on the contrary, describes the free, self-chosen development and destiny of man, and the Divine fostering care, superintendence and guidance previous to it, with respect to it, with it, and subsequent to it.[2]

The history of the fall, in Chapter 3d, is the cardinal point of the second section—the fall as the root of all woe, the occasion of redemption, and the beginning of the history of humanity. It depicts the trial of man's steadfastness, or self-determination, which resulted so disastrously in his commission of sin, arrested his original destiny, and, with the concurrence of Divine grace, conditioned a new development, supported by new means and higher energies. The history of the six-days' creation, however complete, full and well-rounded it may be, *in itself*, with respect to its *own objects*, does *not* suffice that we may fully understand the fall of man, and the displays of human guilt or Divine grace, to which the fall gave rise. The history of this momentous occurrence, demanded a special, a new foundation,

---

[1] Hence also God is called Elohim in the first section — in the second, Jehovah,

[2] For further particulars touching the relation of the two sections to each other, compare my work, *Berträge zur Vertheidigung und Begründung der Einheit des Peutat.* Königsb., 1844, p.45, seqq.

such as Chapter 2d supplies. We there learn that man was formed out of the dust of the earth; a circumstance which vastly increases the guilt and folly of his self-exaltation, by which he would fain be *as* God, *without* God: and also explains how he was, in consequence of the curse pronounced upon sin, to return to the earth again from whence he was taken.

The breath of life God breathed into him, constituted him a personal, conscious, and free being, who, needing development and capable of development, must and could choose for himself, and decide between good and evil, and be responsible for his choice. The garden in Eden, so full of peace and joy, was the place where his trial and fall were to come to pass; the place of happiness from whence he was to be driven after the fall, to eat his bread in the sweat of his face. The command to keep the garden, indicated the existence of a hostile, destructive principle, against the power of which man was thus warned. The tree of life, whose fruit was not forbidden to man in his state of innocence, was, after the fall, guarded by the sword of the cherubim. The tree of knowledge was the chief and most direct medium of his development. The other trees, with their fair and precious fruits, but aggravated the guilt of man in eating of the *only* tree that was forbidden him; for they *all* offered him their bounties, if he would but keep from that mysterious, that fatal tree. The review and naming of the animals introduced the creation of woman, and the latter was the condition of the first and every subsequent development, etc.

§ 11. *The Position and Mission of the first Man.*

We shall now proceed to a more detailed examination of this second and very significant section of the sacred record, so far as it bears upon the end we have in view.

This part of the record dwells with great clearness and at special length on the creation of man, concerning which the first section gives only the most general facts. The chief point here brought into view, is the dualism of man, by virtue of which he may be said to have both a Divine and an earthly nature.

The germs of the varied forms of life, were implanted in the bosom of the earth, by that Spirit which in the beginning swept over the "tohu vabohu" of the primeval earth. Consequently, the special production of the vegetable and animal kingdoms, did not appear as the offspring of *pure*, direct creative acts; but merely as the result of further formative agencies and potencies, brought to bear upon the original germs of life. "Let the earth bring forth!" it was said. And as the lives of plants and animals, from this point of view, appear as the individualized products of the life of the earth, so also does the life of man. But man is the *highest*, and therefore, also, the unique, the representative (einheitliche) product of the earth. Creative energy, which thus far had been employed in different parts of the earth at the same time, producing its countless individual manifestations of life, now concentrated itself in *one* point, to produce the highest form of life, the sublimation of the earth's most noble potencies; and the account could no better and more vividly express this fact, in its concrete, prophetic manner of representation, than by saying, that God "formed man out of the dust of the earth." But man was something vastly more than the highest and most noble manifestation of animal life. The princely form made out of the most refined elements and the most noble potencies of the earth, was, in addition, imbued and filled with a Divine breath of life, whereby man, who on the one hand is of the earth earthy, is on the other the offspring of God,[1] (Acts 17 : 28, 29) and the image of God.[2]

Man was now placed in the garden which God himself had planted in Eden, as his place of abode and employment. He was commissioned to *dress* and *keep* it.

Although we are told that all the creatures which proceeded from the hand of the Creator were *good*, *very good*, it is clear that this perfection could not have been absolute, but merely relative; so that the words are not to be understood as importing that

---

[1] The formation of man out of the dust of the ground, and from the divine breath of life, did not comprehend two processes differing in point of time, so that man was at any time (and were it but for a moment) merely an animate earthly form, like the rest of the animals, differing from them only in grade, but not in nature. But there was, indeed, a distinction in regard to the origin of the elements of which he was formed. Two elements differing "toto cœlo"—the form from the dust of the ground and the divine breath of life from above—met together in the moment of his creation, and the product of the two was Man.

[2] Gen. 1 : 27.

man, and nature, assigned to him as an abode, were immediately advanced, by the creation itself, to the *highest stage* of perfection of which they were capable, and for which they were destined by the Creator. Nay, rather, man was created with *that degree* of perfection which harmonized with the position he at first held, and the mission which was assigned to him. As man was raised, by the Divine breath of life which dwelt within him, from the sphere of mere passive nature, into the sphere of free, personal life, of moral and religious freedom, it is clear that *his* highest stage of development, at least, could not arbitrarily and at once be attached to him, as in the case of a plant. Nay, he must rather, through his own *free choice* and *action*, determine himself and develop himself to those high ends for which he was *destined* and *made capable* by his Creator. In accordance, therefore, with this moral necessity, man was immediately placed under such circumstances as would leave him free to decide for himself, either *for* or *against* the will of God and the destiny originally set before him, so that he might freely enter upon any course of development which he himself should see fit to choose.

But man was not only to find an *abode* in the midst of nature around him, but also a place of *employment* and *activity*. He was to be intimately connected with it, and develop himself in the midst of it and along with it. Consequently, nature itself could not be created in any but a stage of relative perfection; it was requisite that *it also* should stand in need of development and be capable of it; though not on its own account, but on account of man, who, as its priest and mediator, its lord and master, was to conduct it to its ultimate stage of perfection, or to its consummation.

The mission and *end* of man's activities were to be realized by his having dominion over the *whole* earth.[1] But it was necessary that he should *begin to assume* this dominion, in *the place first assigned to him* by his Creator. Therefore, the *first* and temporary task to employ his powers, was the *keeping* and *dressing* of the garden in Eden.[2] This is no *new*, no strange task. The idea of having *dominion*, as we gather it from Chap. 1 : 26, is here merely further represented, in both its *positive* and *negative* phases. The object is still the same as before, though limited by present circumstances. God himself had planted the garden in Eden; it now becomes the duty of *man* to take up the work which God had begun, and advance it to completion. But, doubtless, the dominion of man was not ever to be confined to paradise. Nay, much rather was it to be extended in ever-widening circles, until it compassed the whole earth—appropriating it, and moulding it also into a paradise. Thus was the *beginning* (the dressing and keeping of the garden) to lead to the *end* (man's dominion over the whole earth).

Man was to "*dress* and *keep*," or *guard*, the garden in Eden. *Guard it?* against *whom?* Was there, indeed, an enemy already present, meditating the destruction of the Divine work? The command to keep the garden doubtless intimates the negative

[1] Gen. 1 : 26. [2] Gen. 2 : 15.

phase of man's dominion, just as the command to dress it reveals the positive. But thus far our attention has been absorbed by Divine, creative, good agencies. But were there indeed, besides, neutralizing, bad agencies in existence, which man was to ward off?

§. 12. *The Tree of the Knowledge of Good and Evil.*

There stand out prominently from amidst the innumerable trees which grew in the garden for the pleasure and good of man, two of special note, peculiar both in their kind and also in their design. They are the *tree of life*, in the midst of the garden, and the *tree of the knowledge of good and evil.*

Mysterious and inexplicable objects! Where shall the key be found to the secret mysteries which lie concealed under these names?

Thus much we know, however; the two trees formed a complete contrast with each other. One tree was *called*—and therefore *was*—a tree *of life.* But the other trees also were trees *of life*, in a certain sense. Their fruits, "pleasant to the eyes and good for food," were given to man as his sustenance; and ever as he partook, his physical system was refreshed, repaired, invigorated, strengthened. But *that mysterious tree* was alone, and in preference to all the rest, called a tree of life. The reception of its fruits rendered the continued and undisturbed life of the body absolutely certain. The fruits of the other trees, indeed, restored the worn and wasted powers of life—worn and wasted through the functions and processes of life themselves—but in so feeble and limited a degree as to fall short of preserving for ever the wholesome balance between waste and supply. That our apprehension of the tree of life is not an erroneous one, is proven by Gen. 3 : 22, where, after a judicial sentence appointing death as the unhappy lot of man, all approach to the tree of life is prohibited, "lest he put forth his hand, and take also of the tree of life, and eat, and *live forever.*"

The *tree of the knowledge of good and evil* was wholly different in kind, nature, and design; in all these features, the direct opposite of the tree of life. It was not, indeed, expressly *called* a tree *of death,* but none the less on this account was it such, or at least, capable of becoming such. For thus runs the command of God: "Thou shalt not eat of it; for in the day that thou eatest thereof *thou shalt surely die.*" Still, however, God had planted it in the garden, just as the other trees.

But it was called the tree *of the knowledge of good* and *evil.* Thus was the tree characterized as one by means of which man should attain to the *knowledge* of good and evil—but also as a tree by means of which it was *to be known,*whether man would choose the good—to serve God; or determine himself to and prefer evil—opposition to God. The inability to understand good and evil, and make a distinction between them, is, according to Scripture[1] and experience, a predicate and characteristic of unsuspecting childhood and innocence—but of these in their early, undeveloped stages—which, indeed, favorably contrasts[2] with the consciousness of sin and guilt be-

[1] Deut. 1: 39; Jonas 4: 11; Is. 7: 15, 16. [2] Matt. 19: 14.

longing to mature and developed stages of life, in the present state of the world.

But in view of the original destiny of man, the perpetuation in Adam, or the race, of such a child-like ignorance and characterless innocence, would have been an incompleteness, desirable or allowable under no circumstances. Therefore, according to this view, the tree of knowledge was also a tree of blessings, as well as the tree of life. It was, also, just as the latter, a tree of life, of spiritual life. It was a tree of knowledge, by being the occasion of mental and spiritual activity in the soul of man. But the other manifested its true powers as a tree of blessings and of life, when its fruits were eaten and assimilated through the powers of organic life. It was a tree of life not only *by design*, but also in its own capacity, *by nature*. But the tree of knowledge, on the other hand, was a tree of life and blessings, only so long as man refrained from eating of its fruits. The moment he partook of this tree, it revealed its powers as a tree of death. So we see that it was *by design* only, a tree of life and blessings; but was *in its own proper nature* a tree of death and woe. It was a source of knowledge, so long as its fruits remained untasted; and this knowledge was life. It was also a source of knowledge after its fruit had been eaten; but this knowledge was death.

Man, in the capacity of a creature, could only attain to the knowledge of good and evil, through the fact and subsequent to the fact of its being discovered whether he himself should be holy or sinful, by Him who implanted those qualities in his nature and moral constitution, which made his continuance in *holiness* (for which he was destined and made capable,) or his revolt into *sin*, (which was rendered possible by his moral freedom), a matter of his proper choice and determination.[1] Consequently, we must also retain the second sense of the words which designate this tree; the tree of knowledge was also a *test* by which it was *to be known* whether man would choose good or evil.[2]

But does our understanding of the whole matter, as we have gathered it from the preceding, solve all questions, remove all difficulties, and fathom all mysteries relating to these significant trees? Far from it! Many, very many which crowd upon the thoughtful and inquiring mind, and such, too, as are of no slight import, still remain unfathomed, unsolved. Question on question might here arise, until language itself should fail of terms in which to frame our inquiries. But the sacred account passes by, in all sublime, holy, and child-like simplicity, and unbiassed freedom, the host of questions to which the reflecting and over-curious mind gives rise, just as a child, undisturbed and uninfluenced by the problems of the world and of life which surround it, passes on in its innocent course, as though they were not anywhere to be found, or ever to be grappled with.

---

[1] 1 Cor. 13 : 12.

[2] It was part of the insidious wiles of the tempter that he wholly ignored and obliterated this important and chief sense of the words designating this tree; and, on the contrary, brought forward the other as the only one (Gen. 3 : 5)— for thus only was it possible for him so to magnify the truth of the sense he alone presented, and distort the full sense, from exhibiting it in its one-sided aspect, as to involve a satanic lie.

It here becomes us to lay our hands upon our mouths, and console ourselves with the old proverb:

> "Nescire velle, quæ magister maximus
> Docere non vult, erudita inscitia est."

But, nevertheless, we may still hope that later stages of revelation will lift the impenetrable veil which conceals these mysterious secrets, which enshroud the cradle of the human race—at least we may be confident in the assurance that hereafter, when faith is merged into sight and fragmentary knowledge for ever done away with, along with all the depths of Divine wisdom and grace, these mysteries also shall be fully disclosed to our minds.

What we can satisfactorily gather from the sacred record, is substantially as follows: the tree of knowledge was designed to furnish the occasion and opportunity for the self-determination and decision of man, either for or against the will of God, and which pertained to him and was absolutely indispensable to him as a free, personal being. The tree of life would, in all probability, have completely realized its destiny, only when man had chosen for himself that destiny which God originally appointed for him.

§ 13. *The Formation of Woman.*

Thus was man, at least objectively, prepared to take the decisive step by which he was to pass from a state of child-like, immediate, dependent life, to a knowledge of himself, of the world, and of God: from ignorance of, to the knowledge of good and evil; from a state wherein it was possible to decide either in favor of sin or holiness, to the realization of one or other of these conditions. This was to be the first step in that history which he *himself*, in the capacity of a free person, was to bring about.

But there was still one development wanting, which man was now to experience. This, as he was a free being, indeed lay within the compass of his desires and wishes, but not within his power to effect, since he was a mere creature himself—it lay within creative power alone. It was the creation of woman out of the substance of the first man. This act first introduced the characteristic of sex into human nature. The human being first created was neither man nor woman, still less a compound of the two. That being was just what the person of the resurrection shall be[1]—without sex. But, after the creation of woman, that first human being was thenceforth man—the woman was taken from the man, not the man from the woman.

The cardinal point in the Divine plan with regard to man, was clearly this, that the whole human race, in sorrow and in joy, amidst cursings and blessings, in its undeveloped, as well as in its developed stages, should constitute an organic, generic unity. Therefore was it necessary that man should be created as an individual unit, so that the collective race of man, to as great numbers as were demanded for the fulfilment of his mission upon earth, might proceed from this unit: so that, as says the apostle,[2] God might make "*of one blood* all nations of men for to dwell on all the face of the earth." Hence the necessity of deriving both sexes from the human

[1] Matt. 22 : 30. [2] Acts 17 : 26.

being *first* created. Not only was all humanity to spring from one pair of human beings, but also, in order that in every respect the unity might be preserved, woman was to proceed from man. But as man was created a *free* being, he could be the subject of no kind of development — not even that of having the characteristic of sex introduced into his nature — without his own choice and consent in the matter. It was necessary that he should desire, choose, and will this change. The review of the animals,[1] among which he observed the development that was wanting in himself, and at which time he looked around in vain amid them for a help-meet of his own kind (v. 20), awakened within him this desire. God graciously met his wishes, by taking from him a part of his body, and forming from thence woman. Adam immediately on seeing her said: "This is now bone of my bones, and flesh of my flesh. She shall be called woman, because she was taken out of man."

Upon this creative act of God rests the institution *of marriage,* with its blessing: "Be fruitful, and multiply, and replenish the earth, and subdue it." *Marriage* was the condition and potent beginning of all historical, free, personal developments of man. It was, therefore, necessary that *it* should precede man's free, moral self-determination, either *in accordance with* or *against* the will of God; for the latter conducted him into the sphere of actual history. The decision now about to be made, was to

---

[1] Gen. 2 : 20.

be the decision of the whole race; the triumph of one would be the triumph of all, but also, the fall of one the fall of all.

§ 14. *The Fall.*

All was now prepared for the trial of man — it could, however fraught with woes, be deferred *no longer*. But under the tree which was to be concerned in this melancholy trial, appeared suddenly and unexpectedly, another and a strange being (as a "deus ex machina"), in order also to sustain a rôle, and indeed no insignificant one, in the grand drama which was about to be enacted. It was the *serpent*, the most subtile beast of the field.

The tree of knowledge stood in the midst of the garden, (Chap. 3 : 3). Upon the one hand was the Divine command: "Thou shalt not eat of it," and the admonition: "In the day that thou eatest thereof thou shalt surely die." On the other hand were the allurements of the serpent, and his sadly significant promise: "In the day ye eat thereof, your eyes shall be opened; and ye shall be as gods, knowing good and evil." Between these two, stood man, a free being, endowed with the high prerogative of liberty of choice, with the power to withstand his *sore trial*, which under the circumstances amounted even to a *temptation* — but also left free to fall. He may, nay he ought to conquer; for God had in the creation given him power and ability for a triumph, and had, besides, expressly warned him against sin, and threatened him in the event of its commission. But it was also possible for him to disregard the voice of his Creator, which so graciously warned him and authoritatively threatened him; it was possible for him to fail of being true to the destiny which God had set before him, it was possible for him to choose contrary to the will of the Maker.

But man strangely suffered himself to be ensnared; he yielded where he should have triumphed, he became a slave where he should have been a victor and a conqueror. The tempter succeeded in implanting base and sinful desires in the soul of man—in breathing into him, as it were, another breath, derived from *beneath*, the opposite of that breath from *above* which was breathed into him at his creation. And now the solemn drama, upon whose issue hangs a whole world's history, hastens to its tragic end. The woman looks upon the tree, and sees that it is good for food, and that it is pleasant to the eyes, and a tree to be desired to make one wise. She takes of its fruit and eats; she gives thereof to her husband, and he eats also. "Then when lust hath conceived, it bringeth forth sin; and *sin*, when it is finished, bringeth forth *death*."[1]

God, who so lately condescended graciously to warn man against the commission of sin, has now become a requiting judge. A *curse* lights upon the *serpent:* to be cursed above all beasts of the field, trodden in the dust, hated of all creatures, and bruised by the seed of the woman — this is its well-merited lot. A *curse* lights upon the *woman:* in sorrow is she to bring forth children, her desire is to be unto her husband, and she is to be subject to him. A *curse* lights

[1] Jas. 1 : 15.

upon the *man:* in the sweat of his face is he to eat his bread, until he return to the earth from whence he was taken. Finally, a *curse* lights upon *nature*, in the midst of which man is to have his abode—a curse on man's account: thorns and thistles are to be brought forth by the ground.—Man is driven from the garden in Eden: cherubims with flaming swords cut off all approach to the tree of life,—lest man should put forth his hand, and take of its fruit, and eat, and live for ever.

The trial and decision of man was the offspring of necessity—but not his fall and rebellion. However, what had been a possibility had now become a reality. The deceptive promise of the serpent was fulfilled: man's eyes were opened (Chap. 3 : 7),—but he saw only his misery and nakedness. He was now brought to know good and evil; but with the painful consciousness of having trifled with and lost the one, and of being sunk into the depths of woe by the other. He had become as a god: he had boldly cast off all allegiance to the one God, and assumed sovereignty over himself. He had constituted himself a god, no longer the representative of God; he had become his own master, free as God—but this likeness to God brought not with it the happiness which pertains to the Divine Being, but was fraught with the deepest misery and woe.

Man, by yielding up his will to the will of the tempter, and opposing the will of his Maker, fell into *sin*, and also into *death*, the wages of sin. Whosoever committeth sin is the servant of sin—true freedom is to be found alone in communion with

God, the everlasting source and archetype of freedom. By means of man's freedom was it possible for him to choose sin, but in this very choice of sin he lost all freedom to escape from its power. By no possibility can man redeem himself.

Along with man, and on his account, nature also, in the midst of which he was to live and act, fell under the curse of sin, and the dominion of death.[1] The connection and the relation between spirit and nature, mind and matter, was the ready channel by which destruction and death were poured out over the material world, appointed as the dwelling-place of man.

By means of the unity of the human race, arising from the mode of its propagation, the whole race fell in and along with Adam; for he at this time was still the whole race. If the root was impregnated with poison, it was impossible but that all the boughs and branches of the tree which was to spring therefrom, should be pervaded with the same deadly qualities. All subsequent extensions and diffusions of the human race, could therefore but extend and diffuse sin, and death, the wages of sin, but never check or destroy them.

§ 15. *The Tempter.*

New problems and new mysteries are contained in that portion of the primeval history of the human race which we have just now surveyed. Mysterious and enigmatical was the nature and origin of *the serpent,* which there took so conspicuous a position in

[1] Gen. 3 : 17 seqq; Rom. 8 : 19 seqq.

the foreground of the history; mysterious its sudden appearance, its complete, its inveterate hostility toward God, its connection with and relation to that fatal tree, and no less mysterious the curse it bore off from the scene of action.

Was that, indeed, merely a common serpent, such as may at any time be met with in our fields and forests, and nothing more?

That it was a real serpent, the animal which we call by that name, cannot be doubted for a single moment. Its specific name, the other epithets by which it is designated, and the mode in which the curse pronounced upon it was to take effect, all force us to hold fast the opinion that it was a real serpent.

But was it nothing more? Should not the capacity and manner in which it now appeared, just at the critical moment, its consummate treachery, its well-concealed fraud, its well-applied tactics, point us to some fearful mystery, which for the present stage of revelation was still to be kept close? Should not all this lead us to conclude upon the existence of some personal, spiritual power, to whom it was of the last interest to disturb the designs and the work of God, and bring to naught the counsels of Divine love toward the human race? which made both the tree and the serpent the gladly-found instruments of its despicable designs?

The view imprinted upon the account, of the identity of the serpent and the spiritual agency connected with it—let the real connection between the two have been what it may—was natural enough, and to the mind of Adam, at least prior to the fall, wholly correct; for then his whole manner of thought and perception was direct, immediate, and unsupported by reflection. But immediately *after* the fall, when he had begun to learn about evil, reflection would begin to assert its prerogatives, and busy itself in trying to divine the connection between the outward manifestation and the hidden cause of evil. Thus even at that early time would it be discovered that there had been active in the serpent, or in connection with it, an *evil spiritual agency* or *being*. It must be remembered that very soon, in addition to the tradition of this affair of the serpent, as it appeared to the senses, there would be subjoined a traditional explanation of its nature, and its connection with an unseen, mysterious agency. But whilst both the fact and its attempted explanation were confounded and obscured in the traditions of the heathen nations, the author of Genesis took up the original tradition in its pure form, and without any attempt at unraveling its mysteries; perhaps, for this reason, as *Delitzsch*[1] supposes, that their disclosure would have

[1] The narrator satisfies himself with a statement merely of the outward occurrence, without lifting the veil from the secret; and this he could well do, as the traditions of the heathen themselves supplied more particular though distorted accounts of the matter. He kept the matter veiled because its explanation would not have been proper for the people of his age, so much inclined to heathenish superstition and intercourse with demons. It was from design and in view of the best interests of that age that the narrator remained silent about all but merely the fact as it occurred and seemed to the senses. It may be observed that the Pentateuch very rarely makes mention of demons in other places (Gen. 6: 2; Lev. 16; Deut. 32: 17).

been prejudicial to the interests of that age. "The history would be clear enough and significant enough to every discerning mind without any such explanations."

A *personal* being besides man—an evil being—was, therefore, on the scene of action previous to man's creation. Moreover, as God is so unequivocally called the Creator of the heavens and the earth, and all things therein, it is obvious beyond all dispute, that this spiritual, this personal power, was a *creature* of God; and further, since, according to all Scripture, only that which is holy and good can proceed from the hand of God, that this spiritual being was originally a *holy* one—but now *fallen* from its first estate and high destiny, and *become* evil and sinful, by the abuse of its personal freedom. It is equally obvious, as a necessary inference, that a history of vast power, and pregnant with the most fatal consequences, must have been enacted previous to the creation of man.

Clear and definite views, however, in regard to the origin, progress and end of this history, its mission, its design and its consequences, are not to be gathered from revelation—at least, from that portion of it to which our attention has thus far been confined. But further disclosures are made by subsequent revelations, the investigation of which will soon claim our attention.

§ 16. *Prospect of Redemption.*

The human race had now entered upon a new course of development, which would have hurried it on to the most irremediable destruction, so that it could never have returned and laid hold again on its original and high destiny, had God abandoned it to its own choosings, had he not taken the marred work again into his own hands, and brought about a new state of affairs.

But it was the *will* of God that we should be a new race in Christ: "*He hath chosen us in him, before the foundation of the world.*"[1]

The designs and plans of the tempter seemed completely successful. The promise: "ye shall be as Gods," was fulfilled"—in the deceitful sense intended. But the deceiver was caught in his own snare; he had derided man, the image of God, with the most malicious irony;—God now derides him in return, with the irony of a holy and avenging judge.[2] The tempter unconsciously foretold his own judgment and sentence, in those jeering, equivocal words. For God had, in prospect of the fall, laid the plan of redemption before the foundation of the world; and from this plan, which began to be developed in history immediately after the fall, those words derive a third and a deep sense, which never struck the tempter's mind. Redemption was provided in consequence of the fall. In effecting it God became as man, in order that man, truly and in the proper sense of the words, might become as God.

Man, though fallen, was indeed still capable of being redeemed. He did not engender evil within himself of his own accord; nay, rather, it was forced upon him from without, but still, by a power which

---

[1] Eph. 1: 4. [2] Compare Ps. 2: 4.

he could and should have withstood. His whole being, the whole intricate web of his life, was pervaded and poisoned by sin. But sin was still something foreign to it. His very being had not itself become identical with sin. For there was something still remaining in him, and there still remains in all his descendants something which reacts against sin, opposes it, and finds *no* pleasure in its commission;[1] it rather reproves and chastises the perpetrator on account of his sins. And in spite of all want of delight in God and in his service, which discovers itself in the heart of fallen man, there still dwells there an earnest longing after something of a higher and holier nature, something invisible—a longing which the things of this world can never satisfy. Both his accusing conscience and his longing after communion with God, proceed from the Divine image within him. For this Divine image, however much it has been impaired, clouded and darkened by sin, has not been wholly obliterated and destroyed;[2] and man still continues, notwithstanding the fall, the offspring of God.[3] So long as the faintest spark of the heavenly fire still remains amid the ruins of sin, it may, under proper treatment and with the timely supply of aid, be again fanned into a glorious and heavenly flame.

That voice of longing, those fond hopes of restoration and redemption, are heard, like the echoes of the longing and groaning of the human race, throughout the whole creation which fell through and along with man. For the earnest expectation of the creature

[1] Rom. 7: 15, 16. [2] Gen. 9: 6; Jas. 3: 9. [3] Acts 17: 28.

chimes in with our longings; the whole creation groaneth, and travaileth in pain together until now.[1]

In consequence of the Divine decree, concluded in eternity, and resting on the grace of a merciful God, as well as on man's need and capability of being redeemed, that salvation so long in waiting, began now to be manifested, and to enter into history as the spring of its movements and the regulator of its developments.

But man still retained, even after the fall, his freedom of choice. And as he had freely taken upon himself the commission and guilt of sin, it was also necessary that he should now freely appropriate to himself the offered salvation. As sin was not irresistibly forced upon him, so neither was it fitting that salvation should be. It was possible for him to reject the offered boon, and to persist in that unnatural, perverted course of development, he had so unhappily entered upon, and which would conduct him to final, to irretrievable destruction, as its natural and unavoidable goal. His first decision, as he stood beneath the tree of knowledge, was not an absolutely final one; since, before such an one could be made, it were necessary that the object chosen be fully understood, in all its relations; and no less, that the subject choosing have all his faculties and powers fully developed. These were confessedly both wanting in the case of Adam. The degeneracy of the whole man, which was introduced by the fall, was not, indeed, absolute, hopeless; since he was still susceptible of being regenerated by the power of

---

[1] Rom. 8: 19–22.

God, on the principle of previous incomplete knowledge and imperfect development. But the second decision of man, which is rendered necessary by the offer of salvation, becomes an absolute, a final one, since the above principle will no longer apply—since the restrictions of the first decision are now entirely removed. Faith, which eagerly lays hold of the offered salvation, and unbelief, which persistently rejects it, stand, respectively, at the entrance of the two ever-diverging paths of this final decision.

The mercy of God, who would prepare man for redemption, was abundantly involved in the sentence of punishment which the judicial severity of God pronounced upon him.[1] For all the curses and punishments there inflicted upon him, include, also, benefits and blessings. Though the *woman* was to bring forth children *in sorrow*, still, she was to *bring forth;* and Adam seems to have had some intimation of the blessing involved in this curse, for he, in reference to it, and very significantly too, called his wife, *Eve*—the mother of all living. The former blessing: "Be fruitful and multiply, and replenish the earth, and subdue it," reappears in this curse, with the prospect of its being fully realized, despite the pervading influence of sin. The possibility of salvation depended upon the circumstance of there being evolved from the first man, who potentially contained the whole race, a human race closely and essentially bound together by unity of blood; for redemption was to be brought about, by the Redeemer's taking upon himself human flesh and blood.

[1] Gen. 3 : 16–19.

Had God in righteous judgment recalled the blessing, that "man should increase and multiply," had man remained in his undeveloped unity, he could not have been redeemed.

To labor in the sweat of his face, which was assigned to *the man* as his special lot, was a palliative and an antidote against the power of sin. Thus, too, even his expulsion from Paradise, "lest he eat of the tree of life, and live for ever," and *death* itself, involved both a penalty and a gracious gift. For had man eaten of that tree, his life upon earth, loaded with curses as it *now* was, with miseries and corruptions, would have become eternal, and all possibility of his becoming released from the consequences of sin, would have been for ever set aside.[1] Bodily *death*, on the contrary, which without the intervention of redemption would have been *but* a curse and eternal ruin, now becomes, through that redemption, an everlasting and invaluable blessing. For sinful man attains to the resurrection, through death alone: his body is "raised in incorruption," only on condition of its having previously been "sown in corruption."

---

[1] *Delitzsch* is of the same opinion on this point. He beautifully and appropriately remarks (Gen. 144): "This tree had doubtless the power to completely counteract the mortality (the 'posse mori') of man, and to advance and gradually bring into a most glorious state his corporeal nature. To have eaten of its fruit now, would have established him for ever in his present condition of sad connection with sin, both spiritually and corporeally, and produced, as *Drechsler* very properly remarks, a change in his physical nature, corresponding to the state of his soul, gradually transforming it into an infernal body, the horrible caricature of the glorified body.

The first *express* announcement of a coming salvation, upon which faith might already lay hold, and unbelief destroy itself, was furnished by the curse pronounced upon the tempted:[1] "Cursed art thou above all cattle, and every beast of the field: upon thy belly shalt thou go, and dust shalt thou eat all the days of thy life: and I will put enmity between thee and the woman, and between thy seed and her seed: it shall bruise thy head, and thou shalt bruise his heel."

This curse pronounced upon the serpent has, so far as it contains a gracious promise to man, been very properly styled the proto-evangelium—the first announcement of salvation.

The Biblical account represents the recollections and views of the first pair, preserved by sacred tradition as the venerable relics of a primitive age, in their original character and marked objectivity. The Protoplasts, however, regarded the subtile beast of the field, and the personal, spiritual tempter—whatever connection really subsisted between the two—as strictly identical. The identity of the two seems to be as unmistakable in the *curse* now pronounced upon the serpent, as it was before in the visible appearance and crafty wiles of that animal. The curse, the *whole curse*[2] is *formally* pronounced upon the serpent, *singly and alone.* But the curse was pronounced

[1] Gen. 2: 13–15.

[2] It is but an arbitrary assertion, justifiable in no possible way, to say that the first part of the curse refers to the serpent itself, as the instrument of the temptation, and the second to the devil as the personal agent of it.

for *man's sake* alone, not for the sake of the serpent; it was, accordingly, adapted to the views of *man*, who did not yet discriminate between the visible appearance and the spiritual agency engaged in the temptation. To man, the tempter appeared *as a serpent;* in his view, accordingly, the curse which was directed against *the serpent*, really was a curse pronounced upon the first *author* of sin; and the prospective defeat and destruction of the *serpent*, through the seed of the woman, was regarded as a deliverance from the power and influence of the *author* of sin.

A gracious promise following immediately in the footsteps of the first sin! The Divine Nemesis judging the betrayer through the betrayed, conquering the victor through the vanquished! Divine compassion hastening to pour the healing balm into the fresh and bleeding wound!

Man was not subjected by the fall alone, without any further straying from the path of obedience, wholly to the will of Satan, in servitude and obedience. While sin implanted in him a principle of *opposition to God*, he still retained, ever since his creation, a principle of *opposition to the tempter* also. God assigns to the latter (this is obviously the meaning of the first promise) the victory over the former. Although man had permitted himself to be seduced into a union with Satan, that union was not to be permanent. Not friendship and union were to exist between the two, as the issue of the first decision might lead us to expect; but rather, through the Divine interposition and aid, enmity and continued warfare, which were ultimately to terminate in the complete defeat of the tempter. Eve, the mother of all living, was to bring forth children, and the seed of the woman was to bruise the head of the serpent; i. e. the human race, *as a whole*, was to maintain a contest with the author of sin, and destroy the kingdom which he had established.

The propagation of sin is inseparably connected with the mysterious propagation of the human race—"for that which is born of the flesh is flesh." But the same mystery of generation and birth is also the vehicle and medium of salvation—"for that which is born of the Spirit is spirit."[1]

But man can receive nothing, except it be given him from above.[2] After he had become flesh, through the commission of sin, it was no more possible that the Spirit should be born OF the flesh. Consequently it was necessary that the Spirit should first be born INTO the flesh, in order that it might then unfold itself naturally, and according to the laws of its own generation and propagation. But this birth of the Spirit could be effected only by an act on the part of God—such an act as the implantation of moral and intellectual faculties in man at the time of his first creation. There was then breathed into human nature, a breath of the Divine life, a transcript of his own Being. But there is here wanting something of a still higher and better nature. It becomes necessary that the Divine Being himself, the fulness of the Godhead, should condescend to take upon himself our nature, that he might raise us from the depths of the fall, to our original and high

[1] Jno. 3 : 6. [2] Jno. 3 : 27.

destiny, and conduct us to the goal of our development as it was appointed from eternity. All human destinies or ends, however, depend upon the race being unfolded from unity to plurality; and the unity of the race, in spite of its numbers, is a cardinal principle in God's dealings with it. As, therefore, sin passed from one man over the whole race, so in like manner must salvation be derived and applied from the one to the many. "Therefore as by the offence of one, judgment came upon all men to condemnation; even so by the righteousness of one, the free gift came upon all men to justification of life." "For, as by one man's disobedience many were made sinners; so by the obedience of one shall many be made righteous.[1]

It was necessary that the *new* development, with all its supernatural, its Divine, life-giving powers, should commence at a point in the old, the natural development, specially prepared and adapted for its reception; that it might thence extend itself, through *spiritual* generation and the new birth, over the whole human race. When this point in the old development was reached, when all was prepared, then was it said: "The Holy Ghost shall come upon thee, and the power of the Highest shall overshadow thee: therefore also that holy thing which shall be born of thee, shall be called the Son of God."[2]

From that promise which gave to the "seed of the woman" the final and complete victory over the seed of the serpent, there is carefully preserved to us by sacred history, an uninterrupted series of generations

[1] Rom. 5 : 18, 19. [2] Luke 1 : 35.

of men, stretching through all history down to the time of Christ. These were all characterized by the presence of prophetic powers, and in turn possessed and transmitted this significant promise as a precious token of the Divine good will. They closed in the birth of the second Adam, in whom all promises are fulfilled. He, as the second head of the human race, was to take up again that development which had been marred and interrupted by the fall, and conduct it to its ultimate completion. He also, in the same capacity, was to be the chosen captain and leader of hosts in the contest between the seed of the woman and the seed of the serpent — it was through his Divine power also that the great, the final victory was to be gained.

Thus has this significant promise placed before both the betrayer and the betrayed, a long and severe conflict—one that is to extend throughout the whole history of the world, but whose last, whose decisive issue, in spite of all the varied phases it may assume during its progress, is not left in the least doubt. But we must gain clearer views of the enemy against whom this war is to be waged, before we proceed to contemplate the contest itself, its varied phases and its final issue. We are pressed by every consideration to commence our inquiries at once, just at the point which we have now reached, and they may not well be deferred longer. We shall have to pass over considerable ground in carrying out our design, but the reader will bear patiently with us, as these preliminary researches[1] are not alone important in con-

[1] [The reader will observe that from the close of this section to

nection with the special point before us, but are of equal weight in connection with the leading objects of the whole treatise.

§ 17. *The Morning Stars and the Sons of God.*

In addition to the Hexæmeron in the 1st Chapter of Genesis, and the celebration of the creation contained in the 104th Psalm, the Book of Job furnishes us with a wholly independent description of several points in the process of creation.[1]

God there speaks thus to Job:
"Gird up now thy loins like a man;
For I will demand of thee, and answer thou me.
Where wast thou when I laid the foundation of the earth?
Declare,if thou hast understanding.
Who laid the measures thereof, if thou knowest?
Or who hath stretched the line upon it?
Whereupon are the foundations thereof fastened?
Or who laid the corner-stone thereof,
When the morning stars sang together,
And all the sons of God shouted for joy!
Or who shut up the sea with doors,
When it brake forth, as if it had issued out of the womb!

---

the commencement of the 29th, the author is engaged in the consideration of the angels, and other matters pertaining to the extramundane relations of the world's history. At the latter point mentioned (§ 29), he again takes up the history of the contest between light and darkness, having gained the information necessary to a more complete understanding of this contest.—Tr.]

[1] Job 38 : 3 seqq.

When I made the cloud the garment thereof,
And thick darkness a swaddling band for it,
And brake up for it my decreed place,
And set bars and doors,
And said, Hitherto shalt thou come, but no further;
And here shall thy proud waves be stayed!"

It will be perceived that this description coincides in several points with the Mosaic history of the creation:—with regard to the founding of the earth, the origination of the atmosphere, and the bounding of the seas—all of them there referred to the second and third days' work. But we are also favored with something entirely new, and peculiar to this poetical picture: When God laid the foundations of the earth, the *morning stars* rejoiced together, and the *sons of God* sang their songs of praise to the works of an all-wise and almighty Creator. The morning stars and the sons of God were hence in existence *before* the foundations of the earth were laid. They existed—taking this description in connection with the Mosaic—*previous* to the six days' work.

But what were these *morning stars?* and who were the *sons of God?*

The *morning stars* were, doubtless, the stellar worlds, those glorious spheres of light which ever spangle the vault of heaven. They were called *morning stars*—not *evening stars*—because, in the mind of the prophet, it was morning when God began to lay the foundations of the earth.[1] The

[1] Comp. *Schlottmann:* "By a most beautiful figure the stars are all here called, in respect to the great morning of creation, morning stars." So also *A. Hahn* and others.

voice of rejoicing and exultation with which they celebrated the dawning of creation's first morn, was none other than that silent but eloquent language in which they still declare the glory of God, as sings the Psalmist:[1]

"The heavens declare the glory of God;
And the firmament showeth his handy work.
Day unto day uttereth speech,
And night unto night showeth knowledge.
There is no speech nor language,
Where their voice is not heard.
Their line is gone out through all the earth,
And their words to the end of the world."

We here, as it would appear, meet with a contradiction to the Mosaic account. For while, according to the Hexæmeron, the sun, moon and stars were first placed in the firmament on the fourth day, subsequently to the formation of the atmosphere, the land, and the sea, the Book of Job represents the starry heavens, with all their magnificence and glory, as already in existence when the foundations of the earth were laid—as admiring witnesses of the creative process.[2]

---

[1] Ps. 19: 1 seqq.

[2] *Delitzsch* (p. 73), as also Hofmann (p. 352), is unwilling to recognize any force in the passage from Job, 38: 7, as applied above. "There we behold," says *Delitzsch*, "the accidental poetical connection of the great facts of the creation, which are described by the Mosaic record in their chronological order. The choral songs of the angels and the harmony of the spheres, refer prospectively and retrospectively to the earth, then conceived to

But we have already learned (§ 8) that the fourth day's work was not concerned in the creation and appointment of the stars to be what they are *in themselves*, independent of all connection with the earth, but only in fixing these bodies *in their relation to the earth*—we are told when and how this relation was established. The question touching their first, their real origin, did not there engage the attention of the prophet; consequently we have nothing decisive upon that point. If it be true that we are here to understand that the stars were in existence before the creation of the earth, the discrepancy between the words of the Book of Job and the Mosaic account, is not to be sought in an irreconcilable contradiction

---

be in the act of coming into being, to whatever period of the work of the creation the origin of the angels and the stars must be referred." We willingly allow that there was no special reason why the poet should here strictly observe the chronological order of the different points in the process of creation, and note the particular days to which they severally belonged; nor should we be disconcerted were it even shown (which, however, it cannot be) that there is in this mention of the creation some *inversion* of the regular order. But this is not the question. The passage before us does not in general refer to the creation of the angels and morning stars at all. But of this we are fully assured: *that they were present* when God laid the foundations of the earth and gave to the seas their bounds. And it is from *this* very circumstance that they so oppositely contrast with man, who *was not present* when this took place. "*Where wast thou,*" says God, "when I laid the foundations of the earth, when the morning stars sang together, and all the sons of God shouted for joy?" The antithesis here involved forces us to take the "when" strictly, and to regard a recourse to anthems of praise sung retrospectively, as the offspring of arbitrary interpretation, and a sorry expedient at best.

between the facts upon which the respective accounts are founded, but in a varying apprehension of these facts, arising from their being regarded from different points of view, and which may easily be explained.

Consequently, as we are forced to believe that the stars were created *before* the earth, by the above lines from the Book of Job, and as the Mosaic cosmogony permits us to hold such a view, it must be granted that, according to the Biblical theory of the world, the stars *were* indeed created *before* the earth.

The phrase, "*the sons of God,*" is no less clear and unequivocal than the above expression, "the morning stars." These "sons of God" were doubtless *the angels,* those holy beings which ever surrounded the throne of God, in readiness to execute his commands.[1] They are called angels, from their serving in the capacity of messengers and servants of God; this name is derived from their calling, from the offices they fill—it is their official name. They are called *sons of God* in respect to their nature and being. In contrast to the weak, sinful children of men, the inhabitants of the earth, they are designated by this name, as the high and holy inhabitants of heaven, who sustain and reflect His majesty and glory.[2]

---

[1] Job 1 : 6; 2 : 1; Ps. 29 : 1 [in the original]; 89 : 7; 103 : 21, etc.

[2] It is to be observed here, that the angels are ever called the children or the sons of *God* (Bne Elohim), but never the sons of *Jehovah.* The name *Elohim* designates the Divine Being as the fountain of all life and power, of all majesty, glory, holiness, and blessedness; *Jehovah,* on the other hand, represents Him as the gracious and merciful God, the Redeemer and Saviour, who de-

§ 18. *Spirituality and Corporeality of the Angels.*

Let us now pass immediately to what may be gathered from the Scriptures concerning ***the nature, the position, the mission***, and ***the history of the angels.***

The angels are SPIRITS.[1] This term expresses, first, something ***positive***, and second, something ***negative*** concerning the nature or being of the angels.

The idea of ***spirituality*** is the ***positive*** phase of this term. According to it, the angels are free personalities, endowed with self-consciousness, in opposition to the mere offspring of ***nature***, incapable of freedom and without personality. If, in accordance with the principles of division which universally obtain, we divide all created things into ***spirit*** and ***nature***, we shall have no difficulty in determining to which of the two spheres the angels belong.

The whole Biblical view respecting these beings conforms to this designation of them as ***spirits***, from the most essential peculiarities of their being. They never appear as mere forces of nature, or as unconscious, cosmical life-potencies, although they are, indeed, often revealed as media or bearers of the same.[2] No, they ever appear as free beings, endowed with

nied himself to save man from destruction, and exalt him to partake of the glory of His heavenly abode. (Comp., for further particulars, my work: *Die Einheit der Genesis*, Berlin, 1846, p. 43–53). The sons of *Elohim* are therefore the media and bearers of the divine might and glory: the sons of *Jehovah*, on the contrary, the media and bearers of his *redeeming* grace. In this latter sense Israel is called the first-born son of the Lord (Jehovah), (Ex. 4 : 22).

[1] *πνεύματα*, Heb. 1 : 14.

[2] Compare John 5 : 4.

consciousness and possessed of an independent spiritual existence, whose will is never constrained to accord with the will and designs of their Creator, but is left to choose and decide for itself.

It is the high prerogative of the created spirit to determine itself, and freely enter any course of development it takes. For this reason it was also impossible that the angels should have been placed immediately by the creation, in the highest and most complete stage of perfection which they were capable of reaching, and for which they were designed by their Creator. Nay, rather, they were to reach that advanced goal, through their own strivings and efforts, through the employment of their own highest powers. But they possessed potentially and in the germ, that high heritage to which they were to attain. God ever and always *gives* before he *requires again*, and his demands are ever measured according to his gifts. In harmony with this principle, the capacities with which he endowed the angels, were fully equal to the task of their fulfilling their mission, and being true to their original *destiny*. The freedom of will by which they were to decide for themselves, and enter upon a course of development of their own choosing, involved also the *possibility* of their choosing contrary to the will of God, of their entering upon a course of development other than, and antagonistic to that which had been originally set before them, and leading to wholly different results. This was the possibility of their revolting from their high destiny, of their rebelling against their Creator and Lord; and it was involved in their moral freedom, which was at first of a merely formal character, and not possessed of those characteristics which would be attached to it, so soon as they had realized a condition the offspring of their own choice.

The *negative* phase of the term "*spirits*," by which the angels in general are designated, does not force us to deny all idea of body (σῶμα) in connection with the angels, for there are also spiritual bodies;[1] but merely the idea of a body other than spiritual—a fleshy body, compounded of earthy materials (σῶμα ψυχικόν, σάρξ). "It excludes"—to use the words of an esteemed divine[2]—"it excludes all idea of a life connected with flesh and blood derived from earthly materials, all idea of a form of life holding the same confined relations to place and space as does our gross organism, all idea of dependence upon conditions of life and laws of movement such as we have to do with, without at all denying that the angels have proper bodies, and an outward life conformable to the nature of those bodies. For the Scriptures reveal to us a sphere of corporeal life, in addition to and beyond our own *as it at present subsists*, and which, just as our present life, with its "tabernacle of clay," its gross earthy character, corresponds to our terrestrial system,[3] in like manner, as a faithful transcript of the celestial systems, is adapted to the nature of a pure spirit (πνεῦμα), just as, further, our present body is adapted to the nature of a mere ψυχη.

Angels may be called "πνεύματα," pure spirits, but

[1] σώματα πνευματικά, 1 Cor. 15 : 44.

[2] T. Beck, *Christl. Lehrwissenschaft,* I., 176 seqq.

[3] 1 Cor. 15 : 45 seqq.

not *men*. For the angels possess nothing of a character other than spiritual: their corporeality also is of a spiritual nature, and even their bodily constitution bespeaks the spirit. The corporeality of man, on the other hand, partakes not of a spiritual but of a fleshly character: the dualism of flesh and spirit has not yet in his case been done away with, by his fleshly body having been glorified and transformed into a spiritual body. So long as this dualism still remains, man cannot be called a pure spirit, a *spirit*, without any qualification.

The Bible, indeed, does not expressly treat of the corporeality of the angels. This subject, however, did not fall within the sphere of its objects — nor could it. But, on the contrary, it cannot be denied that the Biblical doctrine of the angels presupposes the fact of their having corporeal forms, and that it furnishes us with many hints from which inferences may be drawn concerning the nature and the constitution of their bodies.

The words of Christ in Matthew 22 : 30, are specially clear on this point.[1] In Matthew it is said: "For in the resurrection they (men) neither marry, nor are given in marriage; but are as the angels of God in heaven." Luke adds: "Neither can they die any more; for they are equal unto the angels; and are the children of God, *being the children of the resurrection*." The unprejudiced reader cannot but believe that corporeality is here indirectly predicated of the angels, who in *one* respect at least, are analogous to what the resurrection bodies of our race

[1] Compare Luke 20 : 35, 36.

shall be—devoid of the characteristic of sex.[1] This is a bodily characteristic, and exerts, indeed, during this mortal life, much influence upon the physical constitution and proportions. But when the body dies, it is immediately done away with. It is *not* in the *incorporeal state*, however, that men are to be like angels. They shall only attain that high honor, *through the resurrection*, when they shall be clothed in new and glorified bodies, which shall indeed possess the characteristics of corporeality, but not those of sexuality, (somewhat as the first created human being was neither man nor woman previous to the creation of Eve). That Christ referred only to a bodily likeness, when he spoke of that high honor which is to be conferred on the children of the resurrection—of being as the angels of God—is placed in the strongest and most unequivocal light, by the closing sentence, "because they are the children of the resurrection;" for the resurrection is not to be concerned in a changing or transformation of the spiritual being, but merely in the reformation or renewal of the body. This remarkable and weighty passage is of still more significance in view of our design, when combined with the Pauline doctrine concerning *spiritual* bodies (σώματα πνευματικά), when speaking of the resurrection.[2] No more shall the resurrection body possess the characteristic of sex, than the spirit (pneuma), to which it is to be *com-*

[1] Compare Meyer on Matt. 22 : 30: Besides it is *clear* from *this passage*, where the resemblance of the angels to the future resurrection body is referred to, that we are not to regard the angels as pure spirits, but as possessing extramundane bodies.

[2] 1 Cor. 15.

pletely conformed. Hence it is that it is called a spiritual (pneumatic) body. It must be assumed, therefore, that *the angels* possess (however different they may be from our resurrection bodies, in some respects) real, spiritual (*pneumatic*) bodies. Thus also have we obtained a proof that the term *spirits* (πνεύματα), by which the angels are designated, by no means excludes the idea of their corporeal nature. It merely excludes the idea of a *fleshly* body, not of a *spiritual* one however.

Supported by the clear, unequivocal import of these words from Matthew and Luke, we can scarcely be in doubt as to the true interpretation of Paul, in 1 Cor. 15 : 40. No one will dispute with us the assumption that Paul was acquainted with the words spoken by our Lord, and that they may have been before his mind, when stating the doctrine of the resurrection at such length and so clearly as is done in the 15th chapter of 1st Corinthians.

The Apostle's course of reasoning in the place in question, is as follows: The Christian doctrine of the resurrection seems to involve an absurdity (verse 35), which must be explained. This the Apostle does, by the instance of a grain of wheat cast into the earth; the grain itself must become corrupt and must decay, before it takes on a new and more glorious form, in the plant to which it gives rise (verses 36 : 37). Next, in order to show us the resurrection body in its two aspects—that it shall be as truly a *body* as our present ones are; but that it shall be a body of *another kind* — he calls our attention to the great and essential differences between the bodily forms to be met with throughout creation. He first mentions (v. 39) different kinds of the terrestrial body (σάρξ): such as the *flesh* of men, the *flesh* of beasts, of fish, and of birds. Thus even upon the earth, we find (amid bodies composed of flesh), those possessed of the most varied characteristics. But still more broad and significant are the distinctions, when we compare these terrestrial bodies with those which do not partake of a fleshly nature — with celestial bodies. "There are also *celestial bodies*, and *bodies terrestrial* (σῶματα ἐπουράνια and επίγεια); but the glory of the celestial is one, and the glory of the terrestrial is another" (verse 40). As the *terrestrial* bodies are evidently those specified in the preceding verse — the bodies of *men* and *other creatures inhabiting the earth*—we naturally and for the best of reasons, refer the expression, "*celestial* bodies," to the inhabitants of heaven, as proving that they do indeed possess corporeal forms. We are forced to this view, however, from their being designated as *bodies* (as σῶματα); for this remarkable word, ever and without exception, not only in the New Testament, but throughout all the Greek classics, designates *only* organic (living) bodies, but *never* inorganic (dead) bodies, that is, bodies in the modern scientific sense of the term. Consequently, verse 40 cannot be explained by the succeeding verse, in which the Apostle, passing into another sphere of the analogy, speaks of there being one glory of the sun, another of the moon, and another of the stars (but does not call these bodies "σώματα"). Nay rather, it must be explained by the verse which precedes it, and in the manner indicated above. And just at that point, as we believe, the words of our Lord touching our future bodily likeness to the angels, may have passed through the mind of the Apostle.

We may think strange of but one point, if the above be the true interpretation of the Apostle's words; and that is this, that he did not distinctly and plainly say, *angelic* bodies, instead of "celestial bodies," as that would have excluded all possibility of mistaking his meaning. But that expression would not have been comprehensive enough for the Apostle's design. Heaven contains other bodies than merely those of the angels. In verses 45–49, Christ is called "the heavenly" (ἐπουράνιος); *his* body is both a "spiritual" and a "celestial body;" and this body of our *Lord*, the direct type of our resurrection bodies, is doubtless comprehended within the scope of the 40th verse.[1]

[1] I must confess that *Hofmann's* opposition to my view (Schriftbeweis, I., 353) caused me to hold it with less assurance for a while. But, after renewed examination, I have again become convinced of its correctness, and find myself forced to reject *Hofmann's* interpretation of the phrase, "celestial bodies," in the 40th verse, by referring it to the sun, moon, and stars, of the succeeding, or 41st verse. The term "σώματα," to my mind, is now as strong and decisive as ever in favor of my view. *Hofmann* asserts, indeed, that σῶμα does not *by any means* designate an organic body, but is antithetically opposed to πνεῦμα. The proof, which I hold to be impossible, rests with him. That σῶμα is not so opposed to πνεῦμα, is evident from the expression "σῶμα πνευματικόν," which would otherwise involve a "contradictio in adjecto." (Σάρξ and πνεῦμα are directly opposed to each other, and a σὰρξ πνευματὶκή would be simply a contradiction in terms). *Meyer*

If this interpretation be received as the correct one, we may find in the same chapter, a hint respecting the constitution of the bodies of the angels. The bodies of men and beasts form the same contrast in connection with the bodies of the angels, as that which obtains between heaven and earth. It is certainly a natural inference to regard the bodies of angels as compounded of celestial material, just as the bodies of men partake of the nature and character of earthly matter (to which the Apostle in verse 47 expressly refers); for the former are called "celestial bodies" in the very same sense as the latter are called "terrestrial bodies." And as, according to the teachings of Scripture, a higher degree of purity and perfection of matter, of splendor

very correctly remarks upon this passage: "Were we, in harmony with the prevailing view, to understand the Apostle to mean the celestial *bodies* (*worlds*), we must attribute to him either our modern scientific mode of speech, or the view that the stars are living beings." (*Hofmann* himself (p. 352) opposes any such idea as the latter, and not without reason). No Grecian philosopher, to say nothing of the Apostle, who adhered to the ordinary modes of speech, would have called the celestial spheres "σώματα." The modern scientific term "bodies" was wholly unknown to the ancient Greeks and Latins. The fact that the term "σῶμα" is applied in the case of plants, in the 37th verse, proves nothing, for plants are also *organic* bodies. I still contend that the "σώματα ἐπείγεια" are the bodies of the *inhabitants of the earth*, (v. 39: men, beasts, fish, birds). The plants, although they are called σωματα in the 37th verse, do not any longer come into consideration. The Apostle mentions two departments of terrestrial σώμάτα, v. 37th: plants, which are σῶμα, but not σάρξ, and v. 39, 40: men and beasts, which are both σῶμα and σάρξ. The transition to a new department, in the σώματα ἐπιγεια, caused him to drop the plants.

and of glory, must be attributed to heaven, than to the earth in its present condition; so also must we regard the "celestial" bodies of angels, as possessing a more refined, ethereal and glorious character, than the "terrestrial" bodies of men.[1] The power of the angels, moreover, as referred to in 2d Pet. 2 : 11, and which doubtless is manifested in connection with a corresponding physical or corporeal constitution, is represented as by far more mighty and influential than that of men.

Were Luther's translation of Psalm 104, 4 ("He maketh his angels winds, and his ministers flames of fire") unqualifiedly correct, we might infer from this comparison of the angels to the winds and to lightning, that the bodies of angels possess all the qualities of lightness and velocity of movement, all that pervading energy and wondrous manifestation of power, which characterize these forces of nature. The translation is grammatically correct; but it seems more in harmony with the connection and the sequence of thought contained in the Psalm, to translate: "He maketh the winds his angels (or messengers), and the fiery flames his ministers." According to this rendering, the Psalmist is not speaking primarily of the angels at all; but of the winds and flames of fire. Still, however, the angels

---

[1] The remarks of J. P. Lange (in his fine treatise: *Die Lehre von der Auferstehung des Fleishes*, in vol. 2, and of his *Vermisch. Schr.*), concerning the law of the embodiment of all finite beings from the material of the *place* where they dwell, and according to the *state* of their moral being, may serve to set this matter in a clearer light.

who are properly the ministers and messengers of God, as the winds and the lightning are less properly called by Luther — the angels, we say, are placed in such relation to these forces of nature, by this second rendering even, as can only be explained by granting that there does exist a resemblance between them, as to their outward appearance, and the manifestations of power they effect through the medium of their physical constitution. And when we observe, on the other hand, that the author of the Epistle to the Hebrews has appropriated to his purpose[1] this verse of the 104th Psalm, giving it the same sense as Luther does — a rendering admissible both as to matter and form (though not in accordance with the special connection in which it stands in the original) — when we observe this, and remember that the Apostle thereby sanctions, if not the translation itself, at least the thought it contains, we are warranted from this point of view also, in holding fast to the points of resemblance before indicated.[2]

---

[1] Heb. 1 : 7.

[2] In connection with the semblance of the bodies of these beings to *fire*, we may be permitted to quote from Beck's *Christl. Lehrwissenschaft*, a beautiful passage, due to the celebrated *Boerhaave*, elem. chem. I. p. 126: Si mirabilis est ignis, in eo sane praecipuum admirabilitatis constituendum videtur, quod subtilitate incomprehensibili ita indagineum eludat, ut et ab aliis pro spiritu verius quam pro corpore sit agnitus. Ipsa ignis elementa ubique, et in corpore solidissimo auri et in vacuo maxime inani Toricelliano habitant, omniaque corpora et spatia aequali distributione et insinuatione obtinent. As to the *wind*, in the same connection, the words of Christ, in John 3d : 8th, are significant: "The wind bloweth where it listeth, and thou hearest the sound thereof, but canst not tell whence it cometh, or whither it goeth."

The mode, also, in which angels have ever appeared upon earth, is in harmony with this view. Thus, Matthew says of the angel which appeared to the women at the tomb of our Savior: "His countenance was *like lightning*, and his raiment white as snow." These words describe, not the likeness to the human form, assumed but for the moment as it were, but rather, that in the appearance which was of a superhuman, of a specially angelic character; not what was involved in a transient appearance, such as the angel then assumed, but that which was a characteristic of, and essentially pertained to his own proper being. The dazzling splendor of his raiment must certainly be regarded as having been the effect of the light proceeding from the bright and glorious body of the angel himself, — this being in accordance with the analogy of what appeared at the transfiguration of Christ.[1] When we keep in mind, in addition to the above, the suddenness with which angels have almost always appeared, and also vanished (ascended), we shall perceive that the marked peculiarities of their bodies, in contrast with those of men as at present constituted, are these — that they are possessed of a more pure and refined nature, more resemble light in appearance and rapidity of movement, and are endowed with powers made adequate to the duties and exigencies of high, immortal, spiritual life.

It is either so expressly stated, or silently assumed as a fact, in all the Scriptural accounts we have of the appearance of angels, that they appeared upon

---

[1] Matt. 17 : 2; Mark 9 : 3.

earth in human form, or at least, in one very similar to it — hence it was that their heavenly nature and origin was so frequently not at first perceived. But this circumstance is far from justifying us in the immediate inference that their form is properly and necessarily one similar to the human. Nay rather, it is not only conceivable, but even more than probable, that they for the time only assumed the human form, in order that they might conveniently and effectually hold intercourse with men. But this inference may legitimately be drawn from the circumstance mentioned: that the bodies of the angels are not so crude and inflexible as ours, nor are they so well-defined and fixed in their outline, but rather possessed of a high degree of fluidity and mobility —that they do not oppose to the wishes of the eager spirit, the clumsiness and inertia of human bodies; but are rather the willing instruments of the spirit, subordinate to all its wishes, and completely adequate to all the wants and exigencies of spiritual life. Nothing of a more definite character touching the corporeal forms of the angels, and of their similarity or dissimilarity to the human form, can be gathered, either by proof or inference, from sources within our reach.

We have already shown that the Bible, so far from forbidding the assumption that the angels are possessed of corporeal forms conformable to the mode of their being, really demands such an assumption, and itself takes its validity for granted. But, apart from the positive teachings of the Scriptures themselves, the idea of an absolutely incorporeal being, is altogether irreconcilable with the idea of a finite creature. But surely no one would presume to dispute that the angels are mere finite creatures.

"Leiblichkeit ist das Ende der Wege Gottes." A creature without any bodily form is wholly inconceivable, since that which is created, as the created, can only work and subsist within the limits of time and space, and since corporeality alone confines the creature to time and space.[1] God alone is an infinite,

---

[1] [We here introduce some interesting remarks by *Isaac Taylor*, which will be found to coincide very closely with those of the author. He says: "We must affirm that Body is the necessary means of bringing Mind into relationship with space and extension, and so, of giving it — Place. Very plainly, a disembodied spirit, or we ought rather to say, an unembodied spirit, or sheer mind, is NOWHERE. Place is a relation of extension; and extension is a property of matter: but that which is wholly abstracted from matter, and in speaking of which we deny that it has any property in common therewith, can in itself be subject to none of its conditions; and we might as well say of a pure spirit that it is hard, heavy, or red, or that it is a cubic foot in dimensions, as say that it is *here* or there, or that it has come, and is gone. It is only in a popular and improper sense that any such affirmation is made of the Infinite Spirit, or that we speak of God as *everywhere* present. God is in every place in a sense altogether incomprehensible by finite minds, inasmuch as his relation to space and extension is peculiar to infinitude. Using the terms as we use them of ourselves, God is not here or there, any more than he exists now and then. Although, therefore, the idea may not readily be seized by every one, we must nevertheless grant it to be true that, when we talk of absolute immateriality, and wish to withdraw mind altogether from matter, we must no longer allow ourselves to imagine that it is, or that it can be, in any place, or that it has any kind of relationship to the visible and extended universe. But in combining itself with matter, by

an absolute Spirit; He only exists above and beyond time and space. A created spirit without a corporeal form to confine it to time and space, to bound its being, and give it a species of form, must either

means of a corporeal lodgment, mind brings itself into alliance with the various properties of the external world, and takes a share in the conditions of solidity and extension. Thenceforward mind occupies one place, at one time, moves from place to place, and may follow other minds, and be followed by others; it may find and be found; it may be detained, or be set at large; it it may go to and fro within a narrow circle; or it may traverse a wide circle; and while, by this same means the material universe is opened to its acquaintance, it is also itself restricted in its opportunities of acquiring knowledge, by its subjection to the laws of gravitation and motion: we may then with some degree of confidence, on these grounds, regard a corporeal state as indispensable to the exercise of active faculties, and to a scheme of government, and to a social economy. That which is finite — a finite mind, for example — must, as we are inclined to think, become subject to some actual limitations, and must undergo some specific relations, before its faculties can come into play, or be productive of effects. . . . . There is reason to conjecture (perhaps stronger terms might be used) that none but the Infinite Spirit can be more than a latent essence, or inert power, until compacted by some sort of restraint. The union with matter, or the coming into a corporeal state, may be, in fact, not a degradation of the mind, but the very means of its quickening — its birth into the world of knowledge and action. The first consequence of this birth is, as we have said, the acquirement of locality in the extended universe. . . The corporeal alliance of mind and matter is, in the present state, and as we may strongly conjecture it will be, the means of so defining our individuality in relation to others, as is necessary for bringing minds under the condition of a social economy. The purposes of such a system demand the seclusion or the isolation of each spirit, or its impenetrability by other spirits. . . Perhaps unembodied

be like God, infinite, omnipresent, and eternal — be God himself; or, since that would be irreconcilable with the idea of its having been created, be dissipated into nothing and utterly lost. Hence, within the province of created life, the possession of a body is the condition of all existence; the corporeal structure is the instrument of all activity of the spirit; it constitutes a tenement for it, gives it a lodgment, and thus enables it to preserve its legitimate boundaries and its identity, — without a body, without a fixed abode, the homeless spirit would be carried everywhither and dissolved into nothing, be utterly lost. Corporeality places a *limit* or a check to the life and activity of the created spirit, and thus prevents them from being infinite, eternal, and omnipresent, like the same qualities in the Divine Being. But corporeality is also a *blessing* and a *beneficent grant* to the creature, since it is through the body alone

spirits (if there be such) may lie open to inspection, or may be liable to invasion, like an unfenced field, or a plot of common land. But although such a state of exposure might involve no harm to beings either absolutely good, or absolutely evil, we cannot imagine it to consist with the safety or dignity of beings like man. . . There is some reason to question whether sheer spirits could (except by immediate acts of divine power) be individually dealt with, and governed, or could be known and employed, or could be followed and detained, or could form lasting associations, and be moulded into hierarchies and polities, or could sustain office, and yield obedience, in any certain manner, if at all. At least it is true that all these functions and social ends are now in fact dependent upon corporeity; and it is only fair to assume that they demand a bodily structure in every case where minds are to live and act in concert with others." — *Physical Theory of another Life*, pp. 25, 26, 38, 39. — Tr.]

that it derives the power, the capacity and the means, for the exercise of its freedom and the pleasure of its will, for the most complete realization of its life. However spiritual and heavenly a nature, therefore, we attribute to the angels, however we exalt them in imagination beyond all connection with such a burdensome corporeal constitution as ours, beyond the restraints and laws of our base "tabernacle of clay," still they are but creatures, and must ever so remain—they too must pay the tribute of corporeality, be their bodies ever so ethereal, pure, and glorious, and however much they elude the grasp of our senses.

§ 19. *Nature, Position, and Mission of the Angels.*

Another point highly significant, and pregnant with the most important consequences in connection with the position and the whole history of the angels, is this: that they were created WITHOUT SEX. Christ himself taught this in express terms as a characteristic peculiarity of the angels, when in reference to the glorified bodies of men at the resurrection he said: "*In the resurrection they neither marry nor are given in marriage, but are as the angels of God in heaven.*"[1] The extraordinary and far-reaching consequences of this constitutional quality or peculiarity, can be fulty understood, only after we have taken a careful and comprehensive view of the whole matter. As a first consequence, it was absolutely necessary that the number of the angels should ever remain just such as God constituted it — the number could neither be increased nor diminished in any other way

[1] Matt. 22 : 30.

than by a direct act on the part of God, — and that so significant a provision as obtains upon earth, and one which conditions and moulds all human history, namely, that man was to unfold himself through the institution of *marriage* from his original unity into a great multitude, should never obtain in the angelic world. A further consequence was, that the bond which connects the single individual to the whole species, could not, as in the case of man, be a bond of *succession*, sustained by the unity of derivation, but merely one of *simultaniety*, conditioned and preserved by their all having the same Creator, a community of nature, of objects to be gained, and of destinies to be fulfilled. So far as their self-determination and the history flowing from it were concerned, this provision was specially and peculiarly important, since it rendered the choice of one part of the species, or of one individual, independent of the choice of all the rest, so that the fall of one could *not* carry with it the ruin of the whole species.

As to the *number* of the angels, we can gather nothing definite from the Scriptures; it is represented as indeterminable, far out-reaching all attempts at computation. The Scriptures when referring to it, groan under the burden of their utterances, since no human numbers are adequate to the task of computation. Daniel beheld in vision the judgment throne of the Lord. His angels, the attendants of his majesty, stood round this glorious throne. "A thousand thousands ministered unto him, and ten thousand times ten thousands stood before him."[1]

---

[1] Daniel 7 : 10.

The Apostle John, in the New Testament, makes use of the same laboring expressions,[1] and a like swelling fulness of terms may be found all through the Scriptures, when speaking of these beings collectively.[2]

There obtain in the angelic world, according to more or less clear intimations of Scripture, various *grades* of position, of dignity, of might, of callings and of destiny. There are angels and archangels;[3] cherubims,[4] and seraphims,[5,6] which are further distinguished by the terms thrones, dominions, principalities, powers, authorities, &c.[7]

Doubtless all the special terms by which the different orders of the angels are designated, denote corresponding specific differences in the nature, being, position, and duties of these heavenly beings. But all further insight into the nature of these differences is forbidden us, since all the angels, so far as their relation to man is concerned, form one vast and general class of heavenly beings, between whom and the

---

[1] Rev. 5 : 11.

[2] See Gen. 32 : 1, 2 ; Ps. 68 : 18 ; Luke 2 : 13 ; Matt. 26 : 53.

[3] 1 Thess. 4 : 16 ; Jude 9.

[4] Gen. 3 : 24 ; Ps. 18 : 11 ; 80 : 2 ; Ezek. 1 : 10 ; Rev. 4.

[5] In regard to the *seraphims*, I agree with *Hofmann* (Schriftbew. I. 328). But none the less do I deviate from his view in regard to the *cherubims*. Compare my *Gesch. des alten Bundes*, vol. I. 2d ed., § 22, 3, where I have developed at length my views touching the nature and position of the cherubims, and their relation to the history of redemption, as derived from a full examination of the sacred Scriptures on this point.

[6] Isaiah 6 : 2.

[7] Col. 1 : 16 ; Eph. 1 : 21 ; 3 : 10 ; 1 Pet. 3 : 22.

inhabitants of this earth, there exists a wide, general, and universal difference. We are permitted to catch a glimpse of the manifold varieties of these mysterious beings, only in order that the glory of God may be to some extent apprehended by our slow and grovelling minds—that glory which ever displays itself in the most varied manifestations of life, and powers of production and adaptation; but never in monotonous sameness, in mere repetition or mechanical imitation of previous creations—that glory, too, which in the midst of all this infinite variety, presides with majestic ease, comprehending the whole within the scope of one grand idea, conceived in the remote counsels of eternity. Nor is it alone with respect to man, that the angels may all be classed together, as one vast and general assemblage of heavenly beings, but also in their relation to God; since, so far as this is made known to us by revelation, we see behind all the differences that may really exist in other respects, one common *vocation*, one general *sphere* of *offices* or *duties* assigned to all orders of these heavenly inhabitants. They all alike belong to and help to form the heavenly host,[1] they are all the ministers who stand round about the throne of God, who execute his commands, who are the organs of Divine power and rule in the visible world; and they all together constitute the vast and jubilant choir, which resounds its heavenly anthems to the praise of the majesty and glory, the wondrous works and ways of God. Not as though God had need of their praise, or had created them merely on his own

[1] Gen. 32: 1 seqq.; 1 Kings 22: 19; Dan. 4: 10–14; Luke 4: 10.

account—nay rather, he created them, constituted them the ministers of his will, set them round about his glorious throne, and opened to them the ravishing visions of his majesty and glory, that they might find for themselves, in a voluntary service and obedience, adoration and praise, an infinite and inexhaustible source and fulness of delight and blessedness.

§ 20. *The Fall in the Angelic World.*

It was necessary that the angelic world, also, no less than our human world, should experience a history, should be concerned in a progress from a beginning to an end, in a development—be it a development in accordance with or contrary to the will of God—of those powers and capacities bestowed upon it at its creation. And it was also necessary that the history of the angelic world should begin as the history of our world—with the realization of a state of freedom—with the trial of the allegiance or self-determination of the angels themselves. It was necessary that they should, in the capacity of free personal beings, decide either in accordance with or contrary to the Divine appointment with respect to them. Neither was it possible that they should have been advanced, immediately at their creation, to the *highest* possible point or degree of perfection of which they were *capable*, so that no further development could have taken place; nor was any constraint allowable on the part of God, either at the commencement or during the progress of the development, which could in the least remove or limit their freedom, so as to *compel* them to be true to their destiny

—to choose *that* course of development which God had set before them.

The trial of their self-determination was conditioned by their position and duties as the creatures of God. Thus was it to be proven whether they would ever seek their greatest, their final happiness, in the service of God, and in unreserved obedience to all his commands; in beholding and celebrating his glory, in reposing beneath its effulgent beams; or whether they would rather choose to seek their happiness—but find perdition—in rebellion against God, in inveterate opposition to the Divine purpose—whether they would rather *yield obedience to* the will of God, and have his *favor*, or desire to be *as* God, and lose *all fellowship* with him.

The angels did not all maintain their allegiance throughout their trial. Part of them misused their freedom, and took occasion by it to rebel against the will and purpose of God. The whole godless movement took its rise in the daring mind of one of them, who was naturally endowed with pre-eminent qualities, and who originally occupied a high and distinguished position in the angelic world; but many others permitted themselves to be drawn away with him into the same daring rebellion and hopeless fall. The natural pre-eminence of this leader, arising from his constitution and the might of his will, still subsists, even since the fall, so that the whole revolted host form under him, as their chief, a regularly organized kingdom of darkness.

We may observe throughout the whole Scriptures a distinction respecting the fallen angels—a distinction between the great (einheitlichen) Prince of darkness, and a multitude of subordinate, but fallen angels. This by no means unimportant distinction is, we are sorry to say, wholly unobservable in Luther's translation of the Bible. The original language of Scripture does not speak of devils, in the plural, at all; it never speaks of but *one* devil, or Satan, of but one prince of darkness; but it does, indeed, frequently speak of demons, of *many demons* (a word which Luther nevertheless ever renders devil, to the obliteration of all distinction between the two classes.)

The fact that the whole rebellion took its rise in the mind of that daring leader, and further, that he occupied, previous to the fall, some superior position in heaven, lies concealed in this well-marked distinction between the devil and demons in general, and is no less clearly revealed in the ascendency which he, the Prince of darkness, is ever represented as maintaining over those (who are now no longer the angels of *God*, but the angels of the *Devil*[1]) who fell with him, as though they were his ministers and subjects. Matt. 12 : 24–26 teaches expressly that they all together, since their fall, constitute a well-organized and concentrated hellish force, under the leadership of Satan.

The Scriptures[2] say nothing as to the reasons and occasions of the fall of these beings, nothing as to

---

[1] Matt. 25 : 41.

[2] Nor does the apocryphal passage, Wisdom 2 : 24, treat of *it*, as J. P. Lange assumes (Dogmat. p. 568). It says: "Death came into the world through the envy of the devil." It reveals the

the manner in which it occurred, or as to the means able to bring it about—nothing as to the outward object with which it may have been connected. Most probably we could not have understood the affair, had an account of it been given, since we are so wanting in all definite information touching the modes of angelic life, the secret nature of these beings, and the peculiarities of their condition and circumstances. Hence, perhaps, the silence of the Scriptures. But however that may be, this much at least is clear, that their fall was not the offspring of the being given to them by their Creator, but proceeded from a fund "of their own,"[1] acquired by a perversion of their powers in the abuse of their moral freedom.

But the revolt in the angelic world was not universal; they did not *all* fall; a great, indeed, in all probability, the incomparably greatest part of these holy beings, remained true to the Divine appointment and "kept their first estate." We are led to this opinion by the cumulative exuberance of expression used in attempting to give us some idea of the number of those "who kept their first estate," in contrast with the absence of all such laboring expressions when the fallen angels are mentioned.

---

devil to us as already envious; consequently, as already fallen. It does not explain by what means the devil fell, but merely how man was caused to fall: not how destruction entered the angelic world, but simply how death entered our human world. And the explanation itself is scriptural enough. For it is clear, from the third chapter of Genesis, that *envy* on the part of the devil, at the high position or destiny of man, lay at the foundation of the temptation.

[1] Jno. 8 : 44.

In consequence of the fact that, in the case of the angels, the idea of species is determined and sustained merely through the oneness of their position, duties and services, but not through natural generation and propagation of kind, the fall of one part of these beings did not, in itself, involve the fall of any or all of the rest. Still, however, the movement which resulted in the fall of some, could not have left the others in a state of indifference, mere idle spectators of the appalling scene. For, in consequence of the fact that a like nature, a like destiny, a simultaneous existence, bound all orders and modes of angelic life together, into an intimate relation, it was impossible but that the determination of one or several of them with regard to the question of their allegiance or relation to their Creator, should force the rest to come to a speedy decision also. The fall of Satan convulsed the whole angelic world, and made it necessary that every individual should take his stand, either on the side of God or on the side of Satan, that he should fall in with the will of God, or with the will of Satan. We cannot imagine that there were there any merely idle spectators, who sided with neither party.

§ 21. *The fallen Angels not capable of Redemption.*

Thus there came to pass a revolt in the angelic world, and it was divided into two hostile, antagonistic parties—good and bad angels. The revolt of the latter, their daring self-determination in opposition to the known will of God, was absolute and final—it left no possibility of a return on their part, nor of salvation from God. Nowhere in Holy Writ do we find the most distant intimation that the fallen angels may be converted and redeemed; their *eternal* condemnation was pre-determined from the beginning.[1] We may not seek the grounds of their condemnation in the will of God, as though God, notwithstanding the possibility of their being redeemed, is *unwilling* that they should be,—this would impeach the holy character of God himself, as seen in the mirror of revelation. All creatures were created and appointed to secure happiness, and God will never allow this design to be frustrated, so long as its realization is yet possible. He has no pleasure in the death of the wicked, but would rather that they should turn to him and joyfully accept the offers of his grace.[2] Had Satan himself been capable of salvation, God would doubtless have provided for him and his subjects, a salvation adapted to their condition. The ground of their hopeless condemnation, therefore, is to be sought only in the angels themselves; it may lie in their *nature*, in their *will*, or in *both* at the same time.

In the case of the angels, as with all free creatures, the grounds of their moral condition or state, of their moral ability or inability, must be sought first and above all in their *own will*. With respect to the incapacities of the *fallen* angels, however, as regards salvation, these grounds are not difficult to trace. They engendered sin within themselves, purely out

---

[1] Jude 6; 2 Pet. 2 : 4; Matt. 25 : 41; Rev. 20: 10, etc.
[2] Ezek. 33 : 11; 2 Pet. 3 : 9; 1 Tim. 2 : 4, etc.

of *their own* will,[1] without any temptation or enticement from without, without any positive occasion or inducement to commit sin; they themselves were the creators and fathers of the evil which now reigns so fearfully within their bosoms. That decision of their minds which was the root of all evil, was thus an absolute, a final decision; the evil which they voluntarily engendered was an absolute evil, consequently a change on their part, or repentance and salvation in their case, is wholly out of the question.

Although, according to the above, the ground of the impossibility of their salvation is to be found first and chiefly in their own *will*, still it may also be that their *nature* is such as not to allow of their being redeemed; not, however, that the strength of the former and primary ground should be at all weakened or destroyed by the latter. Their nature itself may, for aught we know, be such that the decision connected with it, once fixed upon, was an absolute one, one that could never be re-considered.

Besides, the nature of the angels in itself considered, apart from any decisive and final character it may lend to the decisions of their minds, seems to render the redemption of these beings impossible. So far as we can see, salvation is possible only under these conditions; that a new and vigorous life, far exceeding in energy the might of sin and death already existing in the fallen being, a life capable of overpowering and casting out sin and death, should enter the sinful creature—that a supernatural, a Divine life should dwell in the fallen one, in personal

[1] Jno. 8 : 44.

and essential union with it; in order to annul, *in* it and *for* it, the effects of a course of ungodly development, in order to take up the divinely-appointed but neglected development, and conduct it with the creature itself to its highest completion; or, in other words, God himself must become man in order to redeem *mankind:* to redeem the *angels*, he would have had to assume and permanently retain the nature and being of the angels—himself become an angel. This effective assumption of the creature's nature, as we have already seen, was possible in the case of man; but it was not possible in the case of the angels, because they were made by their creation itself a fixed and definite multitude of individuals, not permitting of increase by propagation; because they were not united together by a bond of unity of nature and being founded upon the mode of their origin—their having a common ancestor and head from whom they all derived their descent. Consequently, had God assumed their nature, the want of this indispensable unity of the species, would have prevented the application of the obedience and merits of their substitute, to the special necessities and wants of each individual. Had God become an angel, then would this God-angel have held an individual, isolated position, just as any other created angel; but when he became man, he at once entered into the most intimate relation of both blood and nature with the whole human race, and every particular individual of it. The angels being created without sex, it was impossible that they should all be derived from one, by natural descent. In this lay an *advantage*, as the fall of *one* could not, through the ties of blood and natural descent, include the fall of all the rest, as was virtually the fact in the case of man. But it also involved the *disadvantage* no less, that a *common* redemption from *one* Redeemer, could not be extended to all, on the principle of unity of race—on the principle of substitution and imputation.

We have here advantage confrounted with disadvantage, so that it is still impossible to say that the angels are less favored or made inferior to men. Matthew, indeed, Chap. 22, 30, shows clearly that the absence of the characteristic of sex, is in itself an evidence of a *higher* stage of advancement, since man only at the close of his history, at the resurrection, when he is to be rendered perfect physically, and in all respects, shall attain to *that* condition, or stage of advancement, in which the angels were placed immediately on their creation. The possibility of the angels arriving at a state not admitting of their being redeemed, by obstinate rebellion and perversity of will, is balanced by the possibility of man arriving at a like state, by obstinately refusing to be saved, by rejecting an offered redemption, and thus, like the fallen angels, attaining to a state of absolute evil, of irremediable and everlasting condemnation. The point when a final and absolute decision is to be made, which in the case of the angels must, both from their nature and destiny, have been but a short time subsequent to their creation, cannot and will not be omitted or neglected in the case of man. It merely comes later in his case, conformably to what may be different in his nature, destiny, and development, but is just as unavoidable as that of the angels, and as irreversible in the nature of its consequences.

§ 22. *The Perpetuity of Evil among the fallen Angels.*

But this ungodly self-determination, and opposition to God on the part of some of the angels, did not bring their history to its close. The fallen angels could indeed never return; but they might advance still further on the road to destruction.

It is the prerogative of a free, personal being, not only to determine itself contrary to the appointment of God, but also to continue to exist, after having renounced its allegiance to him, and further, to follow out, wholly unrestrained, the godless course of development it may have chosen, even to its ultimate goal. Both Divine wisdom and justice demand that evil, wherever it has gained a foothold, should be abandoned to its own course. Freedom is not given to the created spirit conditionally, but, as the idea of personality itself demands, *absolutely;* and this quality of its constitution must ever be retained, even though the creature itself be cut off from the eternal source of its being. For the personality of the creature constitutes its likeness to God, so that, so long as God regards himself, it cannot be but that he will regard the personality of the creature. God is strictly just in all his dealings with Satan; and even in his case, respects the personality belonging to the creature. Hence it was neither possible nor desirable that God should annihilate those angels that sinned, nor that he should in the least lessen or otherwise affect their right to freedom and to existence.

Freedom of development must ever be retained by them as an undisturbed possession; but in the case of a finite being, such a boon always brings with it a something else as its balancing and opposite pole — *necessity*. The direction, indeed, which they took, was altogether one of their own choosing; but the goal to which it leads, is a necessary one, and can never be changed. They possessed full power to renounce all allegiance to God; but at the same time, they must abide the consequences of such a daring and wicked act. That eternal condemnation which overtakes the fallen angels, incorrigible in their wickedness and settled opposition to God, is the direct and proper result of God's still respecting the personalities with which they were originally endowed.

Also Divine *wisdom*, no less than Divine justice, demands that evil should be left to an undisturbed development of itself, according to the laws of its nature. So soon as evil came into existence, it manifested itself as an *external* power, a reality, whose inner weakness and futility, when opposed to the Divine will, can be satisfactorily discovered, only when it has completely unfolded itself; when all the germs it contains have been developed; when all its powers have been enlisted in vain; when all the appalling self-deceptions and fearful self-delusions it has practiced, obstinately and of its own accord, are completely unveiled and brought to light. The development of evil is its own overthrow—its every apparent triumph is but a new defeat.

The annihilation of the fallen angels, on the part of God, the abolishment of their freedom, or a forcible restraint of their efforts in opposition to God, would not have been proper or allowable. As they were incapable of salvation, both from their nature and from the character of their will, it was necessary that they should be abandoned to that fate of which they themselves were the authors; and as their revolt from God was at once *absolute* and *decisive*, it was necessary that it should completely unfold itself, and bring to perfection its own proper fruits.

But so soon as this was accomplished, they might, so far as they are objects of the effects of sin, be made to pass through their last judgment. But they are implicated in and take part in another and a no less important affair, which must be brought to a close before their final sentence is pronounced.

We of course refer to their relation to the earth and man, and to the part they play in human history. Here also must that which specially belongs to them—their chapter in the history—be fully unfolded, and their complete overthrow be accomplished, before they shall be ready for the judgment. Compare § 25.

§ 23. *The Abode of the holy Angels.*

The very idea of a created spirit involves the assumption that there exists somewhere in space, a place adapted to the nature and exigencies of spiritual life, a place where the spirit may realize and manifest its life and freedom, and fulfil its peculiar mission.

Heaven[1] is designated by the Scripture in general, as the dwelling-place of the good angels. They ever appear as the *heavenly* host, as the native inhabitants of those blessed heights, to which man casts many a longing and wistful eye, and which he ever fondly recognizes as the place of unalloyed and unfailing happiness and glory. The idea of angelic beings, and the idea of the heavens, are so closely connected, and the correlation of the two is so deep and so real, both according to Scripture and Christian sentiment, that they can hardly be dissociated—one ever suggests the other.

But the word heaven is so general and comprehensive in its application, that we must seek to give it narrower and more carefully defined bounds, before it can be taken as the correlative of the word angel, in the strictest sense.

That significant description in the Book of Job (chap. 38), part of which we have already quoted (§ 17) for a different purpose, sheds some light upon this question.

"Where wast thou when I laid the foundations of
the earth?

* * * * *

[1] [The words *heaven* and *heavens* are used in this section, as indeed frequently throughout the book, as convertible terms: the word *heaven* not being confined to its special designation as a name for the abode of the blessed, but including as well the physical heavens, whether they constitute that abode or not. This usage seems best adapted to giving the author's idea, while speaking in accordance with common modes of expression.—Tr.]

When the morning stars sang together,
And all the sons of God shouted for joy."

Here we have, in addition to the sons of God celebrating the founding of the earth, the morning stars mentioned as joining in the jubilant chorus. But, according to the well-known laws of poetical parallelism in Hebrew poetry, it is necessary that the two corresponding members, "the morning stars," and "the sons of God," should be essentially connected, that they should either be identical in meaning, or at least, be comprehended under one common (einheitlichen) idea.[1]

As we have previously learned, the morning stars are those glorious worlds of light whose undying fires ever light up the vault of heaven. What now

---

[1] I cannot but persist in *this opinion* although *Hofmann* (*Schriftbew*, I. 352,) says: "This passage has been *perverted* for the purpose of showing a connection between the angels and the stars, in the Biblical theory of the world." . . . . "That such was by no means the idea of the poet, appears clearly from Chap. 15 : 15, where "his saints" are put in precisely the same relation to "the heavens," that "the sons of God" are to "the morning stars." But it may soon be seen how inconclusive this argument is. The parallelism of 15 : 15 does not, at all events, consist in the mutual relation of the celestial spheres and the saints of the earth. Nay, rather, "the heavens" must either be regarded (with *Hahn*, 79) as the abode of the saints, (and hence the latter be the angels), or we must (with *Schlottmann* and others,) refer "the saints" to the inhabitants of the earth, indeed, but then the term "heavens" to the inhabitants of heaven, just as "*col haarez*" (the whole earth) is frequently spoken of the *inhabitants of the earth*. That such a metonomy with respect to the heavens and the angels was by no means unusual, is evidenced in the fact that both ideas are designated by one and the same expression, "the hosts of heaven."

is a more natural assumption, since the heavens are so universally represented as the dwelling-place of the angels, than that the inspired and Divinely illumined poet may have regarded the sons of God as the inhabitants of these morning stars?

This argument derives additional weight from the fact that the same view prevails in all the other writings of the Old Testament. For the words, "the host of heaven," designate both the stars of heaven,[1] and also the multitude of the angels who praise the Lord and fulfil his commands.[2]

As to the physical constitution and laws of these celestial and angelic worlds, we need not expect any definite information from Scripture. Revelation must have taken a wholly different stand; have done violence to its nature and object; indeed, must have become a text-book in Astronomy itself, in order to describe the heavenly bodies to us in these relations.

But it does, nevertheless, and in harmony with its design, contemplate and represent their physical nature, in its ethical and religious bearings.

They all bear the marks which characterize every created thing; they were all created out of nothing, according to the will of the Creator, who alone is from eternity and an absolute Being. How mutable also and how incomplete are they all, when compared with the immutability and spotless purity of God! Hence sings the sacred Psalmist:[3]

---

[1] Gen. 2:1; Deut. 4:19; Is. 34:4; Jer. 33:2; Ps. 33:6, etc.

[2] Gen. 32:1, 2; Ps. 103:21; Ps. 148:2; 1 Kings 22:19; comp. Luke 2:13, etc.

[3] Ps. 122:25–27.

"Of old hast thou laid the foundations of the earth:
And the heavens are the work of thy hands.
They shall perish but thou shalt endure:
Yea, all of them shall wax old like a garment;
As a vesture shalt thou change them, and they shall be changed;
But thou art the same, and thy years shall have no end."

And thus the Book of Job (25 : 5):

"Behold even to the moon, and it shineth not;
Yea, the stars are not pure in his sight."

But in all instances, on the contrary, where it is not the leading design of the inspired writer, to set forth prominently the contrast between the Infinite and his finite works, the heavens with all their glittering worlds, are represented as the culminating point of all glory and blessedness, of all order and harmony in connection with created things: and that song of praise which their perfection, their grandeur, and glory, resound to the honor of Him who so created them,[1] surpasses in swelling fulness and majestic harmony all other songs of praise.

How could it, indeed, be otherwise? How could the celestial worlds be the dwelling-place of the angelic hosts, and not correspond in glory to their glorious inhabitants? The body must correspond to the soul inhabiting it, the dwelling-place to the inhabitant.

If we everywhere find the angels represented as

[1] Ps. 19 : 1.

pure, holy beings, who have remained steadfast in the truth, have been true to their calling and destiny, and amidst whom life and happiness, peace and joy, hold undisturbed sway, what is more natural than to suppose that their abode should partake of a character corresponding to these glorious attributes? Every manifestation, evidence, and token of sin, sickness and death, gloom and destruction, of dissension, disorder and tumult, must for ever remain foreign from those holy abodes; every look must be a beam of joy and delight, every tone a hymn of rapture, and every movement be graced with holy love. Countless also as the multitudes of the "heavenly host," must be the number of celestial abodes. The being, mission, and destiny of the angels, appear to partake of remarkably bold, original, and decided peculiarities, of remarkably profuse and varied characteristics. Nature as it surrounds and sustains these heavenly beings, must partake no less of the same marked features and varied adaptation—it must be suited to every condition, development, and exigency of angelic life.

Further, as we have discovered that, according to Scripture, a characteristic peculiarity of the angels consists in the absence of all sexual distinction between them, we are led to expect that this peculiar feature must be mirrored in their heavenly abodes—that every thing which in our world appears as the cosmical transcript of the sexual contrast, must there be wanting: that those blessed abodes, where they neither marry, nor are given in marriage, must also be free from all the physical antagonisms and oppositions, the restless and wearisome play of forces, which constitute such broad contrasts in our world; and that, finally, all cosmical forces must there unite in quiet and combined (einheitlicher) harmony, and in this capacity be completely adequate to perform all their functions and attain all their ends.

§ 24. *The Heavens as the Dwelling-Place of God.*

God exists above and beyond the sphere of all history, yet he still rules in history, and moulds it according to his will: He, the unchangeable, exists above all changes in the created world, yet still does he involve himself in all the changes which his creatures experience, in order to prepare them for and raise them to *his* unchangeable state, to a state of absolute, inalienable completeness and blessedness. He condescends to the low condition of his pupil, grows with him to the height of His manifestation in the creature, so that the latter may be fully prepared to take part in the glory and blessedness of Deity, so that He who is alone holy and blessed, may be *all in all.*[1]

We have contemplated angels and men with respect to their position and mission; we have considered the heavens as the dwelling-place of the angels, the earth as the abode of man; and finally, we have learned the essential features of the relation these two portions of the universe hold towards each other. It yet remains to discover the relation of both to God, and the relation God holds towards them.

The heaven is the throne of God, and the earth is his

---

[1] 1 Cor. 15 : 28.

foot-stool, according to the Scriptures.[1] We likewise are taught to pray: "*Our father, who art in heaven.*" We are told that Christ, who himself was Divine and from the bosom of the Father, passed *into the heavens*, so soon as his work on earth was accomplished, returning to the Father again that he might take possession of the throne of glory. Hence we must infer that the being of God is present in and dwells *in the heavens*, in an eminently proper sense.

The word heaven involves, first, the idea of place, and second, the idea of state or condition. According to the first idea, which is specially prominent in the original Hebrew word (theheight, high), this bringing together of heaven and earth, expresses the contrast involved in the expression, upper and lower. The idea contained in this expression, is in and of itself, indeed, ethically, of no significance, as in a physical sense it is merely a relative idea. But when the soul, created for communion with God, is unable to find on earth what it stands in need of, and what it yearns after, then does it cast its longing glances upward. See we ourselves here surrounded with sins and miseries, then do we there *above* seek holiness and happiness. Thus does this idea come to have an ethical significance (though in itself it possess none), and the ideas of place and of state coincide—place and state become convertible terms. Heaven is not the place of perfect happiness, merely because it is the abode of the blessed, but also because it forms such a wide contrast with the earth, because it is a place so sublimely transcending our earthly abode.

[1] Is. 66 : 1; Matt. 5 : 34, 35.

The earth is the theatre for the developments of sin and death, of discord and destruction; heaven is the blessed abode of holiness, of tranquil peace and eternal joy. Earth is the near, the present, the common, the finite, the sensible; heaven is the distant, the absent, the sublime, the unattainable, the supersensible, the infinite, in the conception of which we abstract in our minds all that is or may be known through the medium of the *senses*, all relations or conditions which obtain here on earth, just as we do in our attempts to form a conception of God.

Thus are the heavens ever represented as standing in a closer relation to God than the earth. God is omnipresent, but he is also wholly separate from sinners; according to the one idea he is upon earth just as really as he is in heaven; but, according to the other, he is distant from and infinitely exalted above it—he is in heaven. Happiness and holiness constitute the being of God; the more intensively these attributes anywhere prevail, there must we suppose God to be more specially present. The earth presents everywhere to our view, the melancholy aspects of sin and of death; heaven is the abode of the angels, where undisturbed harmony and unalloyed happiness ever prevail; hence must we suppose that God is there more concretely,as it were, more powerfully present—*heaven* being *his throne* and *the earth his footstool.*

Moreover, God is not âfar off, he lives in all things, and they exist only in so far as He sustains and preserves them. There are tokens of the Divine presence in every blade of grass: He is the ever Immanent. In him we live and move and have our being, and He it is who gives to all creatures, life and breath and all things.[1] But he is also a God afar off; He is the Infinite One, highly exalted beyond all finite things, separated from them, and different from them. If heaven be to us, in its relation to the earth, the far distant, the sublime, the supersensible, in a measure the infinite, then does it in this respect stand in a closer relation to God than our earth.

But we have not done yet. The greatest intensity of his presence, the greatest power of the actual Divine presence, lies beyond the utmost limits of sense—beyond creation—for here it cannot be contained. Yonder above is the place where God absolutely dwells, in the heaven of heavens, the *most holy place*, from whence Christ came forth, and to which he returned, in order to appear in the presence of God for us;[2] the third heaven, into which Paul was caught up, and where he heard unspeakable words, which it was not lawful for man to utter.[3] There it is that God dwells in light that no man can approach unto.[4]

Beyond all doubt there *is* such a being and such a dwelling of God, infinitely exalted above the bounds of sense and finite creation; for the heaven, even the heaven of heavens, all of which He has created, cannot contain him;[5] and however great and perfect the holiness and purity of the angels and their habitations may be, still this is merely a relative perfection, and cannot bear comparison with absolute, with

[1] Acts 17. [2] Heb. 9 : 24. [3] 2 Cor. 12 : 2.
[4] 1 Tim. 6 : 16. [5] 1 Kings 8 : 27.

infinite perfection. Hence it is said: "Behold, he putteth no trust in his saints; yea, the heavens are not clean in his sight."[1]

§ 25. *Retrospective View of the primeval History of the Earth and Man.*

In the consideration of the primeval Biblical history,[2] we left several significant inquiries unanswered. Since that time we have gathered much new information, from subsequent revelations. May we not, perhaps, find in it the key to a deeper insight into that history?

Although we learned, in the consideration of the fall of man, that the tempter who appeared as a serpent and was cursed as a serpent, must have been a personal spiritual being, we had there to be satisfied with merely raising questions concerning the nature, the being, the position and character of this mysterious personality. But we are now no longer in doubt as to *who* and *what* that tempter was that appeared by the tree of knowledge.

Besides such fact, however, the subsequent stages of Divine disclosures, bear the clearest and most decisive witness to the true character of this being. Christ himself says of the devil, that he is "a murderer from the beginning," since he brought sin into the world, and death by sin. In the Revelation of John (chap. 12 : 9), he is characterized as "the *old serpent*" which deceiveth the whole world. Compare 1st John 3 : 8; 2 Cor. 11 : 3; Rev. 20 : 2, &c.

If it be true that the serpent, by whom man was

---

[1] Job 15 : 15. [1] Gen. 1–3.

betrayed, so soon as he essayed to employ the powers God had given him, was closely and essentially connected with the prince of the fallen angels—whether as his instrument, image, or representative—we have in this history some data toward fixing the *time of the fall* of this prince of darkness. We find him hanging about the cradle of human history, with his malignant opposition to God already fully developed. Hence we may at least conclude, that *his* fall was *previous* to the fall of man; and, as the latter took place when man *first* employed his free will, that it was also *previous to* the *creation of man.* Still further, there is every reason for supposing that this catastrophe in the angelic world happened very soon *after the creation* of the angels themselves. For, as the probation of man, the test of his course in relation to the will of God, stood *at the beginning* of human history, and *first* called into exercise his powers of choice, so also was it with the angels—their history also commenced in a corresponding probation.

As we have inquired concerning the *time* of their fall, so also may we inquire after its *place.* That this catastrophe happened in some particular place in the universe, is involved in the very idea of a creature—a being that can realize and manifest its life only in time and in *space.* We are fully warranted in the assertion, that the fall of the angels must have left corresponding traces of ruin in material nature, in the midst of which they dwelt and where they exercised their power; and that these traces would be the more marked and significant, the more important the position of the rebels, the more portentous and far-reaching in its consequences the catastrophe of the fall. Our authority for this assertion is the intimate and essential connection between spirit and nature, mind and matter.

The fallen angels appear at the commencement of the human race, as confirmed rebels against the authority of God; consequently, the traces of that desolating catastrophe must be sought in times anterior to the existence of man.

Taking up the Divine record, our eyes immediately light upon that enigmatical "*tohu va bohu*," that description of a waste and desolate condition, and of darkness, which reigned over the earth as it appeared for the first time to the eyes of the prophet, anterior to the six days' work.

Have we not here found just what we have been in search of, a waste and desolate condition such as we might expect, and appearing at the very time we should expect it must have taken place?

We were previously compelled (§ 6,) to pass by this puzzling hieroglyphic of primeval history, without being able to interpret it, without being able to fathom and comprehend its origin and true import. But since that time, our stock of knowledge has been vastly increased, by continual drafts upon revelation —perhaps we may now be prepared to grapple with the problem.

We discovered previously, when our inquiring minds dwelt upon this expression, (but without arriving at any satisfactory conclusion), that the words "tohu va bohu," wherever else they appear, designate, beyond all doubt, a positive *devastation* and *desolation*, which has succeeded to previous life and fruitfulness. This very fact involves the probability at least, if not the necessity, that these words should be interpreted in the *same* sense *here* also.

Nor could we formerly conceal from our minds the fact, that the words "*the earth was without form and void, and darkness was upon the face of the deep;*" if interpreted wholly upon their own merits, much more naturally and appositely refer to a devastation of some work of God, which has taken place since its creation, than to a work *not yet* completely finished, one *still devoid* of the higher cosmical attributes—light and life. For a Divine work, as then remarked, though it be not completely formed and advanced to the stage of perfection, must, according to the measure of its completeness and capacity, reflect a Divine harmony and order, light and life.

From that point of view even, we could not avoid regarding it as probable, that this dark, waste, and barren condition of the primeval earth, was the result of a desolating and devastating process brought to bear upon a world originally full of harmony and teeming with life. But while we there inclined to this view as the correct one, the necessary data were still wanting, from which to gather clear views of the origin, nature, and historical bearings of this assumed devastating process.

But we have here found, in the fall of the angels, the data that were previously wanting. We *there* discovered a *waste* and *desolate* condition of the earth, for which we could in no way find an author, but by the present interpretation — *here* we have found a *destroyer*, for whom we cannot otherwise find a corresponding destruction! *There*, darkness enshrouding the wild chaos, a dreary waste, an uninhabited solitude; *here*, a kingdom of darkness, spirits of the abyss, of confusion and destruction. No less do the two coincide with respect to time; for both appear previous to the creation of man, previous to the six days' work.[1]

Since now all the evidences of the two facts, the fall of the angels and the desolating of the primeval earth, so perfectly coincide, we are not only justified in holding the two to be essentially connected, but almost forced so to do, and to regard the "*tohu va bohu*" of Gen. 1, 2, as a consequence of the revolt in the angelic world. And we would still further remark that, with the aid of this assumption, and with it *alone*, many other perplexing inquiries can be satisfactorily answered, and many problems in the history of the human race find their proper and long-sought solution—that an unexpected light bursts forth from

[1] The view here defended is by no means of modern origin. So long ago as in the tenth century, it was said by Edgar, king of England, in confirmation of the law of Oswald: "As God drove the angels from the earth, after their fall, whereupon the latter was changed into a chaos, so has he now placed kings upon the earth, that justice might obtain here below." Comp. Tholuck: *Vermis.Schr.* II, 230. The same view subsequently and continuously received favor, and is now held by very many. And not only Theosophists are addicted to it, and interpreters tainted with theosophy, such as J. Böhme, St. Martin, J. M. Hahn, Fr. von Meyer, Hamberger, and the like; but also, such safe and considerate authors as Reichel, Stier, G. H. von Schubert, Kniewel, Drechsler, Rudelbach, Guericke, M. Baumgarten, Lebeau. A. Wagner, and many others, have expressed themselves in its favor.

this fresh and important acquisition to our knowledge, and is shed over many obscure points in the circle of our religious ideas. This shall be so clearly seen as we proceed with our inquiry, that all doubts and scruples as to the legitimacy and admissibility of our combinations and deductions, which may still for the present be left out of the question, must completely vanish.

Thus we find an earth even in pre-adamite times, and, no less, a history unfolding itself upon and with respect to this earth. The prophet who conceived this primeval history, beheld the earth as "without form and void"—a barren and dreary waste. This waste and void condition was *preceded* by a state of order, light and life, such as it is fitting every Divine work should possess; and it was likewise *followed* by a Divine restitution, in the work of the six days, which called forth light from darkness, order and teeming life from the dreary and barren waste; which constituted the earth as it at present subsists, established order upon, and filled it with abundance of life.[1]

[1] The arguments of Hofmann (*Schriftbew*, I, p. 238, 242) and Delitzsch (Genesis, p. 63,) do not affect the point at issue. I have not asserted that the "*tohu va bohu*" of the second verse of the Bible, can designate *only* a positive *devastation* and *desolation*. Nor have I admitted the correctness of the translation: "And the earth *became* waste and desolate." Nay, rather, I have openly opposed the one as well as the other. And further, I have not wrung my view out of the first chapter of Genesis; but have expressly conceded, that neither he who first conceived this chapter, nor he who subsequently incorporated it into the Scriptures, may have discovered in the "*tohu va bohu*" what I do. My view rests

The *devastation* was a consequence of the fall of the angels, and from this we further conclude that the primeval earth was the habitation of the angels, and the place of their probation — of that part of

---

alone upon a combination of Gen. 1 ; 2, with the facts supplied us in the later stages of revelation. I do not claim for it the authority of revealed truth, nor yet the character of a necessary consequence. It is a hypothesis, a conjecture, and still remains such, claiming only probability, and not certainty. I have become attached to it, because it satisfactorily solves many heretofore inexplicable and connected problems of Scripture and natural science; because it, in my view, places the development-history of the whole Cosmos, under a *single* point of view, etc. As to the further remarks of *Delitzsch* in opposition to my view, they do not affect it in the least. He says: "*The account of the creation speaks, to the unbiassed mind, of the creation of the universe,*" (granted! but of the creation of the universe only so far as it stands related to the earth), "*and not of the mere re-creation of the earth and its solar system.*" (I have not spoken of an actual re-creation being taught in the record, but merely of the enlivening and individualization of a waste and barren chaos. If we may be permitted, from the facts of subsequent revelation, to look upon this chaos as the residuum of a previous creation, now destroyed, then may we properly enough call what is mentioned from the 3rd verse forward, a "re-creation," or as I have designated it, a "restitution or new-creation." In the second edition of this work, I gave up the idea that the solar system was comprehended in this new-creation). *Delitzsch* continues, but to no purpose: "*The cosmological traditions not pertaining to the Israelties, which should here be taken into consideration, say nothing in regard to a chaos caused by the fall of the angels.*" Whether the traditions of the heathen should here be taken into the account or not, I shall not stop to determine. For the present it may be admitted that they should. But what is the consequence? Nothing further than that those traditions, in the times of Moses, knew no more of a chaos produced by the fall of the angels, than did the Israelitish tradition itself; that such knowledge was unknown to

them which rebelled against God, lost their dominion by revolt, and were driven from their abodes upon the earth. In the same degree, obviously, that the fallen angels prior to their fall, had a like being and destiny with the rest of the angels, like capacities, and were included with them under one common idea of species — in the same degree must the mutual abodes of all these beings have been of a like nature. And as there is in general no distinction as to *species* made between the former and the latter, so also must the primeval earth, in its original and uninjured state, have been in general and conformably to the laws of classification, similar in character, nature, and adaptation, to the other celestial worlds.

The *restitution* of the earth, on the contrary, was the offspring of a decree of the Divine mind, by which the plan with respect to this earth was to be saved from being subverted; by means of which God was to raise a whole world of life from the floods of destruction in which it had fallen, banish the destroyer into exile, and place in his stead, *man*, a new inhabitant and lord of the earth. From this we further conclude, that man has taken the place of Satan and his angels, to complete their unperformed task, to finish their discontinued mission, to restore the disturbed harmony of the universe, the

---

all the original traditions, as it could only arise in subsequent times, from the combination of the various facts of revelation. *Delitzsch* then opposes to my hypothesis, one of his own, in which he pretends to retain what of truth belongs to mine, and at the same time to avoid its faults and errors. The most striking point in his hypothesis, to my mind, is that it is in all respects untenable.

marred beauty and proportion of the whole; that *man* has been appointed to conquer and judge the *destroyer*, the great rebel in God's universe. "Know ye not," says the Apostle Paul,[1] "that the saints shall judge the world? *know ye not that we shall judge angels?*"

Man was thus placed in the *very* part of the universe, where he must receive the gaze of all eyes; in that place which was, perhaps, from its original nature and destiny, the most important of all places in space,—at least, from what had already occurred there, and quite as much from what is yet here to pass before the eyes of a wondering world, a place that had acquired a most prominent position, a surpassing importance. All further developments of the history of the universe now depended upon *his* behavior, upon *his* decision and *his* history.

But the rebels who were the authors of that disturbance of the Divine plan, which was now to be allayed, have been shut out from their original abodes, so far as the restitution of the earth has given it a character not conformable to their nature and modes of being. Their element is darkness, waste and desert places; hence when the creative word of God said: "Let there be light!" they fled hastily away—they fled, when at the command of the Almighty, the chaotic confusion was resolved into harmonious order, when the barren and lifeless wastes were made to teem with new and happy life.

But since the beginning, indeed, of the new development in the world, has been brought about, but

[1] 1 Cor. 6 : 2, 3.

not its absolute completion, the fallen angels still remain a power not fully subdued. Ideally (by the decree of God) they may be fully vanquished, but not really (by the act of man). They have been driven from their original habitations, their province has been given to another lord; but their claims to it, though in themselves futile, may still be annoyingly pressed, until their worthlessness is fully exposed; until the voice of history proclaims that their whole project has resulted in a most signal failure; until they receive, after the purifying fires of the last judgment,[1] all that properly belongs to them — the dross that remains — and be consigned with their possessions to the prison of hell, whose adamantine walls stand ever firm.[2]

Their interest in the affairs of earth, their hostility toward man, to whom the province they have lost, but still claim, has been presented; and who has been called to bring about that judgment[3] over them to which they have already, in idea, been subjected, — may all be satisfactorily accounted for from our present stand-point. From hence we see in its proper light, the significance of the earth, as the historical central-point of the universe, where the great contest between good and evil is to take place, where the pending fate of worlds is to be decided: from hence we can see how the whole universe may not be brought into its most perfect state, until the earth has been recovered and perfected. No longer does the intimate connection between heaven and earth, everywhere pre-supposed in Scripture, appear as an

[1] 2 Pet. 3 : 10. [2] Rev. 20 : 9. 10. [3] 1 Cor. 6 : 3.

inexplicable mystery: no longer does it appear as the result of mere accident or arbitrary appointment, that the earth has become the central-point of the universe, the theatre of the most glorious manifestations of Deity, yea, even of the Incarnation of the Son of God. Here, too, may we discover how the incarnation may result to the benefit, not of our poor earth alone, but of the whole universe.

§ 26. *Continuation.*

Enriched with the knowledge we have now acquired, we shall once more take into consideration the Biblical account of the fall of man,[1] hoping that we may now be able to gain a more profound insight into this disastrous affair, than previously, when we had to examine it as a mere fact, standing altogether on its own merits.

Nor does our hope disappoint us. Not only does the temptation, in its form, manner, and substance, now appear in a much clearer light; but no less does the grand mystery of the whole affair—the tree of knowledge, and the serpent—one of them the basis, and the other the instrument of the temptation—become more intelligible to our minds.

It was necessary that man, in the capacity of a free creature, and hence, standing in need of self-determination and self-development, should be placed in a state of probation, should be placed on trial—this is obvious from the outset. But it is not so obvious, why his trial took the form of a *temptation;* why the Divine will, which was to offer an occasion

---

[1] Gen. 3.

for the decision of man, expressed itself, not in a *positive*, but in a *negative* injunction; not in a *command*, but in a *prohibition*.

Arbitrary will is not conceivable on the part of God, least of all, as manifesting itself when he deals with personal, spiritual beings. It must therefore have been necessary from the position of man itself, that his trial should be effected in connection with a *prohibition*, and not a *command*. A prohibition presupposes the existence of evil, be it in the *subject* to whom any thing is forbidden, or be it in the forbidden *object*. But, in the present case, it could not be that evil already existed in man, the subject on trial; partly, because he still remained in the undeveloped condition in which he was originally created, and partly, because, had the case been otherwise, his trial would have been both unnecessary and inadmissible. The evil must therefore have existed outside of man. Yet was everything that God had made, in and upon the earth, *good*, "*very good*."[1] Whence, then, this evil?

The tree of knowledge (compare § 12) was a tree of the knowledge of good AND evil, *not* merely of good OR evil; and man was, in either case, by partaking of or refraining from its fruits, to attain to a knowledge of both good AND evil. Had evil not been in existence, however, man, in case he decided conformably to the Divine will, could not have attained to the knowledge of evil; for to have knowledge of what does not exist, is a contradiction in terms. And wherein lay the necessity that he should,

[1] Gen. 1 : 31.

in either case, attain to the knowledge of that evil which already existed, which, nevertheless, existed only without his own being, yea even as it would appear, wholly without the sphere of his activity—for all upon the face of the earth was good, "very good?"

God had caused the tree of knowledge, just as the other trees, to grow in the garden of Eden.[1] Wherefore, then, did he bid man to beware of His own work? Ah! that tree was a tree of death; for man would become the child of death, the moment he should taste of it,[2]—and still it was necessary, useful, and indispensable, although man was destined, not for death, but for life. The tree was good, for *God* had created it; and still it was, at the same time, pernicious, for it was capable of bringing *death* upon man. How is this to be explained?

God tempts no man to evil,[3] yet still, the trial of man amounted to a *temptation*, and indeed, as we see at the first glance, a direct temptation *to evil.* It is not possible, however, that God should have suggested and occasioned the machinations of the serpent. Nay rather, the tempter derived the occasion for them and the impulse toward them, from his own depraved heart. The co-operation of God may only be regarded as a *permission* recognizing the necessity of the temptation, and, in so far, *favoring* its occurrence. But what were the grounds of this necessity? What special object had the tempter in making *man* the object of his wiles? what impelled him to entice *man* into destruction? Had it been

[1] Gen. 2 : 9. [2] Gen. 2 : 17. [3] James 1 : 13.

merely the general base desire of "the evil one," to have companions in guilt, to drag others down to those depths of woe into which himself was fallen; without any further reasons, without believing that he stood in a special relation to man, it is wholly inconceivable that God should have not only permitted him to display such base passions, but also have opened the way for their successful employment.

But all these and similar difficulties find a ready solution in the fact, that the fallen angels were the previous inhabitants of the earth, and that the earth, which was laid waste in consequence of their fall, has since been restored through the grace and might of God, and assigned to man as a place of abode and spiritual training.

It now becomes obvious what a special object Satan had in enlisting all his powers, to allure man to revolt from God, to trample upon his original destiny, and rush headlong into destruction. His motive was natural enmity, profound hate, envy, wrath and revenge, toward the new aspirant, the favored rival who had received *the* habitation he himself had forfeited, *the* empire he had lost; against him to whom all the glory and blessedness he himself has lost, is to be given, and who is to sit over him in that judgment which is to consign him to the most fearful doom. His were the spasmodic efforts of despair, the utmost endeavors where all was at stake; his was the delusive hope of again getting possession of the lost inheritance, and of escaping completely the great judgment, for which he is now reserved in everlasting chains under darkness.[1]

We now discover how Divine wisdom and justice might, and necessarily should, permit, desire, and bring about the temptation — yea, even in spite of the fact that the fall was foreseen. God had appointed man as the possessor and ruler of the earth, as the restorer of the disturbed harmony of a universe, as the leader in the great and sacred contest of created spirits which was inflamed by the fall of the angels, and as the conqueror and final judge of those first rebels. But it was necessary that man, as a free, personal creature, should make his divine calling his own, by an act of his own free choice; that he should acquire the position to which he was appointed, by his own vigorous efforts; that he should earn his right to the possession of the heirless inheritance, and to the high office of being a judge over the rebels. He possessed the power also, in the capacity of a free moral agent, of forming an alliance with the great enemy of God, and, like him, of attacking the throne of God, instead of falling in with the Divine will and appointment. God is just, also, even towards Satan himself, and did not desire or attempt to prevent him from employing all the resources that he was possessed of, in order to maintain his stand in opposition to Deity. He may not and shall not receive his final doom, until he has tried every possible expedient to regain his former footing, in vain; until he has become fully conscious

[1] Jude 6.

of utter and miserable weakness, which indeed underlies all his apparent conquests.

It may now be further seen why it was necessary that the trial of man's self-determination should appear under the form of a temptation; why man was to maintain his first estate, not primarily in the observance of a command, but in avoiding what was forbidden. As evil was already in existence, and as man could by no means remain indifferent towards it, but rather, his position, his whole existence, mission and destiny were directed *against* it, he must, before all others, by the exercise of his free choice, place himself in a determinate relation to it.

Finally, that strange contradiction is now also explained—that the tree of knowledge should have been created by God, and still be a tree of death and destruction; also, that Satan, after he had forfeited this earthly province, should have obtained a lodgment and a basis of operation in the tree of knowledge—that he should there have been permitted to try his hand at the overthrow of man. Although God had caused this tree to grow in the garden, there must have existed between *it* and Satan, some mysterious relation; there must have existed in it something of a satanic nature, something that was related to and belonged to the great deceiver, something to which he might yet cling, and which he might call his own.

This may readily be discovered. Death and ruin, as cosmical agencies, were brought into the primeval earth, through the revolt of Satan; the earth was made a "tohu va bohu." God, by the six days' work, implanted new cosmical powers of life in the earth which had been laid waste, and caused them to be unfolded in fitting forms. Man was then placed between the two, between good and evil, between life and death. Both were presented to his choice by his Creator — he had but to speak the word, and the decision was made. The cosmical good which God had brought in by means of a re-creation, was concentrated in the tree of life: the cosmical evil which originated in Satan, was concentrated in the tree of the knowledge of good and evil — but not without its being surrounded by the wholesome admonitions and threatenings of God, as a barrier against all approach. *Moral* good or moral evil would be engendered in man, according to the position he should freely take with respect to *cosmical* good and evil respectively, admonished on the one hand by God, and allured on the other by Satan.

But the serpent?. Here is a mystery of the primeval world which we shall not attempt fully to unravel. But the substance of the whole mystery, in a practical point of view, is obvious, — we know the nature, the motives, the leanings, designs and objects of the personal, spiritual principle, which wrought in or through the serpent. It is merely the connection between the bodily manifestation and the spiritual principle, which remains unravelled, unexplained. We may, perhaps, apprehend it according to the analogy of the tree of knowledge, as explained above — that Satan, the serpent, and the tree, belong together, as personal, animal, and vegetable forms of concentrated evil, as embodiments of that evil which sprung from the fall of the angels, was held in check by God, and should have been vanquished and doomed by man.

Man in the beginning of his history was designed to do that which can *now* be effected by the seed of the woman *only* in the fulness of times — to bruise the head of the serpent. This he might have effected by obedience to the Divine commands, by repelling the tempter, by disregarding and resisting his crafty wiles and imposing offers. The serpent and the tree were the last relics of Satanic possessions upon the renewed earth. The sway of the "tohu va bohu" was already broken and held in check by creative might. The serpent and the tree, its last evidences, were to be overcome and banished from the earth by man himself. They were the last, the only footholds of Satan upon the new earth; the only possessions he could still call *his own.* Satan himself would have been vanquished and banished from the earth, so soon as they were overcome; and the task of man, "to dress and keep the garden in Eden," would be reduced to the mere *dressing of it.*

§ 27. *The present Place of the Fallen Angels.*

This is a point of enough significance in a representation of the Biblical theory of the world, to claim our serious attention for a while.

As created beings, subject to the bounds and limitations which belong to all that is finite, the fallen angels must necessarily dwell somewhere in space. There must be somewhere within the wide realm of space, a place which serves as a dwelling-place for them, and conforms to their present condition. But where is this place to be found?

In our attempts to answer this question, we must, in order to guard ourselves against misapprehensions or one-sided views, ever keep in mind, that the fallen angels are *spirits* in precisely the same sense as are their former companions (§ 18) who still preserve their allegiance to God — that they *have* bodies, *spiritual* bodies,[1] but *not* bodies of *flesh* and *blood* —

[1] J. P. Lange (*Dogm.* p. 571,) would look upon the demons as a "host of *disembodied* spirits." I cannot agree with him. The sacred Scriptures do not contain the least intimation in favor of such a view. In accordance with the analogy of human nature, Lange would look upon death as the wages of sin obtaining in the angelic world also, as a law of nature. But this supposition appears to me, not only incompetent, but as in all respects inadmissible. It is clear that it is far from being justified by Scripture. The fundamental differences between the human and angelic natures, as taught by Sacred Writ, forbid us to draw any such conclusion. In all that the Bible says about the nature of the angels in general, and the condition of the fallen angels in particular, it excludes the idea that *physical* or *bodily* death ever obtains in the sphere of angelic being. The words: "The wages of sin is death," so far as they represent a necessary and universal principle, must certainly apply in the case of the angels also; but we can conceive, in respect to these beings as *spirits*, only *spiritual* or eternal death. Physical or corporeal death is not to be imagined in the case of spirits, whose corporeal constitution is from the outset of a spiritual (pneumatic) nature. *Man* is capable of physical death, because his corporeal nature is of a fleshly character, and shall remain thus subject to death, until his body is in the future endowed with the spiritual (pneumatic) character. But the *angels*, whose bodies were originally, at the creation, endowed with a pneumatic character, are not capable of physical death. We are led to the same conclusion from a proper consideration of the significant fact, that the angels are by nature devoid of

and that they move in the unblest sphere which has been allotted to them, with the same ease, rapidity and freedom, as do the good angels, in conformity to their nature, in their own glorious and heavenly abodes.

David Strauss has attempted to cast the reproach upon the Holy Scriptures, that their ideas respecting the present condition and the abode of the demons, cannot possibly be harmonized! "Christ beheld Satan falling as lightning from heaven;[1] but, according to the Apocalypse,[2] this fall of Satan is to take place only in the future; according to 2d Pet. 2 : 4, and Jude 6, the fallen angels are bound with chains, in the darkness of the lower world, reserved unto the judgment; according to Eph. 2 : 2, and 6 : 12, they inhabit the air, and according to 1st Pet. 5 : 8, the devil goeth about (at liberty) as a roaring

---

sex, with its predicates of generation and birth. The like conclusion must be arrived at "a posteriori," from the fact that the fallen angels are *not capable of redemption*, when we reflect that physical death, as the wages of human sin, is not *purely* a curse, but a curse conditioned no less by the Divine counsel of salvation than by human sin — that it is at the same time a *curse* and a *blessing*, for without the intervention of death man could not have been provided with salvation, or redeemed (§ 16). The analogy between the human and angelic natures, which is supplied by Matt. 22 : 30, and 1st Cor. 15, (§ 18), rests upon the juxtaposition of the bodies of the angels, as due to the creation, and the bodies of men as due to the resurrection. If we desire to draw any inference from this analogy, it can only legitimately be this one, that the corporeal nature of the angels since their fall, corresponds (as the wages of their sin) to *that* corporeal constitution wicked men shall receive in the *resurrection of the great day* (John 5 : 29) — an evidence, in the one case just as in the other, of the impossibility of salvation.

[1] Luke 10 : 18. [2] Rev. 12 : 9.

lion." He might have added that, according to Matt. 12 : 43, waste places, and Luke 8 : 31, the abyss, is their place of abode; while according to the Book of Job, Satan appears in the midst of the sons of God, before the throne of the Lord in heaven.

We shall first explain the pretended incongruity, that Satan and his companions should on the one hand be represented as dwelling in heaven, and on the other, as cast out from thence.

The whole contradiction, however apparent it be, rests upon the fact that the word "heaven" is used in several more or less restricted or specific senses, but all related and referring to each other, of which sometimes one and sometimes another is applied, and is to be understood according to the sense and connection of the discourse.

The word *heaven*, first of all (§ 24) designates the great canopy spread abroad above the earth, and enclosing it on all sides. Thus the first idea of the word is a physical one: the idea of *locality*. The use of the word in a plural sense, which prevails in Scripture, is of itself sufficient to show that the locality which is called heaven, is regarded as comprehending several varied divisions. We have first, the heavens composed of the terrestrial atmosphere; and in conformity to this sense of the word, the Bible, as do all nations and all tongues, speaks of the fowls of heaven, the reddening of the heavens, &c. According to common, every-day views and modes of expression, as well as according to the natural appearance to the eye, all regions of space lying beyond our atmosphere coincide; as, for example, Gen.

1 : 8. But, where the object is to bring heaven and earth into strong and distinct contrast, the starry heavens, which are strictly opposed to the earth, are clearly distinguished from the atmospheric heavens, which still belong, physically speaking, to the earth. The former are meant by the Scriptures, when they mention the *heavenly* hosts, whether they be speaking of the stars or the hosts of the angels.

But the word heaven does not designate merely a locality, but also a condition corresponding to this locality. Thus there is connected with the physical idea of heaven, also a symbolic and ethical idea of the same place. The ethical idea arises on the one hand, from the exaltation of heaven above the earth, and hence excludes the lowliness, weakness and poverty which prevail everywhere upon the earth; and springs, on the other hand, from the conceptions we have of the inhabitants of heaven — God and his holy angels — and includes the idea of a supra-mundane Divine *glory* and *blessedness.*

The symbolic idea of heaven, also, arises from the high, the exalted character of that place. The high, the sublime, is, in and of itself, that which ever bears sway and controls. The heaven or heavens surround the earth on all sides, control it, give it rains and fruitful seasons, but also visit upon it the chastisements and judgments of an offended God. Thus in symbolic language, heaven represents a *power* which controls and governs all that is earthly.

The words of Christ concerning the falling of Satan, and also the words of the Apocalypse touching his ejectment from heaven, are to be understood in this last sense — on the principle of symbolic representation. Both passages express the loss of his power, his lordly dominion. This is demanded by the connection in both places, and no less, by the prophetico-scenic character of the description. This is the only possible signification to give to the word heaven in this place, as will be allowed by every honest mind, by all but those who fear not to tamper with Scripture. That the two passages contradict each other, in one representing as past what the other says shall take place in the future, no well-informed and discerning mind can believe. It is clear that the words of our Lord may be easily explained, whether they refer to the first revolt and fall of Satan, or to his subjugation which was now being accomplished through the disciples, to whom the Lord had given power over his kingdom, or finally, to the doom of this powerful spirit of evil, foreseen as taking place at the judgment of the last day.

Further, though the Book of Job represents Satan as appearing among the sons of God, before the throne in heaven, it by no means follows that we are to suppose he dwells in heaven, and among the sons of God. If we strip the scene of its poetical clothing, nothing remains but the mere fact that Satan, at least at that time, had still the right and the power to accuse man before his Creator, and, to the extent of Divine permission, also to tempt and injure him. But this view agrees well with what is said in other books of Scripture.[1] The Book of Job does not speak in general of the abode of Satan at all.

[1] Compare, for example, Luke 22 : 31; Rev. 12 : 10, etc.

Again, let us now consider the passage, Eph. 6 : 12. "For we wrestle not," says the Apostle, "against flesh and blood (against feeble, powerless men), but against principalities, against powers, against the rulers of the darkness of this world, against spiritual wickedness in high places (εν τοῖς ἐπουρανίοις)." The Apostle does not say distinctly, "in heaven;" he chooses, and with evident design, a less determinate expression. But had he used the words "in heaven," his meaning would have been nothing different; it might, indeed, have been more exposed to misapprehension.

But what does the Apostle mean by saying that evil spirits dwell *in heaven*, or *in heavenly places?* Are we to apprehend this expression as having a *local*, an *ethical*, or a *symbolic* meaning?

Certainly not as having an *ethical* one. It is wholly out of the question that the Apostle should have intended to represent those spirits of evil, the rulers in the darkness of this world, as *happy*, as *blessed* beings, from the relation in which he places them toward heaven.

But is the expression *symbolic?* — does it denote their dominion over the earth, their controlling power over men?[1] The connection appears, at the first glance, strikingly in favor of such an interpretation. But we very soon discover how inadmissible a symbolic interpretation would be. We are allowed and compelled to apprehend the expressions used in Luke

---

[1] Thus Hengstenberg (*Die Offenbarung Johannis*, I. 619), who would explain this passage precisely according to the analogy of Luke 10 : 18, and Rev. 12 : 9.

10 : 18, and Rev. 12 : 9, symbolically, from the prophetico-scenic character of the descriptions themselves. But here the case is wholly different. We have no warrant for interpreting this expression aside from its direct literal sense, or symbollically. Besides, such an interpretation would give rise to a tautology, such as could only be made endurable or not apparent, by regarding heaven not as the mere ideal symbol of might, but as the real place where it is collected and treasured up. This would be pursuing the thought, at least; — but we should find ourselves already passed over from the purely symbolic apprehension, to the local one.

Every consideration, therefore, leads us to regard the local apprehension of the expression as the only correct one. — But it may now be asked, are we to understand it as referring to the *lower* heavens, the terrestrial the atmospheric, heavens; or to the *upper* heavens, the heaven of the stars and the angels? — Certainly not to the upper heaven, because that is the abode of the holy angels, and to dwell *there* is nothing short of dwelling in the midst of the most perfect blessedness.

It is objected that the same Apostle, in the same epistle, uses the same expression (ἐν τοῖς ἐπουρανίοις), when speaking of Christ sitting at the right hand of God,[1] and of the dwelling-place of the holy angels. But, if it be granted that the word heaven is used in common language in a twofold sense, it is surely no valid objection to say that the same writer would not have used the word, now in this and now in that

[1] Eph. 1 : 20; 2 : 6.

sense, according to the demands of the case. Thus, when Christ in one place speaks of his Father in heaven, and of the angels of God in heaven, and in another place, of the fowls of heaven and of the heavens becoming red, no one would attempt to convict us of false or arbitrary exegesis, should we in one place refer his words to the atmospheric terrestrial heavens, and in the other, to the abodes of the blessed, far beyond the earth.

Hence, the words of the Apostle in Eph. 6 : 12, were designed to convey this meaning and none other, that the abodes of the evil spirits are to be found in the atmospheric heavens, however strange such a view may appear at the first glance.

And the correctness of this interpretation is proved by the fact that the Apostle had beforehand, in the same Epistle[1] *expressly* called Satan, "the prince *of the power of the air* (τὸν ἄρχοντα τῆς ἐξουσίας τοῦ ἀέρος), the spirit that now worketh in the children of disobedience."

All attempts, even the most recent,[2] to explain away the view contained in these words, that Satan

---

[1] Eph. 2 : 2.

[2] Hofmann (*Schriftbew*, I. 402 seq.) regards τοῦ πνεύματος as in apposition to ἀέρος. He holds that the Apostle called the "spirit that worketh in the children of disobedience," *contemptuously*, "ἀήρ," and afterwards explained himself by the term "πνεῦμα." And that this could the more easily be done, as ἀήρ and πνεῦμα are ideas of similar etymological signification. But as the use of the word ἀήρ (air, atmosphere, etc.) is constant and determinate without exception, so that it never signifies *breath*, *wind*, to say nothing of its being used figuratively in the sense of spirit; as, further, the etymological signification of the word πνεῦμα (*wind*, *breath*) is so little observed in the language of the New Testament,

and his legions dwell in the air, have signally failed. They have all been frustrated by the incontestable fact that the word "*air*" (ἀήρ), cannot denote anything else than the air which surrounds the earth,

---

that no one should think of attending to it in the course of reading, unless in a case such as John 3 : 8, where the context clearly indicates that we must take the word in its literal and etymological significance; and as, finally, the genitive τοῦ πνεύματος applies freely and without constraint to the whole conception, τῆς ἐξουσίας τοῦ αέρος (so that it stands in apposition to the chief and governing genitive τῆς ἐξουσίας, and not to the subordinate and unimportant genitive τοῦ αέρος)—as all this is plain to the mind of the reader, it is clear that we cannot but understand the Apostle to speak of Satan and his powers dwelling in the air. In addition, we have the fact that the same view signally prevailed in the minds of the Jewish Rabbis (comp. Meyer). The Apostle, who was educated under the care of these men, could not have written the words ἐξουσία τοῦ ἀέρος, without calling to mind the rabbinic view on the matter he was discussing. If he regarded the view a false one, he certainly should so have chosen his words as to exclude it altogether from his own writings. But his words in reality contain an express acknowledgment of that view. It is the more striking that *Hofmann* wholly ignores this fact, since he concedes to the book of Enoch, which is nothing better than the rabbinic legends, such an important influence upon the Epistle of Jude, and the Second Epistle of Peter. The two cases are precisely analogous. Just as Peter and Jude drew from the traditions of the book of Enoch, only what they knew to be true, passing by all its fables and puerilities, so also Paul, under the guidance of the Spirit, as they were, retained and made use of only such facts due to his rabbinic training,as stood the test of the Spirit of Wisdom from above. "Thus much," says *Meyer*, "is clear enough, amid the confused trash of rabbinic traditions, that the province of the demons is, according to these sages, in the air; and we find the Apostle Paul agreeing with them. Hence we have no right to deny that he may have received this idea in his rabbinic schooling, and afterwards made use of it in his

the atmosphere, the region of the clouds, the lower stratum of the air (in opposition to the *ether*, as the upper and purer air of heaven): that the word is not used in a single instance, either in classical or Biblical writings, in a different sense.

If we now proceed to the explanation of the passage, Luke 8 : 31, where the *abyss* (the regions of the lower world,) is mentioned as the proper abode of demons, we shall discover, on a clear insight into the passage, that by the abyss is not meant the present abode of the fallen angels, but rather, their future prison, to which they shall be irrevocably consigned, only at the day of judgment.[1] The petition of the demons, that the Lord would not send them into the deep must be explained by the preceding words: "Art thou come to torment us *before the time?*" They were tormented by the fear that the Lord might now consign them to that prison of despair and death, which they well know awaits them at the last day. "Evil spirits are not," says Hofmann, "confined where death alone bears sway, but they work in the living also, in order to lead them to sin and drag them down to death."

Further, granting that the view of Matt. 12:43, which represents the demons as walking through *dry places* seeking rest, is contained even in other parts

---

epistle; while at the same time it is by no means allowable to attribute to Paul the same curiosity which was connected with this view or axiom in the minds of the Jewish sages, since he says nothing more than that the devil and his powers inhabit the air."

[1] Rev. 20 : 3–10.

of Scripture;[1] we do not feel ourselves warranted in giving all these passages a figurative meaning; neither can we discover in them any substantial contradiction to Eph. 2 : 2, and 6 : 12. There is no reason to suppose that if demons inhabit the regions of the air, they cannot inhabit the dry or waste places of the earth also—both predicates in respect to them may be alike true at the same time. And though it be true that the devil, according to 1st Pet. 5 : 8, goes about as a roaring lion, seeking whom he may devour, and that he walks to and fro upon the earth, amid the habitations of men—all this does not exclude the idea that he should dwell in the regions of the air, and in the waste places of the earth. This alone necessarily follows, that he is not, together with all his power and influence, banished to and confined in these regions and places.

Paul himself, in other places, calls the Satanic power of the air, the spirit which worketh in the children of disobedience, and the evil hosts of the heavenly places, the rulers of the darkness of this world. And, according to the words of our Lord in Matt. 12 : 43, demons pass to and from waste places upon the earth generally, at will.

It yet remains for us to incorporate the results we have just now attained, with our previous knowledge of the history and condition of the fallen angels.

Satan possessed a foothold upon the renewed earth, and a place where to commence his direful operations, only in the tree of knowledge, and in the serpent, the instrument of the temptation. He made

---

Lev. 16 : 10; Is. 34 : 13, 14; Rev. 12 : 9.

use of *them* as means of a betrayal by which he hoped to acquire, in the earth from which he had been driven, a new province where to exercise his dominion. He succeeded; but still, not so fully as he had hoped.

The fall brought man under the power of him by whom he was betrayed, and the latter became the ruler of the darkness of this world, which, as the consequence of sin, again cast its brooding shadows over all things; Satan became the prince, yea, even, the *god* of this world.[1] But this dominion was prevented from being absolute and universal by the intervention of the plan of salvation; for the evil one finds his proper subjects *only* in the children of disobedience, he rules *only* in the darkness of this world, and *only* the darkened minds of the unbelieving acknowledge him as their prince and God. The "tohu va bohu," which had been reduced by the six days' work, again broke forth, however, in the form of thorns and thistles growing from the sin-cursed earth; of fatal poisons in the animal and vegetable kingdoms; of deserts and barren solitudes upon the face of the earth; of careering storms and pestiferous miasmata in the atmosphere. But the new and glorious character the earth received in the six days' creation, with its bright and warm suns, its fruitful seasons, its green valleys, its liveried beauty, its teeming life and showers of blessings, still predominated. Thus it appears that the leagued hosts of Satanic power were not again permitted to take complete possession of the earth; not even after the fall of

---

[1] 2 Cor. 4 : 4.

man, and the wide-spread ruin arising therefrom — no, not even to find in its waste places that rest which they seek.[1] Hurled from heaven, the place of glory and blessedness, shut out from the society of the holy angels, their former companions, and still strangers upon earth, as the theatre of a salvation in which they have no part, they take a position between heaven and earth—they make the air their abode. The earth was their original home; they have *old* claims upon it, on account of the "tohu va bohu" out of which it was formed; they have acquired *new* claims upon it, by means of the sin and ruin they have caused, to mar the glory of our human world. To make good these claims, to enlarge and extend them, is their one fond hope, the object of their leagued endeavors.

§ 28. *The Universal History of the Cosmos.*

The Holy Scriptures clearly place the development and destiny, the aim and end of *the whole creation*, in this striking, this unique (einheitlichen) point of view, that they are all comprehended under one single world-plan, and that all history is governed, animated and sustained, by this same all-controlling plan. They supply us with a drama of the development of a world, in its most general and essential features, in which all the *creatures* formed and disciplined, so well as the *Creator* forming and disciplining them, take an active part; in which there is given, either by God himself, or through the medium of a free choice on the part of the creature, to the finite *spirit*,

[1] Matt. 12 : 43, 44.

and also to physical *nature* where it is to manifest its activity—to the *angel* of heaven as well as to *man* upon earth,—a special part to sustain—to each one his particular rôle; but this ever in conformity to the nature and measure of his Divine vocation. They point us to "*one God, the Father, of whom are all things*, and we in him: and *one Lord Jesus Christ, by whom* are all things and we by him."[1] They open to us a view into the *counsels* of the Divine will, from whence in the beginning *all things* proceeded, by which *all things* were ordained from eternity, according to the good pleasure of his will, so that he might, in the fulness of times, *gather together in* ONE, ALL *things in Christ, both which are in heaven, and which are in earth;* even in him,[2] and to him;[3]—so that *every name that is named, not only in this world, but also in that which is to come*, may be united in everlasting harmony and fulness under him the one head, and in such a manner that one may not be made perfect without the other,[4] so that when all is finished, *God may be all in all.*[5]

Thus do the Holy Scriptures furnish us with a history, now by hints, and now more at length—ever according to man's special need of knowledge and capacities for it, for his religious wants are always first looked to—a history which, taking within its wide grasp the whole universe, and placing its total development in such a point of view, that it may be seen as the offspring of a single Divine decree, and to tend to one final goal—a history which may thus with

[1] 1 Cor. 8 : 6. [2] Eph. 1 : 10. [3] Rom. 11 : 36.
[4] Heb. 11 · 40. [5] 1 Cor. 15 : 28.

special emphasis be called a *universal history*, a history for the full and thorough knowledge of which we must await the clear vision of the eternal state, when our present fragmentary knowledge shall be for ever done away.[1]

This history, according to its elements and fundamental features as derived from Scripture, may be significantly divided into *four* grand periods, or *ages of the world* (αἰῶνες).

The *first* period, which may be called the *primeval age*, comprehends the creation of the whole universe, together with its original inhabitants, the angels; and also the development and partial fall of the latter, by which at least *one* of the happy and glorious worlds of the beginning, was engulphed in the floods of destruction, ruined and laid waste, made a "tohu va bohu."

The *second* period, which we shall call the *past age*, includes the restitution of the earth which had been destroyed on account of the sin of the fallen angels, the creation of man as the lord and inhabitant of the new earth, and also the trial of man which resulted in his opposition to God, and thus produced a new schism in the unity of the universe, a fresh discord in the harmony of the spheres.

The *third* period, called in Scripture ὁ αἰὼν οὗτος, and which we shall, in harmony with its terms, call the *present age*, comprehends the redemption of man through Christ, and the renewal of the earth which has been marred by fall of the human race. In Christ we are about to see the unfulfilled mission of

---

[1] 1 Cor. 13 : 9, 10.

the second age of the world, most fully and gloriously realized, and the twice disturbed Divine plan of the world fully carried out.

The *fourth* period, the *future age* of the Scriptures, ὁ αἰὼν ἐκεῖνος, ὁ αἰὼν μέλλων, is to be the eternal Sabbath of all God's creatures who have remained steadfast to him, and of those who have been redeemed — in it they shall enter into His everlasting rest. That time is to be one with eternity; in it all developments shall have been fully unfolded, all changes brought to rest, and all histories finally and eternally closed.

We have already considered the chief points in the history of the two first ages of the world, so far as they may be gathered from revelation — we have considered them, both in themselves, and in their relation to the whole. We have also become acquainted with the third age of the world, in its fundamental features, with respect to its mission and tendencies (§ 16); but have been prevented from pursuing our acquaintance further, from the necessity of understanding this age in its extra-mundane relations. We shall now take up the thread of our representation, just at the point where we were previously compelled to abandon it for a time.

§ 29. *The Interest of the Angels in Earthly Developments.*

The universal history of the Cosmos was by no means brought to an end by the fall of the angels, which closed the first age of the world. A jealous God would not endure the idea that a world which he had created, should be hopelessly abandoned to that wasting ruin which the fall of its inhabitants had brought upon it. Hence he renewed it in the work of six days, and gave it new inhabitants. Thus there began a new act in the great drama of the world.

The fallen angels have borne the sentence of condemnation in themselves ever since the day of their daring revolt. But they still play a part in the new world; their history is intimately interwoven with the history of man, the new inhabitant; and only when history is fully brought to a close will the history of Satan be completed. The latter will still continue to press his claims, to push his projects, to vainly war against the plan of redemption, until he sees himself completely overthrown by its most triumphant and enduring success. So long as there still remains upon earth anything accessible to him, which he can take pleasure in destroying; so long as there still remain here any beings not fully delivered by the redemption of Christ from this evil world, from the darkness of unbelief and alienation from God, of which he is the prince;[1] so long as he may still find the means and opportunity of accusing men before the throne of eternal justice;[2] yea, so long as it remains perhaps abstractly possible that he may, by the most extreme, subtle and persistent endeavors, deceive even the elect[3] — so long must the tremendous curse to which both he and his fol-

[1] Eph. 6 : 12.
[2] Job 1 : 2; Zech. 3 : 1; Luke 22 : 31; Rev. 12 : 10.
[3] Matt. 24 : 22 24.

lowers are subjected, and which hovers above them like a gathering and an appalling storm, delay to hurl its shivering bolts upon the incorrigible offenders.

This waiting for the judgment of the great day, keeps the *good angels* in a state of eager expectation; for neither can the end of *their* history, *their* unending Sabbath of holy rest and perfect blessedness, be brought about, until the contest between good and evil, between light and darkness in the minds of created beings, in which they play so important a part, is brought to a triumphant and glorious issue. The disturbed harmony of the universe must first be restored, all elements of opposition to God removed, and all discords resolved into one harmonious, majestic, universal and unceasing anthem of praise to a great and good Creator.

Those words in the Book of Job (38 : 7), to which we have already several times referred, now appear to us in a new and clearer light; we now begin to perceive why the restitution of the earth should have filled the sons of God, the inhabitants of the morning-stars, with joy and delight, and inspired them with adoring and triumphant songs.

We now also perceive how deep and thrilling is their interest in man and in human history; why they are ever ready to guard and further[1] the interests of God's kingdom upon the earth; rejoice at its advancement,[2] sorrow over its want of success,[3] to co-operate with man in his wrestlings with the powers

[1] Ps. 91 : 11, 12: Heb. 1 : 14. [2] Luke 15 : 10.
[3] Matt. 18 : 10; 1 Cor. 11 : 10.

of darkness,[1] and to participate in all the sorrows and joys, the struggles and triumphs of the human race. Their calling, as ministers and messengers of God, to protect and defend the interests of his kingdom, is not adequate to the explanation of their sympathy with man—the deep interest, the joy and delight they manifest at every success of the plan of salvation, does not arise wholly from the blessedness they should experience in the service of God — in the execution of his commands and the maintenance of his decrees—were there no real and close connection between this plan of salvation, these decrees, and themselves, had the former no influential reference to the nature and position of the latter. No, they have, besides, a special, a personal interest in earthly affairs; the history of man is also *their* own history; every success of the plan of salvation here upon earth, brings *their* history nearer to its triumphant conclusion; but, also, every obstruction of this plan retards the progress of *their* history.

§ 30. *Participation of the Angels in the Preparatives to Salvation.*

The first promise to man placed in prospect a long and arduous struggle between the seed of the woman and the seed of the serpent. The final issue of the contest was, indeed, not left in doubt; for the head of the serpent was to be bruised, the seed of the woman was to conquer, but not without receiving many wounds and suffering many reverses.

---

[1] Jude 9.

Satan seemed, however, for a while victorious. He had again become a *power* upon that earth which he had forfeited, and though it had been renewed, he had succeeded in marring it a second time; he had become "the prince of this world,"[1] yea, even "the *god* of this world,"[2] and his angels "the rulers of the darkness of this world."[3] He was at least permitted to appear in the garments of truth, and promise to those whom he would fain induce to serve and obey him: "All this power will I give thee, and the glory of them: for that is delivered into me: and to whomsoever I will, I give it."[4] He was permitted to appear as "the accuser of our brethren,"[5] hypocritically appealing to the justice of God, and demanding that the same curse which had already lighted upon himself, or was held in sure reserve for him, should be visited upon them (when they would not own him as their master),—for, were they not, too, all of them, sinners against God?

He acquired a wide province to his dominions in the worship of nature as practised by the heathen,[6] and but too often did he succeed in planting the growing seeds of idolatry, in the *heart of that nation* which God had chosen as the bearer of the yet undeveloped plan of salvation. But he was yearly reminded in the services of the Hebrews, that a great Atonement had been found, which was both unobjectionable and all-sufficient, and before which Satan

---

[1] John 14 : 30. [2] 2 Cor. 4 : 4. [3] Eph. 6 : 12.
[4] Luke 4 : 6. [5] Rev. 12 : 10; Job 1 : 9; Zech. 3.
[6] 1 Cor. 10 : 20, 21.

himself, the great accuser, must stand silent and confounded.[1]

But the angels of God were far from being listless beholders of the developments and contests that were now taking place upon the earth. They constitute the heavenly host, the powers of the upper world, after whom God names himself "the Lord of hosts" (JEHOVAH SABAOTH), and "the captain of the host of the Lord."[2] They ever surround the throne of the Almighty, ready to be sent to minister to them who shall be heirs of salvation; prepared to protect and defend the righteous, and keep them in all their ways, lest they dash their foot against a stone.[3]

As their destiny neither was, nor indeed could be, to settle the changing fortunes of earthly affairs, by taking a decisive step in the great contest, in reliance upon their own unborrowed strength, but rather, to be messengers and ministers of Him who alone has power to conduct to a final and triumphal issue the battle which was being waged—we may well suppose that their deepest interest and most active participation would not be manifested, until

[1] On the great day of atonement two goats were brought forth, and determined by lot, the one "*for the Lord,*" and the other "*for Azazel*" (a name for Satan). The sins of the whole people were then typically atoned for, by means of the blood of the first goat; after which the sins *already atoned for*, were laid upon the head of the other goat, which was then sent into the wilderness to *Azazel*, in order that he might learn what had happened, and become conscious that by virtue of the atoning grace of God, he no longer had power over the people of Israel. For further particulars, compare my work: *Das Mosaische Opfer*, Mitau, 1842, p. 266–302.

[2] Josh. 5 : 14, compare 6 : 2.

[3] Ps. 91 : 11, 12.

the great Captain of salvation placed himself personally at the head of the ranks, until the preparatory stage of salvation had been passed, and the point reached where it was to be accomplished and offered to a waiting world.

But, even in that preparatory stage, they were not mere intent beholders of what was passing, for it is said that "the law"—our school-master to bring us to Christ—"was received by the disposition of angels,"[1] or, according to the Apostle Paul: "ordained by angels in the hand of a mediator."[2] This is further proven by the fact that angelic beings, who move unseen through our material creation, sometimes (when they wished specially to strengthen the faith of some one, or supply needed comfort,) embodied themselves so as to be apprehended by the senses, either in dreams or during waking hours; as when Jacob, during his flight from the land of promise, beheld in a dream, the angels of God ascending and descending between heaven and earth[3] as the active and untiring bearers or media of Divine agency—or when, as he returned, a double host of the angels of God met him;[4] or, finally, when upon the prayer of Elisha, the Lord opened the eyes of the desponding servant of the man of God, so that he saw the mountain to be full of horses and chariots of fire round about Elisha.[5]

---

[1] Acts 7 : 53. [2] Gal. 3 : 19; Heb. 2 : 2. [3] Gen. 28.
[4] Gen. 32 : 1, 2. [5] 2 Kings 6 : 17.

§ 31. *Christ the Second Adam.*

But when the fulness of time was come, God sent forth his Son, made of a woman, made under the law, to redeem them that were under the law, that we might receive the adoption of sons.[1] The Word became flesh and dwelt among us, and all whose spiritual eyes were opened to discern majesty arrayed in the garments of lowliness, might still behold "his glory as the glory of the only begotten Son of God, full of grace and truth."[2] The eternal Word by whom the heavens and the earth were created, appeared in the world, to save the world and conduct it to its ultimate consummation. The first born of all creatures, the image of the invisible God, the prototype of man created in the image of God, became man; the Lord of glory appeared in the form of a servant, and became like to us in all things, sin excepted. And as formerly, when God renewed the earth which had been laid waste through the fall of the angels, the morning stars sang together, and all the sons of God shouted for joy; so also now the advent of Christ, the second Adam, the Redeemer of men, was celebrated by the praises of the angelic hosts.[3] The celestial worlds joined in the celebration —the star in the east, the sign of the new-born king, came and stood over the lowly stable in Bethlehem, where the matchless wonder of worlds had taken place.[4] Christ took part in the history of the world, in the history of the universe, as the second Adam, the restorer of the human race, in order to accom-

[1] Gal. 4 : 4, 5. [2] John 1 : 14. [3] Luke 2 : 10–14. [4] Matt. 2 : 2 seq.

plish finally and gloriously the counsels of the grace of God, which had been devised in eternity, but twice disturbed by the revolt of his creatures.

He was to do away with the false, the unhealthy development upon which man had entered, which ended only in sin and death; and he was to restore all that it had destroyed. He was to take up again the missed or neglected development God had appointed for man, which was to lead to unalloyed and endless perfection and blessedness; to bring this earth to its final and perfect state, and to resolve the whole universe again into one harmonious and glorious whole.

In order to accomplish these high ends, he entered the organism of the human race by being born of a woman,[1] as a new and holy member of this race, and furnished with a fulness of life that might never be diminished or exhausted. He, the new and healthy member, bears all the infirmities and sicknesses of the whole organism — he heals all its diseases from his inexhaustible life-giving resources. Thus he becomes the *head* and the *heart* of the total organism, and as formerly our sins and infirmities brought suffering and sorrow upon him, so also now since his death and victory, his grace and healing power pervades the whole organism—the power of the victory

[1] Christ, from being born of a (however devout and pious her character, still) *sinful* woman, was *no more* affected with the general sinfulness of humanity, than the noble scion assumes the ignoble nature of the wild fruit-tree into which it is grafted. Although nourished by the sap of the wild, the ingrafted shoot bears not the fruit of the wild tree, but that of its own noble species.

is felt by all. The new life-blood flows from him into the body of humanity, pervades and quickens into renewed life by its wondrous power every single member of the whole body, so far as its influence is received, through the channel of a bond uniting that member with the great Head. But all who are not united in community of life with the head, shall perish and be cast away.[1] As we have all sprung from Adam by natural generation, and thus partake of Adam's sin and guilt, so also are we to obtain Christ's righteousness and holiness by being spiritually begotten of him. We must abide by him, the captain of our salvation, who endures great conflicts and gains great victories *for* us and *with* us; we must follow him through contests and victories, that we may finally be exalted with him to that glory which he has acquired by his own matchless power.

He took the place of the first Adam, the place of the whole human race; he did that which we should have done, but could not do, since we are sinners; he *suffered* where we should have suffered — should have suffered eternally. He obtained *eternal* redemption for us. For by his *death* he has acquired a *merit* which receives, from the Divinity of his nature, infinite worth and eternal validity, and thus removes the immeasurable guilt of our sins;—by his *resurrection* he has brought to light a fulness of *life* and immortality, which, flowing from his Divine nature, is adequate to heal every disease, to supply every deficiency, and to give victorious strength to those that are ready to faint. He has by *his own*

---

[1] John 15 : 4–6.

death, taken away the sting of death; for the sting of death is sin; — his *resurrection* has opened the way for our resurrection, for:

> "Shall Head and members part in twain,
> And never be rejoined again?"

His *ascension* is the pledge of our future exaltation; he, by *sitting at the right hand of God* with regal power, completes our salvation, and conducts heaven and earth, angels and men, to that state of absolute perfection decreed in the counsels of eternity. Thus is the human race finally to attain that position in the universe for which it was originally destined.

When the Lord of glory became man, he laid aside his Divine form;[1] but when he *ascended* again to heaven, he assumed anew the full glory of the Divine character. He appeared in the likeness of sinful flesh, amidst the sinful inhabitants of this earth;[2] but he arose from the grave with a glorified body, and he now sits in the same glorified human form — flesh of our flesh, and bone of our bone — on the right hand of power. While he tabernacled upon the earth, his Godhead took part in the lowliness, the griefs and sorrows of human nature, by means of the personal union of the two natures; now as he sits at the right hand of God on high, his human nature partakes of all the infinite attributes of the Godhead. He who calls us brethren,[3] rules the world: the man Jesus it is, who is the judge of the quick and the dead.

The Redeemer, even during his earthly life *in the form of a servant*, gave, in his wondrous works, the

---

[1] Phil. 2 : 6. [2] Rom. 8 : 3. [3] Heb. 2 : 11.

beginnings, the types and pledges of that full redemption which he as the *glorified* Son of man is to bring about, at the winding up of earthly affairs. His miracles pertained essentially to his character and work as the second Adam, the restorer of all things. Man by the fall lost that dominion over physical nature and the creatures of the earth which God at the first bestowed upon him — the proper relation between spirit and nature was thereby disturbed. Sickness, pain, misery and death, coupled with many disturbances in the economy of nature, entered the life of humanity. It was the mission of the Son of man, to recover and exercise this lost dominion, to do away with all the consequences of sin, and restore completely the proper relation between spirit and nature, between mind and matter. This can, indeed, be effected in its full extent and completeness, only at the end of this economy of affairs, at the close of the third age of the world — when the new life which Christ has implanted in humanity, has thoroughly pervaded and transformed the whole race. But it was both possible and fitting that the first fruits of this restoration should already be visible, as types and pledges of its final and full accomplishment. Hence he appeased the raging storm of the sea, with the potent words, "Peace, be still!" as a sign that he would in future heal all the wounds and convulsions of nature; hence he showed his absolute control over the sustaining properties of food and drink, by turning the water into wine, and satisfying with a few small loaves five thousand men that hungered. Hence he healed all manner of sicknesses, and called the dead back into life again, as a sign that he was about to entirely abolish the power of death; and hence, finally, he destroyed the fearful power of the prince of darkness over men, which was specially observable in the possessed of his time, in order to show that he was come to destroy all the works of darkness.

And as he ascended into heaven, so in like manner shall he come again,[1] in the clouds of heaven, with Divine majesty and glory, that he may judge the quick and the dead, and conduct both heaven and earth to their state of ultimate perfection. He has gone away, as he said he should, "*in order to prepare a place for us.*[2] When he returns again, then shall that place which is being prepared, receive his followers into eternal rest and blessedness.

§ 32. *Co-operation and Opposition of the Angels in the Life of Christ.*

The life of the Redeemer upon earth, was the central and turning point of the whole history of the human race, and—on account of the peculiar position of man with respect to the universe—of the history of the whole universe also.

Hence we have here, swelled to great power and concentrated, the efforts and strivings of angels and demons, of the seed of the woman and the seed of the serpent; the one sympathizing and co-operating with the holy Redeemer, the other hating and opposing him. On the one hand we see enmity and hate, an enlistment of all the powers of darkness

[1] Acts 1 : 11. [2] John 14 : 2.

to injure the Lord's Anointed and prevent the completion of his glorious work. This satanic warring against the Prince of life was persisted in from the manger to the cross! Herod's thirst for blood, the cruel persecutions of the high priests, the treachery of Judas, the wild and murderous clamor of the multitude, Pilate's fear of man, the temptation through hunger and the offer of worldly advantage in the wilderness, the temptation of sorrow in the garden of Gethsemane — all these were enlisted by Satan against the holy Redeemer; — "For of a truth, against thy holy child Jesus, whom thou hast anointed, both Herod and Pontius Pilate, with the Gentiles and the people of Israel, were gathered together; *for to do whatsoever thy hand and thy counsel determined before to be done!*"[1]

The first, the chief, and the most decisive onset of the prince of darkness, was *the temptation in the wilderness.* It was the same in form, substance and aim, as the the temptation of the first Adam. The latter, as we have already seen, was necessary and indispensable. But since the first Adam did not withstand the temptation, the second Adam had to be subjected to the same trial. As the false, the unnatural development of the human race, by which it was involved in death and ruin, began by *subjection* to the power of the tempter, so also was it necessary that the new development which was to lead to the redemption, restoration and perfecting of the human race, should commence with the *conquest* of this arch traitor. "And when the devil

[1] Acts 4 : 27, 28.

had ended all the temptation, *he departed from him for a season.*"[1]

But it was necessary that the whole weight of human sorrows should come before him in the form of a temptation, as well as thirst for earthly gain and glory, so that he might be like to us in all things, and be tempted in all things like as we are, though without sin.[2] Hence Satan was permitted to tempt him anew and in another point—to try if he could not induce him to abandon his vocation as the Redeemer of men, by presenting to his mind the fearful burden of sorrows which awaited him.

This temptation was first presented to him under the form of tender love and sympathizing regard, by the words of his disciple Peter: "Be it far from thee, Lord: this shall not be unto thee!" But the Lord was not to be deceived; he well knew how to distinguish between the weak and erring love of a disciple, which served as a covering to the designs of the tempter, and the satanic origin and bearing of the words spoken. Hence he said: "Get thee behind me, Satan; for thou art an offence unto me!"[3]

But it was in the garden of Gethsemane that this same temptation displayed openly its full, its matured power. And when the Redeemer now came forth victorious from this temptation also, and prepared to face all the terrors of death, with courage undaunted, Satan himself hastened, in impotent rage, to bring these terrors quickly to pass, in order to—undo himself and destroy his own power. He had put it into the heart of Judas Iscariot, one of the disciples, to

[1] Luke 4 : 13. [2] Heb. 4 : 15. [3] Matt. 16 : 22, 23.

betray his Master;[1] Satan entered into this son of Simon after he received the sop from the hand of his Lord.[2] And now the whole multitude, inoculated with his infernal rancor, and raging like wild beasts against *him* who in pitying love left his throne in heaven to save *them*, cry out in fanatical rage: "Crucify him, crucify him."

But, on the other hand, we behold the holy angels showing the liveliest interest in the Redeemer. Heaven was again opened, and the angels of God ascended and descended upon the Son of man.[3] Angels announced to the elect that the time was near for which,

> "Patriarchs, and holy Seers,
> Had hoped for many weary years."

And when the hour had arrived, when the ever-during wonder of the world's history had come to pass, in the lowly manger of Bethlehem, the angelic choirs praised in joyful, swelling anthems, the wondrous, the boundless grace of God. Angels also kept watch over the little child, prepared the way for his escape into Egypt, from the bloody Herod, and brought him safely back after the danger was past.[4] When the Redeemer came forth victorious from the temptation in the wilderness, "behold, angels came and ministered unto him;"[5] and when he had gloriously endured the fearful agony of Gethsemane, "there appeared an angel unto him from heaven, strengthening him.[6] And when he had now

[1] John 13 : 2. [2] John 13 : 27. [3] John 1 : 51.
[4] Matt. 2 : 13 19. [5] Matt. 4 : 11. [6] Luke 22 : 43.

destroyed the power of death, had victoriously burst its bands and brought life and immortality to light, angels were the triumphant announcers of this victory of life over death; and they, too, lingered behind, when the Lord ascended to heaven, to announce to the bereaved disciples that he should come again in glory.

§ 33. *Ascension of Christ, and Progress of the Contest till His Return.*

The death and resurrection of the Redeemer brought his earthly work to a close. He had now procured salvation, and the means (means of grace) by which it might and should be applied to all who would not obstinately harden themselves against the grace of God. His work accomplished, the Lord ascends to heaven, returning to re-assume the glory he had with the Father before the foundations of the earth were laid.

The warring of light against darkness was by no means yet ended—no, it still as ever went forward, and the earth was still the battle-field. Christ the Lord did indeed ascend to heaven, but not thereby did he withdraw himself from the conflict. He still remains, since his ascension, the captain of salvation, the leader of the hosts of light—this now in its proper and comprehensive sense.

The substance and significance of the *ascension of Christ* is comprehended in this, that he again took upon himself the "form of God"[1]—the eternal Divine existence and mode of life—which he had re-

[1] Phil. 2 : 6.

nounced in becoming incarnate and like to us in all things, in order that through death he might destroy him that had the power of death.[1] That form of life which is peculiar to the Divine Being is *this*, that He, the Everlasting One, is at once as infinitely exalted above time and space (his transcendence), as he constantly and everywhere, with his essence and the energy of his will, pervades and controls, preserves and sustains both time and space (his immanence).

Hence it is clear that the ascension of Christ was not merely *a departure* from the earth; it was rather, at the same time, both *a departure* and *a coming.* It was a departure, in the sense that he returned to his Divine transcendence—in the sense that he was no longer *bodily visible* among his followers. But it was no less an all-pervading and plenary coming, in the sense that he now returned to his Divine immanence, in order that he might fulfil the promise, "lo! I am with you always, even to the end of the world;[2] and also, the promise, "where two or three are gathered together in my name, there am I in the midst of them."[3]

He who had humbled himself as the servant and minister of all, now resumed the sceptre of universal dominion as his sovereign right, and became Lord over all and blessed for evermore. God the Father placed him "far above all principalities and power, and might, and dominion, and every name that is named, not only in this world, but also in that which is to come; and put all things under his feet, and gave him to be the head over all things to the church,

[1] Heb. 2 : 14, 15. [2] Matt. 28 : 20. [3] Matt. 18 : 20.

which is his body, the fulness of him that filleth all in all.[1]

Hence his saying: "All power is given to me in heaven and upon earth."[2] Hence might he well assure his disciples: "It is expedient for you that I go away;[3] and again, "I go to prepare a place for you," and "in my Father's house are many mansions."[4]

To prepare a place for us. He shall bring, when he returns in glory, that place with him, "prepared as a bride adorned for her husband."[5]

But he has not left us "comfortless"[6] upon this poor earth, still groaning under the curse; comfortless amid all our sufferings and griefs in this "body of death;"[7] while he, in his Father's house above, where are found the throne of glory and the abodes of the blessed, prepares a place for us; while he, as the ruler and judge of the world, conducts all things to their ultimate consummation, and thus provides and secures for man, that glory which awaits him in eternal life. For, being the ruler of the world, he is at the same time the head of the church, the first born among many brethren; he pours out his Spirit upon all flesh, sends the Comforter to lead us into all truth, and to prepare us *for our place* as he (Christ) prepares *it for us.*

Conflicts await us, severe conflicts; for the Lord came not to send peace upon earth, but a sword[8]—not to send *that* peace which would be more disgraceful than the most hapless conflict, but a sword for

[1] Eph. 1:21–23. [2] Matt. 28:18. [3] John 16:7. [4] John 14:2.
[5] Rev. 21:22. [6] John 14:18. [7] Rom. 7:24. [8] Matt. 10:34.

such a contest as may gain for us a true and lasting peace. He himself is still, as ever, the sovereign leader, the great champion in all the warrings between the hosts of light and the hosts of darkness; and he commands us to take our weapons from the armory of his Spirit—to "put on the whole armor of God, that we may be able to stand against the wiles of the devil;" to take to ourselves "the breast-plate of righteousness, the shield of faith, the helmet of salvation, and the sword of the Spirit, which is the word of God.[1]

Since Satan has been foiled in all his attempts to prevent the *accomplishment of the work of salvation*, he now bends all his endeavors to prevent or hinder the *appropriation* of that salvation. Therefore "he goes about as a roaring lion, seeking whom he may devour[2]— but "resist the devil," say the Scriptures, "and he will flee from thee." [3]

But the angels of God, on the other hand, are ever ready to protect and defend the elect against the powers of darkness. And although that visible, sensible manifestation of themselves and their power, which was still common even in the days of the Apostles,[4] ceased, as did all outward miracles in general, as soon as the gospel and the Church were immovably fixed upon the rock of eternal salvation — still, by no means did they then cease to be actively and efficiently present amid our earthly affairs—with a presence which can be discovered by the eye of faith alone. For they are "all ministering spirits,

[1] Eph. 6 : 11–17. [2] 1 Pet. 5 : 8. [3] James 4 : 7.
[4] Acts 8 : 26; 10 : 3; 12 : 7, etc.

sent forth to minister to them who shall be heirs of salvation."[1] And "there is joy in the presence of the angels of God, over one sinner that repenteth."[2]

But in the pregnant future, when the development of the world has reached its ultimate goal, and the Lord visibly returns with great majesty and glory, then shall these bright beings surround him as the radiance of his own glory, ready to execute his will and bring the great judgment to pass.

Till then, the harvest, the judgment of the great day, the tares sown by the hand of the evil one, grow uneradicated amid the good wheat of the field. The opposition, the antagonism ever becomes more marked, more striking, the conflict more desperate. All the powers of darkness now become enlisted for one last despairing effort, as the decisive hour draws speedily on. *Anti-christ*, the highest development and embodiment of all the powers of darkness, the exact counterpart of the true Christ, now appears at the close of the drama—He, "the man of sin, who opposeth and exalteth himself above all that is called God, or that is worshipped; so that he, as God, sitteth in the temple of God, showing himself that he is God; that wicked one, which shall be revealed, whose coming is after the working of Satan, with all power, and signs, and lying wonders."[3]

This is the *Messiah* of Satan's sending, possessing all the attributes of the spirit of the abyss: this is the *Redeemer* commissioned by Satan himself, to redeem men after a Satanic manner—to free them from all Divine laws, to withdraw them from the new-

[1] Heb. 1 : 14. [2] Luke 15 : 10. [3] 2 Thess. 2 : 3–10.

creating influences of the Spirit, and introduce them into the liberty of the children of Satan.

But the very fact of the mystery of iniquity having reached in Anti-christ its highest development, makes way for the speedy coming of the final judgment. As soon as the man of sin, the son of perdition has revealed himself in all his impotent madness, then "shall the Lord consume him with the breath of his mouth, and destroy him with the brightness of his coming."[1]

§ 34. *Return of Christ and Renovation of the Heavens and the Earth.*

Sudden, and unexpected "as a thief in the night,"[2] shall be the coming of the great day of the Lord: "as the lightning cometh out of the east and shineth even unto the west, so also shall the coming of the Son of man be."[3] Sudden, inevitable and irremediable destruction shall come upon all the enemies of God, and they shall not escape.[4] The appearance of the now coming judgment of the world, shall be fore-shadowed by fearful signs in heaven and upon earth. All creation shall be seized with a sudden, strange, and indescribable woe. Terror and despair shall take hold of the godless; even the hearts of the good shall seem to fail them, for fear, and for looking after those things that are about to come upon the earth. The anxious expectation of the creature shall be resolved into appalling fear, and the groanings of creation into fearful quakings; for every new thing in this sinful world is brought to

[1] 2 Thess. 2: 8. [2] 1 Thess. 5 : 2. [3] Matt. 24 : 27. [4] 1 Thess. 5 : 3.

the birth, not without the long since imposed tribute of pain and sorrow. Thus through the whole compass of creation: "upon earth there shall be distress of nations, with perplexity; the sea and the waves roaring; men's hearts failing them for fear, and for looking after those things which are coming upon the earth; for the powers of heaven shall be shaken;[1] but the Spirit and the bride (the church of Christ) say: Come! . . . Even so, come, Lord Jesus."[2]

"Immediately after the tribulation of those days, shall *the sun be darkened*, and *the moon shall not give her light*, and *the stars shall fall from heaven*, and *the powers of heaven shall be shaken.* And then shall appear *the sign of the Son of man* in heaven; and then shall the tribes of the earth mourn, and they shall see the Son of man coming in the clouds of heaven with power and great glory. And he shall send his angels with a great sound of a trumpet, and they shall gather together his elect from the four winds, from one end of heaven to the other."[3]— "The Lord himself shall descend from heaven with a shout, with the voice of the archangel, and with the trump of God; and the dead in Christ shall rise first: then we which are alive and remain shall be caught up together with them in the clouds, to meet the Lord in the air; and so shall we ever be with the Lord."[4] "The day of the Lord will come as a thief in the night; in the which the heavens shall pass away with a great noise, and the elements shall melt

[1] Luke 21 : 25, 26. [2] Rev. 22 : 17–20.
[3] Matt. 24 : 29–31. [4] 1 Thess. 4 : 16, 17.

with fervent heat, the earth also and the works that are therein shall be burned up."[1]

The Apostle *John* also beheld in sublime *vision* the developments of this last, this great day: "Fire fell down from God out of heaven, and devoured the enemies of God. . . . And I saw a great white throne, and him that sat on it, from whose face the earth and the heaven fled away; and there was found no place for them. And I saw the dead, small and great, stand before God; and the books were opened. . . . And the dead were judged out of those things that were written in the books, according to their works. And the sea gave up the dead which were in it; and death and hell delivered up the dead which were in them. . . . And whosoever was not found written in the book of life was cast into the lake of fire. And I saw a new heaven and a new earth. . . . And he that sat upon the throne said, *Behold, I make all things new.*"[2]

Thus shall "the earnest expectation of the creature," which has for so many weary ages waited "for the maifestation of the sons of God," finally reach its long and earnestly desired object; for "the creature itself also shall be delivered from the bondage of corruption, into the glorious liberty of the children of God."[3]

Nature was created with the capability of being developed and also needing development; it fell to the lot of the created *spirit* to conduct this outer world, this material creation, to its highest development, to its ultimate and absolute state of perfection.

---

[1] 2 Pet. 3 : 10. [2] Rev. 20 & 21. [3] Rom. 8 : 19–21.

We see this end now in part reached. A curse, followed with devastating ruin, had first been brought into this *terrestrial* region, by the fall of the angels; and, again by the fall of man. The *celestial* worlds, the dwelling-places of the angels, had also suffered through this double catastrophe; not positively, but *by privation*, for by it the consummation of their high and perfect development, their harmonious connection and absolutely perfect state, was hindered. *Man* had taken the place of the exiled angels; he was to fill up the void, to check the disturbance, and restore the universe to its wonted harmony. But instead of so doing, he himself fell as the angels did before him, and thus dragged the earth a second time into devastating ruin. He thus became absolutely incapable of fulfilling his mission. Hence *Christ*, the second Adam, came in the stead of man, to renew and complete what man had destroyed and failed of accomplishing—to make all things new—the heavens and the earth. He was to finish the work of man, in conducting this world to its perfect state, and thus establishing harmony between our and all other worlds. But this could no longer be done in the method originally designed—by a quiet, gradual organic development—for this method had been disturbed and destroyed by the entrance of sin. No: a new development was demanded, one which should complete its energetic course and perfect itself, only by breaking out in the appalling catastrophe, the consuming and purifying fires of the last day. But "*a new heaven* and *a new earth*," purified from all dross and defilement, shall proceed from those *flaming elements*

—"a new heaven and a new earth wherein dwelleth righteousness."[1]

We must at present, however, instead of permitting ourselves to be further carried along in our narrative, by the rapid stream of startling events to be disclosed on that inconceivably grand and majestic day—instead of proceeding immediately to the consideration of the changes which that day shall produce in the condition of *free, personal* beings, stop a moment, to review with a hasty glance the coming changes and developments in material nature, which is devoid of *personality*. We have already spoken of these changes and developments; it remains to take into more detailed and careful consideration, the difficulties they individually present.

---

[1] 2 Pet. 3 : 10–13 — comp. Is. 65 : 17; Rev. 21 : 1.

[2] The different points in that great day of the future cannot be separated and arranged in chronological order. It is scarcely to be expected that such an order will be observed in the realization of the events predicted, but rather that all will take place at once. The appearance of the Lord, the resurrection of the dead, the transformation of the yet living, the purification of the earth, the judgment, the infliction and execution of the sentence, will all be the work of an indescribably glorious and solemn moment, pregnant with the weal and woe of a whole eternity. Just as the sun, appearing in all its magnificence and glory, produces a thousand different effects at the same time and by the same power — here causing a germ to be developed, a bud to unfold, fruit to ripen, the well-watered plains to abound — there, the nipped blossom to fade, the uprooted tree to wither, and the dry fields to be scorched—so also shall the coming of the everlasting, uncreated Sun, in all his majesty, produce all at once, by the power of his holiness, according to the different objects affected thereby, the various effects of attracting and repelling, of cheering and appalling, of blessing and cursing, of purifying and consuming, of crowning with blessedness and filling with woe.

The passages of Scripture we have quoted on the last two or three pages, evidently describe a grand catastrophe by which the world as it at present subsists, in its present condition, relations and connections, is to be brought to an end. From one point of view, this catastrophe may be regarded as a destruction of the world, as its complete overthrow. But Prophecy represents the end of the present economy as the commencement of a new order of things; it places, side by side with the destruction of the present world, the rise of a new and more glorious one. Only when we keep in mind both sides of this question, and allow each aspect its proper weight—when we have succeeded in combining both into one *well-proportioned* (einheitlichen) view—only then, may we give ourselves credit for rightly apprehending the import of these prophetic touches. But this is no difficult matter. We are doubtless to recognize in the fires of the last day, not a destructive, but a purifying process, just as ore is cast into the furnace not to be consumed and annihilated, but that the true metal may be separated from the dross, and come forth pure as gold seven times refined. The world itself shall not cease to exist; its present faulty, marred and imperfect condition, merely, shall pass away. That existence of the world which conforms to its original creation and destiny, shall not cease, but merely the existence of whatever belongs to it contrary to its first creation and destiny—the whole web of adverse and destructive influences and agencies, all incongruities and bad properties attached to it and implanted in its elements by the false and godless development chosen by its inhabitants, together with all that is faulty and imperfect which has been prevented from being set aside on account of the presence of sin.

That the sin-crushed earth, filled with death and woe, that the twice ruined earth, with its lone solitudes and dreary wastes, with its storms and convulsions, its poisons and pestilences, its scathing heat and deadening frosts, with its lawless and wildly raging elements, with its countless perverted ends and agencies — that this earth must, before it passes into its state of ultimate and lasting perfection and becomes the happy abode of man redeemed, be submitted to a purifying and renovating process, and that this process can only take place and be perfected in the form of such a tremendous and awful catastrophe as is to be realized on the last day, must be of itself apparent to every thoughtful mind.

But that the *heavens*, the lofty abodes of those glorious beings which have been true to their destiny, have kept their first estate and ever persevered in unswerving obedience to God, that the heavens stand in need of the renovating powers of such a catastrophe; — that the stars which shine with such immortal radiance from the sweeping arch above us, filling every beholder with the deepest sense of unalloyed purity and unchangeable stability, of the most blessed harmony and undisturbed peace, shall fall from heaven and lose the position they have held for thousands, yea, perhaps, myriads of years — that the powers of heaven, which force themselves upon our minds as the types and representatives of all stability and perfection, shall be shaken and moved; — that the starry canopy yonder above, with its countless sparkling gems, shall grow old like an earthly garment, and pass away that it may be renewed in greater magnificence and perfection — this all does not strike the mind as quite so plain and reasonable.

Many interpreters, in order to avoid the not insignificant and seemingly insurmountable difficulties attending a literal apprehension of the text of the prophecy here concerned, have sought to avail themselves, by way of remedy, of strange, false, artificial and unnatural interpretations.

The legitimacy in general of a literal interpretation of these passages, has been contested, and they have been held as symbolic descriptions of subjective conditions in this human world. It is indeed not to be denied that, in the poetic language of the Old Testament, the obscuration of the light of the heavenly bodies, of the sun, moon, and stars, is the image and comparison under which the extinction of mortal life or hope in individual persons, is represented;[1] and still more frequently, the image to denote calamities and judgments visited upon whole states or nations.[2] It is not to be denied that more particularly the light of the sun is used as the symbol of Divine revelation, and the light of the moon to represent human knowledge, wisdom, or culture;[3]

---

[1] Eccles. 12 : 2; Jer. 15 : 9.

[2] As, for example, Is. 5 : 30; 13 : 10; 34: 4; Jer. 4 : 28; Ezek. 32 : 7, 8; Amos 8 : 9; Mich. 3 : 6.

[3] Rev. 12 : 1.

and further, that the stars are used to represent the ministers of God's Church upon earth—the shining lights in the firmament of the Church.[1, 2] The application of the argument, however, drawn from a prophetic, figurative mode of speech, is not admissible in the present case. And for this reason: Every thing that is intended to be represented in Matthew 24th, which is the chief passage, under the figure of the sun and moon becoming dark, and the stars falling from heaven, has been spoken of previously and in another place, in plain, direct and unmistakable words, as something entirely different from and not at all necessarily connected with those signs in the heavens. It is also further to be observed that the doctrine of a real destruction of the present earth, followed by the appearance of a new heaven and a new earth, pervades all Scripture, and is often mentioned in such connection and with such clearness, that it becomes wholly impossible to interpret the passage in a figurative or symbolic sense.[3]—We have just the same reasons for believing that the signs of the coming of the Son of man, the falling of the stars and the obscuration of the sun and moon, are to be real, sensible appearances in the heavens, as

[1] Compare R. Stier: *Die Reden des Herrn Jesu*, vol. II., p. 562.

[2] Dan. 8 : 10, 11; Rev. 1 : 20.

[3] In addition to the passages already mentioned, compare, also, Joel 3 : 3, 4; Haggai 2 : 6, with Heb. 12 : 26, 27; Ps. 102 : 26–28 and Is. 34 : 4, with Rev. 6 : 12–14; Matt. 5 : 19 seq., etc. Comp., on the necessity of a literal interpretation, *J. P. Lange: Leben Jesu*, II., 3, p. 1273 seq., and particularly the very excellent work by J. A. L. Hebart: *Die zweite Zukunft Christi, eine Darstellung der gesammten bibl. Eschatologie.* Erlang., 1850.

we have for believing that the star in the east at the birth of Christ, and the obscuration of the sun at his death, were real, outward, visible appearances. The *truth* with regard to the contested figurative apprehension is just this, that those appearances in the heavens leave us room to assume the existence of corresponding facts upon earth and among men, since heaven and earth, spirit and nature — mind and matter—form one closely connected and related whole. But such an assumption is altogether unnecessary, for these passages themselves teach plainly and in direct terms, the real existence of such corresponding facts (earthquakes, famines, pestilences, wars and rumors of wars).

Another misinterpretation of these passages virtually (but undesignedly) reduces the reality of the occurrences foretold into a mere illusory pretence or show, and resolves their objective actuality into mere subjective perceptions.[1] The heavens, it would be maintained, are not to be renewed really, in and of themselves, according to their own proper nature and constitution, but shall merely present themselves as being so renewed, to the perceptions of men: they shall not themselves be changed, but merely the medium through which we view them. It is, to say

---

[1] Thus J. P. Lange, *Verm. Schriften* II., p. 249: "In the same sense that the creation of the heavens is involved in the creation of the earth, in Gen. 1st, so that the pre-existence of the stars is not contradicted by the fact that they are first noticed as they appeared in relation to the earth, on the fourth day, when the atmosphere became purer and more homogeneous, so also may it be spoken of the *new heavens* in connection with the new earth.

the least, probable that we shall be enabled, by the transformation and re-moulding of the earth and its atmosphere, as well as by the perfecting of our powers of vision, and the increase of our mental capacities which we may reasonably hope hereafter to experience — that we shall be at once enabled by all these, to see the form, the splendor and glory of the heavens, much more distinctly, and with a more comprehensive range of vision than at present. And just *here* lies the *truth* contained in this false apprehension of the passages before us. But it explains but one aspect of what is foretold—the *renovation* of *the heavens* — and this one but half way. The grand, the real difficulty—that the heavens shall grow old, be changed, and vanish away, as is so distinctly and unmistakably foretold — still remains altogether unexplained, and still calling for solution.

A third, and no less objectionable interpretation, would limit the purifying and renovating process of the last day, to our earth with its planetary heavens, just as the fourth day's work of the Hexæmeron is sought to be confined to the creation of the solar system. With respect, therefore, to the point before us—the destruction of the heavens—it is said we are to understand the Scriptures as speaking of our planetary heavens alone. But as we previously found the view respecting such a limitation of the fourth day's work to be inadmissible (§ 4), we now find *this* kindred view much more objectionable still. For the advantage that we would be enabled, by assuming the latter, far more easily to grasp the catastrophe which is to take place in the heavens, will never justify us in arbitrarily limiting the application of words which treat of the *whole* created heavens, and which are in themselves so clear and unmistakable. But still, this view, also, may contain its *measure of truth.* For it is possible, yea, even probable, that by means of the close connexion, by means of the articulate organization which obtains in our solar system, the effects of the catastrophe which twice involved the earth in ruin, may have extended themselves more or less to the neighboring members of the system; that the surges of that disastrous flood may have reached even to the outermost boundary of this special province of the universe. If such be indeed the fact, we can easily perceive that the final catastrophe of the judgment day, must also assume a different character here, from what it will present in the worlds contained in the heavens of the fixed stars—worlds which have not been immediately affected by blight of sin and death. We can also further perceive, why the transformation and renovation of the lower, the *planetary* heavens, which are to be placed in somewhat the same category as the earth, should be specially thorough and attended with astonishing displays of power.

But we must not, however, give up this point, that the *whole* heavens, with all their hosts of stars, *are to be subjected to* an effectual change and transformation of their whole complexion and arrangement, of their mutual relations and references, but specially of their relations and references towards the earth; and that hence, in spite of all the completeness which now obtains in the heavens, and in spite of all the excellence and blessedness of their present inhabitants, there is still *demanded* such a change and transformation as we have mentioned—a renovation and perfecting of the heavens themselves.

In accordance with this view, we are compelled to assume that the present perfection of the heavens is not an absolute but merely a relative perfection — that flaws and defects still cleave to the heavens, so that, as says Job, "the stars also are not pure in *his* sight" (25 : 5). But this must be allowed at the outset, that this imperfection consists not in the *abstraction* of a degree of perfection originally possessed and received at the creation, but merely in the *privation* of that degree of perfection in the heavens which it was designed they should attain — that they were not subjected to ruin through the sin and revolt of their inhabitants (as was the case with the earth), but were, by some circumstance or other, stopt or impeded in that course of development by which they were to arrive at a state of absolute perfection; so that they might now reach that goal, only through the powerful coming of Christ to close up the affairs of this world and renew both the heavens and the earth.

These delaying circumstances are involved in the fact that the whole universe is one organic whole, so that one part of it cannot be brought into an absolutely perfect condition, while another part still remains imperfect—that disharmony was introduced into the music of the spheres, first, by the fall of part of the angels, and again, by the fall of man; and that the good angels also, with the worlds they inhabit, while waiting for the judgment of the great day, are kept in a state of anxious expectation or delay. The more important the original position and the abode of the fallen angels, the more significantly and influentially, thereupon, man and his history were destined to be involved in the further development of the universe (§ 19), so much the greater must have been that disturbance which introduced such a discord into the harmony of the whole, which caused such a breach, such a chasm in the integrity of the whole; and so much more oppressively must its attendant sinister influences have borne upon the holy angels and their blessed abodes. But now, when at the end of this world's course the judgment is set, when Christ the Lord shall separate the good from the bad elements, by the purifying fires of the great day, when he shall burst the hampering bonds of the development, renew and rejuvenate in the creature the divine powers of life—then, sudden as a flash shall the hidden and retarded perfecting process of the universe burst on the sight, in the form of one universal catastrophe, and then shall all the relations of the heavens to themselves and to the earth be changed and newly established.

The *objective* point in the prospective transformation and renovation of the heavens is this: the heavens shall really be changed in their whole complex cast and character, so as to form a new and a different heavens. But this change may, indeed, as all will allow, involve a *subjective* point also—the whole aspect of the heavens as they appear to the eye, as well as the impression they produce upon the mind, may be very different, since the powers of human vision are to be increased and the capacities of the mind exalted; — while, on the other hand, it may very easily be the case that the catastrophe will assume a more mild and peaceful character in the worlds of the upper heavens, than in the lower ones of the solar system, which stand so closely related to our earth, and which hence have perhaps been more or less involved in the ruin brought upon the planet earth.

The conflagration of the world, regarded as a *purifying* process, is to separate all the good from the bad elements in the world; it is to purify the true and noble metal from all admixture and defilement of dross. All the elements in the world which Satan may with right call *his own*, all dross and impurities which are not capable of being renewed and ennobled, shall be returned to him as his peculiar possession. And they probably shall constitute the eternal abode of this arch-fiend and all his followers —that abode called figuratively by the Apostle John, "a lake of fire and brimstone,"[1] described by Christ, as a place of "outer darkness, where there shall be weeping, and wailing, and gnashing of teeth,"[2] and by Peter, as a place of "the mist of everlasting darkness."[3]

Thus are the heavens and the earth to be thoroughly purified by the fires of the last day, and at length reach a state of high, complete, and everabiding perfection, in which all members shall be organically united with the whole, and where universal peace and harmony shall prevail. Thus is the earth

[1] Rev. 19 : 20; 20 : 10. [2] Mark 8 : 12. [3] 2 Pet. 2 : 17.

to become, in accordance with its original destiny, the central and culminating point of the whole universe, the throne of the most immediate presence of God within the sphere of the created. For, says St. John the Divine:[1] "And I saw a new heaven and a new earth; for the first heaven and the first earth were passed away; and there was no more sea. And I John saw the *holy city*, the new Jerusalem, coming down from God out of heaven, prepared as a bride adorned for her husband. And I heard a great voice out of heaven, saying, Behold, *the tabernacle of God is with men*,[2] and he will dwell with them, and they shall be his people, and God himself shall be witn them, and be their God. . . . And I saw no temple therein: for the Lord God Almighty and the Lamb are the temple of it. And the city had no need of the sun, neither of tne moon, to shine in it: for the glory of God did lighten it, and the Lamb is the light thereof."

§ 35. *The Judgment and the Eternal Consummation.*

A similar separation between the good and the bad—the godly and the godless—shall take place in the world of spirits, through the great judgment of

---

[1] Rev. 21 : 1 seq.

[2] As to the signification of the expressions here used: "the new Jerusalem," "the holy city," "the tabernacle of God with men," compare my *Lehrbuch der heil. Geschichte*, 6th ed., § 201, 2 Anm. It is there supposed that the symbolico-typical signification of the tabernacle, of the temple, and the holy city, as the place where God dwells with his people, here attains its highest, most comprehensive, and glorious fulfilment, its most complete realization.

26

the last day. "The hour is coming, in the which, all that are in the graves shall hear his voice, and shall come forth; they that have done good, unto the resurrection of life; and they that have done evil, unto the resurrection of damnation."[1] All who have died shall, hence, come again to life—the ungodly, that they, too, may attain their unchangeable state, their consummation—in eternal damnation. They, not having been united to Christ, cannot have their bodies changed and fashioned like unto Christ's glorified body. No: they must receive bodies conformable to their spiritual condition; bodies which shall be to themselves media of pain and condemnation, just as the bodies of the righteous, "fashioned like unto Christ's glorious body,"[2] shall be to them media of enjoyment and blessedness.

"Flesh and blood cannot inherit the kingdom of God; neither doth corruption inherit incorruption."[3] Hence the necessity that those who still remain alive at the end of the world, should be subjected to some great and sudden change, in order that they may be advanced to the perfected state of those who shall rise from their graves. Paul lifts the veil from this mystery:[4] "We shall not all sleep, but we shall all be changed, in a moment, in the twinkling of an eye, at the last trump; for the trumpet shall sound, and we shall be raised incorruptible, and we shall be changed." The terrors of death,[5] the abhorrent dread of corruption, and the transport of being glorified, are all here concentrated and combined in the very moment of the great change.

---

[1] John 5 : 28, 29. [2] Phil. 3 : 21. [3] 1 Cor. 15 : 50.
[4] Verses 51 & 52. [5] Rom. 5 : 12.

As to the nature and constitution of the glorified bodies of the righteous, we may learn much from the accounts the evangelists have given us, of the appearance of the risen Redeemer, since we are justified upon the authority of the promise in Phil. 3 : 21, ("he shall change our vile body, that it may be fashioned like unto his glorious body;") in referring to the risen bodies of the saints, the qualities and characteristics which belong to his glorified body. In this connection, we may specially mention, the unexpected appearance of our Lord to his disciples, when the doors were shut, his frequent sudden appearance to them, and his equally sudden disappearance from their sight, and also, that he was customarily and without any apparent design on his part, invisible to mortal eyes, etc. We may hence regard these as natural peculiarities of the glorified body; that its material composition is of the most refined and exquisite character, and of a nature so spiritual and ethereal, as not to be grasped by our senses as they are at present constituted; that it is highly raised above the cramping conditions and circumstances of the present life of the body; that it is altogether free from the bonds and impediments necessarily belonging to "this tabernacle of clay;" that it serves with alacrity and unconditionally obeys every motion or command of the in-dwelling spirit, and that, even in the ranging flight of thought, the mind may still be attended by the willing and easy services of the body.

The Apostle Paul, in particular, gives us, in 1st Cor. 15th chapter, still further and more explicit information on this point. "It is sown," says he, "in corruption, it is raised in *incorruption:* it is sown in dishonor, it is raised in *glory:* it is sown in weakness, it is raised in *power:* it is sown a natural body, it is raised a *spiritual* body." Finally, we must here once more remember the words of our Lord: "In the resurrection they neither marry nor are given in marriage, but are as the angels of God in heaven."

Perhaps we may succeed in drawing from these known circumstances of glorified human bodies, some adequate conclusions as to the nature and circumstances of the earth, after it shall have been purified and renewed. We may, with good reason, as we think, suppose that the prospectively new earth is to be purified, ennobled and glorified, in an analogous manner and to the same degree, as our corporeal frames; and that our future glorified bodies shall bear the same relation to the material of the then glorious earth, which now exists between our present bodies and the elements of the earth as it now subsists.

The judgment of men virtually takes place in the resurrection itself, since the bodies they then individually receive, have already enstamped upon them, the marks which characterize the results of the judgment in each particular case. But to the prophet's mind, as to the human mind generally, the single points in this great closing scene, which the exalted Son of Man shall carry through all together, as though they called for but one exercise of his power, must appear separately and in due order. Hence the judgment is represented by the prophet, as something different from and subsequent to the resurrection.

The true nature of the last judgment is best learned from the parable of the division of the sheep from the goats.[1] "Come ye blessed of my Father, inherit the kingdom prepared for you, from the foundation of the world;" and "Depart from me, ye cursed, into everlasting fire, prepared for the devil and his angels."

But as the final catastrophe in the material creation, which is devoid of all personality, is to be widespread and general, is to stretch itself over the whole universe, affecting both the heavens and the earth, so also the last judgment is not to be confined alone to men, but is to embrace the angels also, in its solemn, all-disposing, and closing process. As the judgment in the case of the redeemed will be *no* judgment, inasmuch as they retain no sin for which they might be judged, and yet still *a judgment*, since it will deliver them from all the evils inseparably connected with sin and death, so also shall it be in the case of the angels of God. Thus, then, may it be explained, how the latter are represented by the Scriptures, as being, on the one hand, objects of the judicial process, and on the other, as subjects actively engaged in carrying on this process. It is said of the *angels*, Matt. 13 : 49, "They shall come forth and sever the wicked from among the just;" and *the saints* are represented as being helpers and co-workers with Christ in the work of judgment, as being those whom he is not ashamed to call brethren, to whom also, as mem-

[1] Matt. 25 : 31 seq.

bers of his body, he will give an interest in all his glory.[1] Christ himself, speaking to his disciples, says: "Verily, I say unto you, That ye which have followed me in the regeneration, when the Son of Man shall sit in the throne of his glory, ye also shall sit upon twelve thrones, judging the twelve tribes of Israel"[2]—and the Apostle Paul appeals to the Corinthians: "Know ye not that the saints shall judge the world?" . . . "Know ye not that we shall judge angels?"[3]

Thus do we see Satan's project, so long and ardently pushed, brought to a fruitless end; the final sentence of judgment pronounced and executed. We see man also, whom he had dazzled with the delusive prospect of becoming, by rebellion against God, *as God*, made, through the boundless grace of God in Christ, partaker of all the glory and blessedness of the Deity. For God has become, for time and eternity, *as man*, so that man might for eternity become *as God.*

Christ in anticipation of this time says, in his intercessory prayer:[4] "The glory which thou gavest me I have given them, that they may be one, *even as we are one. I in them, and thou in me*, that they may be made perfect in one; and that the world may know that thou hast sent me, and hast loved them as thou hast loved me," &c. Paul says, Rom. 8 : 17, that as children of God we are also "*heirs of God* and *joint-heirs with Christ;*"—the Apostle John affirms that "we shall be like Him,"[5] and Peter

---

[1] John 17 : 20 seq. [2] Matt. 19 : 28. [3] 1 Cor. 6 : 2, 3.
[4] John 17. [5] 1 John 3 : 2.

speaks of "exceeding great and precious promises, through which we may be *partakers* of the *Divine nature.*" [1]

The great judgment shall close the present and introduce the future age of the world. The grand characteristic of this future age shall consist in this, that time shall then be absorbed into eternity and become one with it. Time shall not cease to be time, any more than the creature shall cease to be a creature; for time and the creature are correlatives which may never be separated—neither of them can exist without the other. But time, by merging into eternity, shall partake of all the attributes of eternity, just as the humanity of Christ, since his exaltation to the right hand of the Father, partakes of all the attributes belonging to the Godhead of the Son, with which this humanity is personally united; and just as we also, through the mediation of this humanity, shall become partakers of the Divine nature. Thus shall all historical developments, all changes, be brought forever to a close. The creature shall have reached a state of the fullest and closest communion with God, the state for which it was originally destined (and beyond which no higher development is possible or conceivable); or—in case it have persistently refused the saving grace of God—a state of absolute separation from God (such as cuts off all possibility of a return to him or re-union with him).

---

[1] 2 Pet. 1 : 4.

§ 36. *Retrospective Glance at the Position of the Angels.*

We shall close the present chapter with a retrospective glance at the position of the angels in relation to that of man.[1]

We are accustomed, without special thought on the subject, to look on the angels as beings of a superior nature, as holy and blessed spirits, who surpass us as much in power and glory as heaven does the earth. And this view is undoubtedly the correct one, so long as it proceeds from the contrasted present conditions of angels and men. For the Scriptures give to man in his present state, where he is subject to the curse, and groans beneath the burden of his sins, a position far below that of the angels, whom they set up on high as principalities and powers, as mighty champions of God, who esteem it their highest honor to execute his commands and exercise toward him adoring love; as the heavenly hosts, from whom the king of the whole universe does not disdain to borrow one of his names (Jehovah Sabaoth).

But whether, hence, this present superior might and dignity, is necessarily possessed by the angels, so that it shall outlive all developments and changes;

[1] *The position of the angels in the economy of the universe* has seldom been sufficiently regarded. The old Protestant theology, however well it did in frowning down upon angel-olatry, nevertheless does not seem to have arrived at a clear and unprejudiced apprehension of the Biblical doctrine of the angels. As to our view of the relation of men and angels, it is essentially agreed with by Molitor, *Philos. der Geschichte*, II., p. 115, obs.; Ebrard, p. 57 seq.; Martensen, *Chr. Dogmatic.*, Kiel, 1850, p. 153 seq.

whether it be grounded in their original nature, in the very essence of their being as given them at their creation, and shall hence outlast all developments and manifest itself in eternity—this is another question, and one to which upon the ground of Divine revelation, and in opposition to generally received notions, we must oppose a most decided negative.

It cannot, indeed, be denied, on the one hand, that the nature with which the angels were endowed at their creation, was relatively superior to that of man—that it was a nature already unfolded through the creation itself. This admission is founded on the fact that the angels, being created without sex, undoubtedly possessed from the first, all the advantages arising from a provision so unlike that in the human economy — advantages which man was to attain only at the end of his development. But this was not an advantage absolutely, but merely relatively, and was more than balanced by corresponding advantages in the human race, arising from the possession of sexuality (compare § 16–18).

But man, on the other hand, was created in the image of God, as his deputy and representative, and was from the very beginning destined for a calling far above that of any angel. From this original position, from this high dignity, he fell into an estate of sin, misery, and death. But for the very reason that his original position was so all-important, not only with respect to this world which was assigned to him as a place of abode and a place where to employ his activities, but also in respect to the whole universe — for this very reason did God himself take his place, become man here upon earth, in order to redeem man, and, with man redeemed, reach the pre-determined goal. Were man indeed the least of all creatures, in respect to his original position, the wonderful fact, that God became man, that he took human nature into personal union with himself, and that he shall henceforth ever remain God-man — this fact alone is enough to raise man to a position of dignity, honor and significance, far above that of any other created being.

And is it possible that we can still cherish doubts as to the new, unparalleled and sublime position to which man is to be raised by redemption, when we reflect that he is to be adopted as a son into the family of God, to become an heir of God, and a joint-heir with Christ;[1] that he is destined to become one with the Father through the Son, as the Son is

[1] ["If human nature had, in its native construction, lacked any capital element — intellectual or moral — that is possessed by higher orders, it could not have admitted of such an alliance as it has (with Divinity). . . . Is it asked on any side, what do we mean — what do we pretend to, when we speak at large of glory, honor, immortality; or of a crown of life, or of being constituted kings and priests unto God; or of sitting on thrones to exercise powers of judgment, even over superior natures? We reply at once, that we pretend to whatever is involved in the union of the members with the Head — that Head being divine; and we expect whatever may fairly be presumed when it is said of all believers, that they shall be 'like Him,' and near Him (as his kinsmen), who is the 'brightness of the Father's glory, and the express image of his person." . . "In the scheme of redemption, the original purpose of the Creator, when he said: 'Let us make man in our image,' is at once expounded and authenticated, and it is seen that nothing great or illustrious was to be denied him."—Isaac Taylor, *Saturday Evening*, pp. 317–347.—Tr].

one with the Father, and to be made a partaker of the Divine nature? when we consider that the office of judging the world, yea, of judging angels themselves, is to be entrusted to the saints?

Angels, on the contrary, are never represented as being the offspring of God, as bearing the image of God in that eminent sense applicable to man — an image that rendered possible, yea, even shadowed forth and realized beforehand, as it were, the incarnation of God. They are never spoken of as the rulers and judges of the world, as co-heirs with Christ, and brethren of our Lord; or as partakers of the Divine nature. No: they were created as *messengers of God*, as *ministering spirits*, sent forth to *minister* to *them* who shall be heirs of salvation.

The second chapter of the Epistle to the Hebrews, bears direct testimony to the correctness of the views we have advanced. The Apostle there, by the application of the 8th Psalm to Christ, infers the superiority of his human nature over the nature of the angels. But all that holds good of the human nature of Christ the Son of man, the second Adam, also holds good of all believers, for they are all created anew in him, and shall bear the image of the heavenly, just as they have borne the image of the earthly.[1] To whatever sublime height the human nature of Christ is raised, by means of its personal oneness with the Godhead, above angels and archangels, equally high shall the faithful of the New Covenant, the members of the body of Christ, be raised hereafter (when they shall be made perfect) above all angels and every living creature.

---

[1] 1 Cor. 15 : 49.

CHAPTER FIFTH.

ASTRONOMICAL INVESTIGATIONS AND RESULTS.

"Non propterea abjicienda est doctrina certa et utilis vitæ, de multis rebus etiamsi multa ignoramus, præparemus etiam nos ad illam æternam academiam, in qua et integram plysicen discemus, cum ideam mundi nobis architectus ipse monstrabit."

MELANCHTHONIS *Initia Doctr. Phys. præfat.*

WE have been conducted by the Spirit of Prophecy, through a realm of knowledge hidden from mortal eye; but one which is ever in painful remembrance and ardent hope, claimed and greeted by the restless longings of the human mind, created in the *image* of God and *for* God, as its true, its rightful possession. We now hasten to explore another, yet a related sphere, which, though lying at a distance so remote, has been forced to disclose itself to the bold and piercing glance of man. Ourselves untravelled in its labyrinthine paths, we shall seek the hand of safe and practised guides, who will point out and explain to us the wonders of a region which has but lately been reclaimed from the depths of space, and added to the province of human knowledge.[1]

[1] We do not of course design to give *instruction* in regard to matters of astronomy in the present chapter. Our object is merely to place in connection before the reader, in a general way, such facts and views pertaining to this domain of science, as may serve to establish and unfold the Biblical theory of the world, or such as may stand in alleged contradiction to that theory, in order thereby to gain a basis for the discussion of the succeeding

§ 1. *The Sun.*

The *Sun*, the mighty king of day, first attracts our attention. Two all-controlling agencies constitute the sceptre of his dominion—gravity and light. His volume is so enormous as to be capable of furnishing material for the composition of almost one million and a half such globes as ours. Were all the planets and moons of the solar system thrown together into *one mass*, they would not constitute a body of more than the five-hundredth part the volume of this vast central sphere. The proportion is somewhat different when gravity is taken as the principle of comparison. The sun, with little over one-fourth the density of the earth, still surpasses it in weight 345,936[1] times;

chapter. We may, however, recommend the following treatises, as sources of information proper in regard to astronomical science. The works of J. H. Mädler (*populäre Astronomie*, 4th ed., Berlin, 1849; *Nachträge* thereto, Berlin, 1852; *Astronomische Briefe*, Mitau, 1846): the work of J. Lamont (*Astron. und Erdmagnetismus*, Stuttg., 1851): of John Herschel (*Outlines of Astronomy*, Lond., 1849, 3d ed., 1850): of Humboldt (*Kosmos*): the works of G. H. von Schubert (*Die Urwelt und die Fixsterne*, 2d ed., Dresd., 1839; *Lehrbuch der Sternkunde*, 3d ed., Erlang., 1847; *Naturlehre*, Calw., 1847; *Geschichte der Natur.*, vol. I., 3d ed., also under the title, *Das Weltgebäude, die Erde und die Zeiten des Menschen auf der Erde*, Erlang., 1852.

[1] It is, therefore, "certain beyond doubt, that no creature belonging to our earth is possessed of strength enough to move its limbs or walk upon the surface of the sun, as upon the earth; since the force of gravity is some 28½ times more powerful there than upon the surface of our globe. The greater and more dense the world to be inhabited, the stronger must be the bodies of its inhabitants. The most Herculean frames of the earth, were they transported to the sun, would at once reveal themselves as the

and the combined weight of all the other bodies belonging to the system, about 700 times. This vast excess of gravity in the sun, binds all the lesser masses of its vassals so irresistibly to itself, within its own control, that were they all to appear in conjunction on *one* side of the sun, and there expend their united powers of attraction upon that great sphere, their influence would scarcely visibly affect it. But still their nature and position are not wholly of a passive and subordinate character; they possess likewise, independent, individual life-powers; spontaneous forces never to be suppressed, in addition to mere receptive capacities. Were it not that the inalienable and unconquerable power of the proper and independent movement of the planets away from the sun, balances the preponderating attractive force of the central body; were it not that the centripetal force is opposed by the centrifugal;[1] mass would be hurled against mass with appalling and destructive power. These terms, borrowed from

most helpless and pitiable weaklings."— Mädler, *Astron. Briefe*, p. 236.

[1] ["A planet moves in its elliptical orbit with a velocity varying every instant, in consequence of two forces: the one tending to the centre of the sun, and the other in the direction of a tangent to its orbit, arising from the primitive impulse given at the time it was launched into space. Should the force in the tangent cease, the planet would fall to the sun by its gravity. Were the sun not to attract it, the planet would fly off in the tangent. Thus, when the planet is at the point of its orbit furthest from the sun, his action overcomes the planet's velocity, and brings it towards him with such an accelerated motion, that at last it overcomes the sun's attraction, and, shooting past him, gradually decreases in velocity until it arrives at the most distant point, when the sun's attraction again prevails."—Tr.]

the vocabulary of mechanical science, are scarcely adequate to note the proper secret nature of the mysterious reciprocal action which here takes place, much less to exhaust its whole significance. Here, too, we meet with a mysterious sphere of dynamic life-forces, where proud science is brought to a stand by a "Hitherto shalt thou come, but no further." We can indeed behold the manifestations of the secret life-forces, which appear in the material frame-work; but these forces themselves, the animating soul, we cannot fathom. Though *Kepler*, the physiologist of the heavens, with prophetic powers of vision, has permitted us, in his three laws, to catch some glimpses of the secret vital relations of our solar system; and though the incomparable *Sir Isaac Newton*, following in the footsteps of Kepler, cast the treasures his predecessor had dug from the mines of knowledge, into current coin, by embodying them in his celebrated laws of gravitation, and thus rendered them tangible and fruitful to science; what does it all amount to, but merely to open to the human science a new theatre of effort and inadequate attainment? — how little, really, with all this advance, is the innate thirst of the human mind after knowledge satisfied![1]

Still more mysterious and equally unfathomable is the other sovereign power of the sun—its radiant,

---

[1] Kepler's *laws* are as follow: "The planets revolve about the sun in ellipses (mostly varying but very little from circles), having the sun in one of their foci. 2. If a line be drawn from the centre of the sun to any planet, this line, as carried forward by the planet, will sweep over equal areas in equal portions of time. 3. The squares of the periodic times of the planets are as the cubes of their mean distances from the sun." From these Newton

ardent, and all-enlivening light. The nature of light is still a problem, which, just as all the secret courses of the processes of life, has never been solved. Its solution is, indeed, perhaps not within the compass of human attainments. The system of emanations, which formerly obtained in connection with this problem, is now generally abandoned on the part of science. And, doubtless, the theory of undulations will also have to give way before that later and more profound theory, according to which light originates from the co-incident activity of cosmical contrasts, induced through a galvanic excitement of latent elementary light or light-ether. "Were not your eye adapted to the sun, how could you behold the sun?" How could the sun light up the earth, were not the nature of the earth adapted to receive the light—were it not impressible and excitable by light? To the masculine exciting agent, corresponds a feminine excitable object; to the imparting agent, a receiving object; the former remains an exciting and imparting agent, only so long and so far as it is opposed by a corresponding object, capable of being excited and of receiving, which from this very circumstance must partake of the same nature. Thus much, however, is satisfactorily known; that the

---

deduced the law of gravitation, according to which attraction decreases in proportion as the square of the distance increases. For a more complete understanding of the laws of Kepler, and their relation to the laws of life in general, compare Schubert, *Die Urwelt*, sec. IV., but particularly his *Ahndungen e. allg. Gescht. d. Lebens*, in the second section of the volume: also Hugi, *Grundzüge einer allgem. Naturansicht*, vol. I., Solothurn, 1841, p. 64 seq., 192 seq.

atmosphere of the sun is the source of light to the planets of our system, and that this atmosphere surrounds the sun, a dark body in itself, to the height of from 500 to 600 geographical miles. "We inhabitants of the earth behold a vast expanse of some 600,000 million square miles, (the extent of superficies the sun presents to us,) contracted to a small disk of one foot in diameter, and though its concentrated rays strike our eyes with such dazzling brilliancy, there have not been wanting astronomers, who have maintained that were this light equally distributed over the enormous body of the sun itself, its effects might not be so blinding[1] there, but moderate and beneficent."[2]

---

[1] ["In measuring, photometrically, the light of the three different structures of the sun, Sir William Herschel found that the light reflected *outwards* by the clouds of the inferior stratum, was equal to 469 rays out of 1000, or less than one-half of the light of the outer stratum; and that the light reflected by the opaque body of the sun below was only seven rays out of 1000. Hence he concluded that the outer stratum of the self-luminous or phosphoric clouds, was the region of that light and heat which are transmitted to the remotest part of the system: while the inferior stratum, which is obviously of a different character from the other, is intended to protect the inhabitants of the sun from the blaze of the stupendous furnace which encloses them. In confirmation of this view, the faint illumination — the *seven rays out of a thousand* — is a proof that the light of the outer stratum, and consequently its heat, must be extremely small on the dark body of the luminary which we see through what are called the solar spots, which are now universally admitted to be openings in the luminous stratum, and not opaque scoriæ floating on its surface." *More Worlds than One*, Brewster, p. 98. No mention is here made of the *true outer* or third stratum of the solar investment.—Tr.]

[2] Schubert, *Urwelt*, p. 22.

Thus we see that the difference between the sun and the planets is by no means certainly so great as it is ordinarily taken to be. The solid central body of the sun seems to be of a planetary nature, and at the very place where the difference appears to be most marked, in the atmospheres of the sun and the planets respectively, even there "the distinction is no greater nor more pervading than that existing between two beings of the same species and internal constitution, but differing in sex, one being masculine and the other feminine. For the atmosphere of planets also, and, still more, that of the comets, partakes, under certain circumstances, of the attributes of an independently self-luminous substance, giving out light without the intervention of any agitation from without. Indeed, the quality which we call transparency is, in a certain respect, nothing more than an attribute of a substance which is co-luminous and self-luminous through merely negative excitement from without."[1] "The contrast," says another student of nature,[2] "in which the sun and the planets stand to each other, as bodies that give and bodies that receive light, appears not to be a complete and absolute one, any more than do many other contrasts in nature. It cannot be said that the planets do not possess any proper power in themselves of developing light. The northern light of the earth, the remarkable circumstance that sometimes when our skies are without a moon, the clouds become luminous through some influence from above,

[1] Schubert, *Urwelt*, p. 21.

[2] Perty, *Allgem. Naturgesch.* I., p. 222.

the illumination of the dark side of Venus, the total eclipse of the moon in which it does not become altogether invisible, though receiving no light at all from the sun, and perhaps, also, the so intense light of Jupiter and Vesta—all these indicate this remarkable fact. Thus it appears, therefore, that as the sun contains a dark body, something of a planetary nature, so, also, each planet possesses something of a solar nature; but as the solar principle predominates in the sun, so likewise does the planetary principle in the planets." Indeed, *Hugi*, who holds gravity to be a polar relation between the centre and circumference, a tending of the individual members to the centre of the whole; and light to be the directly opposite working—a stretching of the grand centre of the whole upon the individual members of the periphery, —expresses himself to this effect, in his above-mentioned ingenious original work (page 44); that very probably the gravity of the planets in relation to the body of the sun, manifests itself in the form of light, so that, conversely, the effect of the sun in producing light upon the planets, appears upon the sun itself, under the nature of gravity, or as an outward striving towards the planets by which it is encircled.

[Some additional facts and views in regard to the sun may not be without interest. This great central sphere, as well as the planets which are dependent upon it, is possessed of an axial rotation. "Its period of rotation is 25 days, 7 hours, and 48 minutes. The axis upon which it revolves is very nearly perpendicular to the plane of the earth's orbit, and the motion of rotation is in the same direction as the motion of the planets ronnd the sun: that is, from the west to the east." The remarkable phenomena of the solar spots happily furnish us with the means of arriving at the period of rotation in this case, which otherwise would, in all probability, present insuperable difficulties to the astronomer. The spots are all found to revolve in the same time —something over 25 days. "The only circumstance of regularity which can be said to attend these remarkable phenomena is their position upon the sun. They are invariably confined to two moderately broad zones parallel to the solar equator, separated from it by a space several degrees in breadth. The equator itself, and this space which thus separates the macular zones, are absolutely divested of such phenomena." . . . . "The prevalence of spots on the sun's disc is both variable and irregular. Sometimes the disc will be completely divested of them, and will continue so for weeks or months; sometimes they will be spread on certain parts of it in great profusion. Sometimes the spots will be small, but numerous; sometimes individual spots will appear of vast extent; sometimes they will be manifested in groups, the penumbræ or fringes being in contact."

"The duration of each spot is also subjected to great and irregular variation. A spot has appeared and vanished in less than twenty-four hours, while some have maintained their appearance and position for nine or ten weeks, or during nearly three complete revolutions of the sun upon its axis. . . . . The magnitude of the spots and the velocities with which the matter composing their edges and fringes moves, as they increase and decrease, are on a scale proportionate to the dimensions of the orb of the sun itself. When it is considered that a space upon the sun's disc, the apparent breadth of which is only a minute, actually measures 27,960 miles, and that spots have been frequently observed, the apparent length and breadth of which have exceeded 2′, the stupendous magnitude of the regions they occupy may be easily conceived. The velocity with which the luminous matter at the edges of the spots occasionally moves, during the gradual increase or decrease of the spot, has been in some cases found to be enormous. A spot, the apparent breadth of which was 90″, and into which our earth might have dropped without grazing its edges, was observed by Mayer to close in about 40 days. Now, the actual linear dimensions of such a spot must have been 41,940 miles, and, consequently, the average daily motion of the matter composing its edges must have been 1050 miles, a velocity equivalent to 44 miles an hour."

These spots are now generally supposed to be excavations in the luminous envelope of the sun, though they have also been supposed to be vast scoriæ or masses of incombustible matter floating upon the surface of the sun. *Sir J. Herschel*, who has devoted much attention to the subject of solar spots, believes the rupture in the luminous investment of the sun, giving rise to the phenomena of solar spots, to result from the action of agencies somewhat like the trade-winds and anti-trades, hurricanes, tornadoes, water-spouts, and other violent atmospheric disturbances upon our earth, induced by somewhat similar conditions of axial rotation, equatorial accumulation of atmosphere, unequal temperature, and the like, in connection with the sun itself. These agencies must, of course, be proportionate in extent and power to the surpassing size of the sun. It will be observed that the region of the spots in the sun corresponds to that of greatest atmospheric disturbance upon our globe. The possibility of such a production of solar spots may perhaps be better understood after considering the sun's atmosphere as a whole.

We may be permitted, in this connection, to make some extracts from a very interesting paper on the physical consitution of the sun, by Arago, part of which may be found in the *Annual of Scientific Discovery*, 1853, p. 135 seqq.

After briefly reviewing the phenomena of the solar spots, and the peculiar radiance, less luminous than the rest of the orb, with which they are surrounded,—the penumbra, M. Arago says: "This penumbra, first noticed by Galileo, and carefully observed by his astronomical successors in all the changes which it undergoes, has led to a supposition concerning the physical constitution of the sun, which at first must appear altogether astonishing. According to this view, the orb would be regarded as a dark body, surrounded at a certain distance by an atmosphere, which might be compared to that enveloping the earth, when composed of a continuous bed of opaque and reflecting clouds. To this first atmosphere would succeed a second, luminous in itself, and which has been called the *photosphere*. This photosphere, more or less removed from the interior cloudy atmosphere, would determine, by its circumference, the visible limits of the orb. According to this hypothesis, spots upon the sun would appear as often as there were found in the concentric atmospheres, corresponding vacant portions, which would permit us to see exposed the dark central body. Those who have studied with powerful instruments, professional astronomers, and competent judges, acknowledge that this hypothesis concerning the physical constitution of the sun, supplies a very satisfactory account of the facts. Nevertheless, it is not generally adopted; recent authorities describe the spots as scoriæ floating on the liquid surface of the orb, and issuing from solar volcanoes, of which terrestrial volcanoes are but a feeble type.

"It was desirable, then, to determine, by direct observation, the nature of the incandescent matter of the sun; but when we consider that a distance of 95,000,000 of miles separates us from this orb, and that the only means of communication with its visible surface, are luminous rays issuing therefrom, even to propose this problem seems an act of unjustifiable temerity. The recent progress in the science of optics, has, however, furnished the means for completely solving this problem.

"None are now ignorant that natural philosophers have succeeded in distinguishing two kinds of light, viz., natural and polarized. A ray of the former of these lights exhibits, on all points of its surface, the same properties; whilst, with regard to polarized light, the properties exhibited on the different sides of its rays are different. . . . . Before going further, let us remark, that there is something wonderful in the experiments which have led philosophers legitimately to talk of the different sides of a ray of light. The word 'wonderful,' which I have just used, will certainly appear natural to those who are aware that millions and millions of these rays can simultaneously pass through the eye of a needle, without interfering one with the other. Polarized light has enabled astronomers to augment the means of investigation by the aid of some curious instruments, among others, the polarizing telescope, which . . . furnishes a very simple means of distinguishing *natural* from polarized light.

"It has long been believed, that light emanating from incandescent bodies, reaches the eye in the state of natural light, when it has not been partially reflected or strongly refracted, in its passage. The exactitude of this proposition failed, however, in certain points. A member of the Academy has discovered that light emanating under a sufficiently small angle, from the surface of a solid or liquid incandescent body, even when polished, presents evident marks of polarization; so that in passing through the polarizing telescope or polariscope, it is decomposed into two colored pencils. The light emanating from an inflamed gaseous substance, such as is used in street illumination, on the contrary, is always in its natural state, whatever may have been the angle of its emission. The means used to decide whether the substance which renders the sun visible is solid, liquid, or gaseous, will be nothing more than a very simple application of the foregoing observations, in spite of the difficulties which seem to arise from the immense distance of the orb.

. . . . . "Observations made any day of the year, looking directly at the sun, with the aid of powerfully polarizing telescopes, exhibit no trace of polarization. The inflamed substance, then, which defines the circumference of the sun, is gaseous. We can generalize this conclusion, since, through the agency of rotation, the different points of the surface of the sun come in succession to form the circumference. This experiment removes out of the domain of simple hypothesis, the theory we have previously indicated concerning the constitution of the solar photosphere.

. . . . . "The constitution of the sun, as I have just established it, may equally well serve to explain how, on the surface of the orb, there exist some spots not black but luminous. These have been called faculæ, others of much smaller dimensions and generally round, have been called lucules. . . . . By experiment it was found that a gaseous incandescent surface of a determined extent is more luminous when seen obliquely, than under perpendicular incidence. Consequently, if, like our atmosphere, when dappled with clouds, the solar surface presents undulations, the parts of these undulations which are presented perpendicularly to the observer, must appear comparatively dim, and the inclined portion must appear more brilliant; and hence, every conic cavity must appear a lucule. It is no longer necessary, in accounting for these appearances, to suppose that there exist on the sun millions of fires more incandescent than the rest of the disc, or millions of points distinguishing themselves from the neighboring regions by a greater accumulation of luminous matter.

"After having proved that that the sun is composed of a dark central body, of a cloudy-reflecting atmosphere, and of a photosphere, we should naturally ask if there is nothing besides. If the photosphere terminates abruptly and without being surrounded by a gaseous atmosphere, less luminous in itself, or feebly refracting? Generally, this third atmosphere would disappear in the ocean of light with which the sun appears always surrounded, and which proceeds from the reflection of its own rays upon the particles of which the terrestrial atmosphere is composed." M. Arago then proceeds to mention observations made in connection with eclipses of the moon, in regard to the existence of a *third* and *outer* stratum of the solar investment, and arrives at the conclusion that the actual presence of such a stratum is scarcely any more a matter of doubt. Sir J. Herschel, years previously, maintained the existence of a gaseous atmosphere of vast height above the second or luminous stratum of the solar investment.—As to the physical cause of the sun's heat, it may be remarked, however, that philosophers widely differ; as there are great difficulties to the hypothesis of combustion, involving such extensive chemical change, many incline to the view that it is produced by electric or electro-magnetic action.—TR.]

§ 2. *The Planets and Satellites.*

Let us now turn our attention from the king of day to the attendants of his majesty. Among the PLANETS, so far as they are known to us, "there obtains this common and universal character: that they, being in themselves more or less dark bodies, stand in need of the vivifying light of the sun; that they move around their common central sphere, in orbits which mostly deviate but little from the form of a circle, and lie in a plane which very nearly coincides with the plane of the sun's equator;" that they all turn on their axes, and that through the inclined position of these axes to the planes of the orbits of the several planets, there is produced a change of seasons and a lengthening and shortening of day and night; and that, finally, "they are all composed of matter which appears not to vary very substantially from the material of which our earth is formed (ranging from the solidity of metal to the lightness of water)". But the variety of their individual conditions is by no means limited by this general similarity. "In spite of the unity of their plan, and an obvious striving towards the same grand idea, uniformity is still avoided. In each of these revolving globes, though they are but partially known to us, we meet with some peculiarities which belong respectively to the particular individual only. Nature has nowhere repeated herself. In every heavenly body, both great and small, we behold an individual independent in itself — between them all, however, there exists at the same time, a harmony that is simple, complete, and ever-abiding."[1]

The similarity as to the constitution and arrangements of nature, is most marked in Mercury, Venus, the Earth, and Mars, the four planets nearest the sun; and the dissimilarity increases proportionally with the distance.

Mercury is a body very similar to our earth, with a mountainous surface, and surrounded by an atmosphere. The length of its day is much the same as that of ours. Its year, however, contains but 87 days, and is divided into seasons of very unequal length. Its diameter is but some 3000 miles, on which account the power of gravity upon its surface, in spite of a somewhat greater density (its proportion

[1] Mädler, *Astr. Briefe*, p. 129.

to the earth being in this respect as 6 to 5), is very materially less than with us. Our pound there weighs but 7¼ ounces. From Mercury the sun appears, at a distance of but 37,000,000 miles, as a disk two feet seven inches in diameter; and the planet, consequently, receives seven times the amount of light and heat that our earth does.

VENUS, that brilliant star which was called by Homer of old, the most beautiful one in all the heavens, does not deviate very much from the earth, in respect to size, density, and power of gravitation. Its day's length is nearly the same as that of ours; its year, however, about one-third shorter than ours. The self-illumination of its dark side is perhaps a manifestation analogous to our northern light, though it is far more intense. The planet contains mountains of considerable size, and is surrounded by a very pure and clear atmosphere.[1]

[1] [From the careful observations of Schröter, as well as those of Beer and Mädler, in regard to the rotation of Venus, it may now be considered as fully settled that this planet turns upon its axis in a period of about 23 hours 15 minutes. De Vico still more lately has arrived at results altogether in harmony with those of Beer and Mädler. Schröter conjectures from indications he has observed, that the southern hemisphere of this planet is more mountainous than the northern. The direction of its axis of rotation has not been satisfactorily determined. "If, as it is generally supposed, it be inclined to the plane of the planet's orbit at an angel of 75°, the sun must at some time be vertical to all points not within 15° of the poles, and as the utmost limit at which the sun is vertical marks the tropics, the latter must be within 15° of the poles of the planet, or 75° on each side of the equator, and so include 150° of its surface. By this arrangement the sun is vertical twice a year to all places on the planet Venus, except those situated within 15° of each pole, producing a most remarkable vicissitude of sea-

The distance of our EARTH from the sun, amounts to about 95,000,000 miles. Its satellite, the *Moon*,[a] 50 times less in size, and 80 times less in weight, is distant from it 240,000 miles. In spite of the intimate relation existing between the moon and the earth, their mutual physical conditions and arrangements are very different and unlike. In this connection may be mentioned especially, the complete absence of water and of any atmosphere in the moon, the highly peculiar volcanic and kettle-shaped excavations of the surface, the coincidence of its axial rotation with its revolution round the earth, &c.

The most remarkable agreement with the physical constitution and relations of the earth, may be observed in the next and smaller planet, MARS.[1] Its

---

sons. During one-half of Venus' year — that is, sixteen weeks — the sun continues at one pole without setting, while the inhabitants of the other pole are involved in darkness. In this respect Venus resembles our earth, for each pole has a night of half a year. But unlike the earth, the inhabitants at Venus' equator have two winters and two summers in every year."—*Familiar Astronomy*, Bouvier, p. 353.—TR.]

[1] [Mars revolves on its axis in 24 hours, 37 minutes and 10 seconds, the direction of the axis being at an inclination of 28° 27′ to the plane of the planet's orbit. Beer and Mädler, who have devoted much time to observations in connection with this planet, suppose, in accordance with the author's remarks, that the whiteness observed at the poles of this planet is occasioned by snow and ice. To the same observers, changes in appearance were manifest in other parts of the planet, "but through those changes the permanent features of the planet were always discerned; just as the seas and continents of the earth may be imagined to be distinguishable through the occasional openings in the clouds of our atmosphere, by a telescopic observer in Mars."—TR.]

a [For matter connected with letters, see additions to the sections.—TR.]

red color would lead us to conclude at least that its atmosphere is quite as dense as that of our earth. Both dark and light spots, which are not subject to change, may be observed upon its surface. "The first appear to be seas, and it may be worthy of remark that, just as it is upon the earth, the greater mass of the waters is collected upon the southern hemisphere; while the northern contains a preponderance of dry land." Besides, the neighborhood of its two poles is rendered conspicuous by a specially brilliant white color. As these clear white zones yearly increase and decrease, in a regularly recurring manner, according as winter or summer is present at the pole concerned, we are left fairly to conclude that they are composed of snow and ice. The day's length in Mars is much the same as ours; but on account of its greater distance from the sun, its time of revolution round that body is almost twice as long as that of the earth; its light and heat are of course much less intense than with us. Though not varying much in density from the earth, gravitation is but one-half as strong upon its surface. Mars, just as the two planets nearest the sun, has no *satellite.*

With respect to the ASTEROIDS, whose number has been so rapidly[b] swelled of late years, that it now amounts to 42, observation has gathered little of importance. The reason of this ill success is their distance and their extreme smallness — the diameter of Vesta, for instance, is thought to be about 260 miles — and observation has been forced to confine itself pretty much to the investigation of their wonderfully intricate, deviating, and extremely elliptical orbits.

JUPITER is the largest of all the planets.[1] Its bulk is 1414 times that of the earth, and almost the one-thousandth part that of the sun. Its distance from the sun is about 495,500,000 miles; and, seen from it, the sun presents a disc of but $2\frac{1}{3}$ inches in diameter. The degree of light received by this planet is 27 times less than that received by the earth. Its density is equal to that of the sun — 4 times less than that of the earth. But, on the other hand, the power of gravitation is much greater upon its surface than with us, a pound there weighing $2\frac{1}{2}$ pounds. From all this it appears how very much its physical constitution and relations differ from those which obtain upon our globe. Its light is at least twice as intense as it would be upon the earth, were the

[1] [The human mind is filled with wonder in contemplating the grand scale on which magnitude, motion and distance are displayed even in our own planetary heavens. Jupiter rotates upon his axis in 9 hours and 56 minutes, producing an astonishing rate of motion at his equatorial surface, when we remember that over 1400 such globes as the one we inhabit would be required to make a mass the size of this huge planet. But still more astonishing is his prodigious orbital motion as he sweeps round the sun at such an immense distance. He moves at a speed sixty times greater than that of a cannon-ball, or 700,000 miles per day, 30,000 per hour, and 500 per minute. Sir William Herschel considers it probable, from his observations, that an analogy to the axial-rotation of our moon in relation to the planet it accompanies is to be found in the adjustment of the Jovian moons — that they rotate once upon their axes in the time of their respective revolutions about the planet, thus ever presenting the same side to that body. There are indications that this is a general law in regard to the satellites of the solar system.—TR.]

earth equally remote from the sun. It is surrounded by a very dense and high vapor or atmosphere. Parallel with the direction of its equator, broad stripes may be observed extending across its disc. These have been regarded as cloud-formations; but they must differ very much from corresponding appearances in our atmosphere, since that great girdle of clouds which passes across the disc of that planet near its equator, has experienced no material change in form and extent for the last 200 years, though the other stripes have suffered various modifications and divisions. On account of the diminished density of this planet, which at its surface is equal to but one-half that of our water, sedimentary deposits, seas, and such like, must, if they exist there at all, be of a wholly different and peculiar constitution. Jupiter has *four satellites*.

The most interesting system, and the one most planetary in its character, is presented by SATURN, with its rotating *ring*,[f] and its *eight satellites*. Its mean distance from the sun is about 906,000,000 miles, its year's length 28½ of our years, the length of its day 10½ hours. It surpasses the earth in size about 770 times, and its mean density is eight times less than that of the earth. Its outer crust, therefore, does not probably possess a specific gravity quite equal to that of cork-wood. The light this planet receives from the sun would be, according to the usual calculation, 90 times less than that of the earth; but it is, in reality, only some 20 times less, owing to a superior capacity in this body for receiving light. These circumstances show, indeed, a striking difference between the physical economy of Saturn and that of the earth; but it becomes still more remarkable when we examine that strange and mysterious ring, which encircles and revolves around the equator of the planet, at a distance of about 19,000 miles. The edge of this arching ring, with a thickness of not much over 135 miles, is turned towards the planet; the ring is about 29,000 miles broad, and extends away from the sphere lying in its plane, and which it encircles, like a great disc with a piece taken out of its centre. The ring is besides not a simple one, but "consists of several concentric rings, of unequal breadth, completely detached from each other by intervening void spaces."

Uranus is distant from the sun about 1822 millions of miles. It revolves round that body in 84 years; the time of its rotation on its axis is still unknown. The light of the sun, which reaches our earth in 8′ 7″, does not reach this planet under 2 hrs. 35′ 42″. From it the sun presents a disc of but $\frac{3}{5}$ of an inch in diameter. Its bulk is about 82 times that of the earth; its specific weight 6 times less than that of our globe. It is surrounded by a very dense atmosphere, which perhaps possesses sources of light and heat in itself, since the brilliancy of this planet is at least four times what it should be according to calculation. Its axis is so much inclined towards its orbit that the two fall almost in the same plane; the length of its day and night is consequently almost wholly independent of the rotation of the planet itself. It has at its poles, in turn, both daylight and summer for a period of 42 terrestrial years; these are then followed by a wintry night of equal length. *Six satellites* have been discovered holding their courses around this planet.

NEPTUNE, the most distant of known planets, in whose discovery mathematical analysis has won its highest and most brilliant triumph, describes its orbit around the sun in 164 years, and is distant from that body about 3000 millions of miles. Its size is very nearly the same as that of Uranus. The rays of the sun are, when they reach that planet, 1300 times less intense than when they strike on earth. Two satellites have been discovered in connection with Neptune.

How diverse and unaccustomed may not the constitution of nature be at such a distance!

That there may still be planets unknown to us, existing without the orbit of Neptune, but nevertheless controlled by our sun, cannot, particularly since the discovery of Neptune, be reasonably contested. A planet, though a hundred times more distant than Uranus, would at least have no occasion to apprehend disturbance from the nearest of the fixed stars: (—the surpassing distance of these latter bodies will hereafter claim our closer attention). If we apply the analogy of increasing distance which obtains all the way from the sun to Neptune, to merely the most extreme limits of the *known* solar system (the aphelion of the comet of 1680), there will still be room for four undiscovered planets beyond the orbit of Neptune, the most distant of which must be 620 times the distance of the earth from the sun (58,500 millions of miles), and require 15 thousand years to complete one revolution.

---

* [" The entire surface of the visible hemisphere of the moon is thickly covered with mountainous masses and ranges of various forms, magnitudes and heights, in which, however, the prevalence of a circular or crater-like form is conspicuous. . . . . . Uniform patches, of greater or less extent, each having an uniform gray tint more or less marked, formerly supposed to be large collections of water, have now been proven to be regions diversified like the rest of the lunar surface, by inequalities and undulations of permanent forms. They differ from the other regions only in the magnitude of the mountain masses which prevail upon them. . . The more intensely white parts are mountains of various magnitude and form, whose height, relatively to the moon's magnitude, greatly exceeds that of the most stupendous terrestrial eminences; and there are many characterized by an abruptness and steepness which sometimes assume the position of a vast vertical wall altogether without example upon the earth. . . . Circular ranges of mountains, which, were it not for their vast magnitude, might be inferred from their form to have been volcanic craters, are by far the most prevalent arrangement. These have been denominated, according to their magnitudes, *bulwark plains*, *ring mountains*, *craters*, and *holes*. *Tycho*, the most remarkable of the *ring mountains*, is distinguishable without a telescope when the lunar disc is full. . . . The area which it encloses, and which is very nearly circular, is 47 miles in diameter, and the inside of the enclosing ridges has the steepness of a wall. Its height above the level of the enclosed plain is 16,000, and above that of the external region, 12,000 feet. There is a central mount, height 4700 feet, besides a few lesser hills within the enclosure. *Craters* and *holes* are the smallest formations of the circular class. *Craters* enclose a visible area, containing, generally, a central mound or peak, exhibiting, in a striking manner, the volcanic character. *Holes* include no visible area, but may possibly be craters on a scale too small to be distinguished by the telescope. Formations of this class are innumerable on every part of the visible surface of the moon. . . . Among the most remarkable phenomena presented to lunar observers, are the systems of streaks of light and shade, which

radiate from the borders of some of the largest *ring mountains*, spreading to distances of several hundred miles around them. . . . Herschel, the elder, suggested for their explanation streams of lava; Cassini imagined they might be clouds; and others even suggested the possibility of their being roads! Mädler imagines that these *ring mountains* may have been among the first selenological formations; and, consequently, the points to which all the gases evolved in the formation of our satellite would have been attracted. These emanations produced effects such as vitrification and oxydation, which modified the reflective powers of the surface."—*Handbook of Astronomy*, Lardner, p. 208, 209. "Dr. Scoresby, in an account he has given of some recent observations made with the Earl of Rosse's telescope, says: With respect to the moon, every object on its surface of one hundred feet was now distinctly to be seen, and he had no doubt, that, under favorable circumstances, it would be so with objects sixty feet in height. On its surface were craters of extinct volcanoes, rocks, and masses of stones almost innumerable. He had no doubt that, if a building, such as he was then in, were upon the surface of the moon, it would be rendered visible by these instruments. But there were no signs of inhabitants such as ours, no vestige of architectural remains, to show that the moon is, or ever was, inhabited by a race of mortals similar to ourselves. It presented no appearance which could lead to the supposition that it contained anything like the green fields and lovely verdure of this beautiful world of ours. There was no water visible, not a sea, or river, or even the measure of the reservoir for supplying town or factory; all seemed desolate." —Tr.]

[t] [Well may the author speak of the rapid increase of the number of the *asteroids!* When he penned those lines, some five years ago, they numbered 18, so far as they were known; now they number 42, and there is no reason to suppose that the unwearied eye of the astronomer may not succeed in descrying many similar small masses, moving across the heavens in the general path of planetary motion. Early in the present century, when only three of these bodies had been discovered, the sagacious Dr. Olbers ventured the conjecture that they all had a common origin; being, as he supposed, the fragments of a large planet revolving between Mars and Jupiter, which was rent asunder by some tremendous

catastrophe in the unknown past. Many astronomers have subsequently sympathized with this view. Le Verrier supposes the sum of all the asteroids cannot exceed one-fourth of the bulk of the earth; but, granting the hypothetical planet to have been of that size, a vast number of additional fragments may still be coursing their way unseen in the region of the broken world, as it would require 400 bodies, equal in size to the largest of the asteroids, to make up one-fourth of the earth's bulk.

Within a few years, an interesting paper has been produced by an astronomer of our own country, Prof. Alexander, of Princeton, in regard to the size, form, rotation, distance, etc., of the original asteroid planet, some of the views of which we here present, as contained in the *Annual of Scientific Discovery*, 1856.

"By a skilful use of evidence, Prof. Alexander has arrived at almost a certainty that, in the space between Mars and Jupiter, once revolved a planet a little more than 2·8 times as far from the sun as our earth. The equatorial diameter was about 70,000 miles, but the polar diameter only 8 miles! It was not a globe, but a wafer, nay, a disc of a thickness of only $\frac{1}{9000}$th of its diameter. Its time of revolution was 3·698 days, say 3 days, 15 hours, 45 minutes. The inclination of its orbit to the ecliptic was about 4°. It met a fate that might have been anticipated from so thin a body whirling so furiously, for its motion on its axis was $\frac{1}{16}$th of its velocity in its orbit, say 2477 miles per hour. It burst as grind-stones and fly-wheels sometimes do. We have found 42 of its fragments, and call them *asteroids*. When it burst, some parts were moving 2477 miles per hour faster than the centre did, and some as much slower; that is, some parts moved 4954 miles per hour faster than the others. These described a much larger orbit than the planet did, and the place where it burst was their perihelion. Others described a smaller orbit, because they left that point with a diminished velocity—it was their aphelion. Some flew above the orbit of the planet and had their ascending node. Others flew below, and it was their descending node. They seemed to go almost in pairs. Two went very far out of the plane of the orbit, so that they pass the limits of the zodiac, and it is found that the ascending node of 18 corresponds nearly with the descending node of 17. Thin as the planet was, it had not cooled so much at the time of the explosion but that some of the fragments could assume a spherical form. Three or four independent processes for finding the place of the planet agreed in their results surprisingly. He interpolated it as a lost term in a geometric series, from Mars to Saturn, for the first approximation. He compared it with Saturn and Jupiter, and with Mars and Jupiter. He found where a planet would be dropped off in the successive cooling and contracting of the solar system. And he compared its orbit for size and ellipticity with those of the asteroids, etc. . . . It is curious to see how the history of this planet verifies the theory of La Place, that a heavenly body must be either nearly a sphere, or a disc, and that the latter must be unstable." — TR.]

[e] [Recent observation has made it certain that another and a partially transparent ring exists within the space circumscribed by the ring or rings of Saturn heretofore known to exist. This ring was, as it would appear, discovered almost simultaneously by Prof. Bond, of Boston, and Mr. Dawes, of England. Dr. Galle, of Berlin, had years previously noticed some indications which are now supposed to have been connected with this ring, but their true import was not then understood.

"By observations made by Mr. Lassell, of Malta, it appears that the new ring is transparent to such a degree that the body of the planet can be seen through it. The following is the language of Lassell: Perhaps the most remarkable phenomenon, which I now notice for the first time, is the evident transparency of the obscure ring; both limbs of the planet being distinctly seen through it where it crosses the ball, quite through to the inner edge of the inner bright ring. To my apprehension, I cannot better describe the entire aspect of the obscure ring than by comparing it to an annulus of black crape stretched within the bright ring, which, when projected against the black sky, as at the curve, would, from its reflecting some light, appear of a dark-grey shade; and when projected on the ball, would, from the transmission of a portion of the reflected light of the ball, appear of a much lighter grey. What the precise nature of this marvellous appendage can be, would be an interesting subject of speculation, exhibiting, as it were, a connecting link between nebulous and solid matter.

"Mr. J. P. Bond maintains that Saturn's ring is in a fluid state, or at least does not strongly cohere." He is led to this conclusion from the changes observed in the rings, from the difficulties in supposing numerous small solid concentric rings near each other, and the like. Peculiar circumstances may require the separation of either the inner or outer portion of the ring as commonly seen, giving rise to the subdivisions sometimes seen. This separation may be necessary for the preservation of the equilibrium of the ring or the parts of which it is composed.

"Prof. Peirce has undertaken to show, from purely mechanical considerations, that Saturn's ring cannot be solid. He maintains unconditionally that there is no conceivable form of irregularity, and no combination of irregularities consistent with an actual ring, which would serve to retain it permanently about the primary, if it were solid. He maintains that Laplace's statement of the sustaining power of an irregularity, was a careless suggestion, which was dropped at random, and never subjected to the scrutiny of a rigid analogy. Moreover, the fluid ring cannot be regarded as one of real permanence without the aid of foreign support. This support he finds in the action of the satellites. The satellites are constantly disturbing the ring, and yet they sustain it in the very act of perturbation.—*Recent Progress of Astronomy*, Loomis, p. 116 seq.—Tr.]

§ 3. *Shooting-Stars.*

There have been added by more recent investigation, to the planets proper — the dignitaries, as it were, in the widely extended realm of the sun — a countless number of smaller planetary masses, which encircle his majesty in thickly crowded millions. Their presence is betrayed alone by the fact that they meet the earth in their mysterious career, and then, at the boundaries of our atmosphere, assume a glowing brightness, through some process not yet understood — perhaps an electric one — and finally, overcome in many cases by the attraction of the earth, lose their independence and fall to its surface. These are the so-called Shooting-Stars, together with *fire-balls* and *meteoric stones* (aërolites), which all with scarcely a doubt belong to the same category.[1]

The height of the shooting-stars—the point at which they commence or cease to be visible — varies from 18 to 160 miles. The relative velocity of their movement is from 18 to 40 miles in a second — much the same as that of the planets nearest the earth (Mercury 28 miles in a second, the Earth 19), though somewhat greater. "They fall either singly and not very frequently — sporadically — or in numbers amounting to thousands. The latter cases (Arabian writers compare them to swarms of locusts) are periodic." The most noted of these periodic occurrences is the so-called November phenomenon (from the 12th to the 14th of Nov.), as also the stream of St. Laurentius (from the 9th to the 14th of August). The latter is so called, because of its taking place during the festival of this saint (Aug. 10th), whose "fiery tears" were long since represented in old church-calendars of England, as regularly recurring phenomena.

Alex. von Humboldt first called attention to the periodicity of these phenomena. He was led to remark it particularly, from the unparalleled phenomenon of shooting-stars which was observed in North America, on the 12th and 13th of Nov. 1833. On that occasion they fell from one region of the heavens, thick as snow-flakes—at least 240,000 were seen in some places in the course of an hour.

The final result of *Humboldt's* inquiries, to which

---

[1] Compare, particularly, Humboldt's *Cosmos*, and Mädler's *Astr. Briefe*, p. 335–343.

most scientific minds assent, is this: "The different meteoric streams, each one of which consists of myriads of small cosmical bodies, probably cut the orbit of the earth. We may imagine them as forming a closed ring, and pursuing the same common orbit."

Meteoric stones are, indeed, composed of elements such as are to be met with upon the earth (particularly pyrites, magnetic ore, iron and nickel); "but scarcely ever in such combinations as obtain in bodies belonging to our globe."

§ 4. *The Comets.*

Before leaving the realm of the sun, we must cast a glance at another class of his vassals—the COMETS. They at times approach in their sweeping career much closer to this mighty sovereign sphere than is ever dared by any of the planets; and then pass off in their extremely elliptical orbits, to the very outer limits of the solar system, absenting themselves for centuries, and even for thousands of years. The province of the solar system, of which Uranus was still the outer planetary sentinel, so far as knowledge until *recently* extended, had been enlarged at least forty times through the far-distant adventures of these erratic bodies — enlarged truly, but to a vast "terra incognita." In spite of the roving nature of these bodies, they still obey, as do all the worlds of the universe, the great laws of cosmical movement discovered by Kepler. Thus, the comet of 1860, which ventures 44 times further away from the sun than the extremely remote Uranus, and completes one revolution only in 9000 years, moves at the rate of 248 miles in a second, or with thirteen times the velocity of the earth, when at its perihelion (only 144,000 miles from the sun's surface); but when at its aphelion, scarcely 10 feet in a second. But it must be confessed such extremes of motion are not to be found in the case of any other comet.

The physical constitution of the comets differs very widely from that of the planets. "It would indeed be carrying the matter too far, to deny them all materiality, and, consequently, all substantial reality; but still, observation has taught us that our accustomed ideas of physical bodies, appear to entirely fail of application to them. In spite of a diameter of many thousands, yea, of hundreds of thousands of miles, they are quite transparent, and possess no power of refracting light. Our air when most rarefied would not be so completely impassive and devoid of power to produce effects. The nuclei of these bodies even, are probably much rarer than common air, so that our conceptions of heavenly bodies as solid masses, lose all application in this connection. This idea is favored by the fact that they experience very sudden and momentous changes in their aspects, which certainly proves a remarkable volatilization and mobility of their parts. The end they fulfil in the great plan of the universe is in all probability beyond the power of our minds to discover."

"That comets are not solid bodies appears from the fact that they experience such great and sudden changes. Neither can they partake of a fluid or gaseous form, for in either case the rays of light would then be refracted. But what, then, are these bodies? We can but confess our ignorance, and say that as the earth furnishes us with nothing analogous, it is altogether impossible to profess any definite knowledge on this point. Perhaps they are composed of extremely minute, diffused, dust-like particles."[1]

It has been satisfactorily determined by actual experiment, that the light of comets is not an inherent light, but that it is derived from the sun. — The orbits of these strange wanderers lie in all directions about the sun — they pass from east to west, as well as from west to east. The number of the comets has never been determined. Although but some 500 of them have been closely observed, doubtless many thousands may still be speeding their ways through the remote regions of the solar system, entirely withdrawn from all human observation for the time.

[The great comet of 1843 presents one of the most remarkable of these phenomena on record, and may serve to give the mind some idea of the wonders connected with the life and experience of a *comet*. It approached the sun so closely as to become *red hot* (according to Loomis), and retained a peculiar *fiery appearance* for some days after its perihelion. It absolutely almost grazed the sun, and whirled around it at such a prodigious rate, that in two hours it swept over more than 1,000,000 miles of solar surface. Sir John Herschel computed the heat it must have received from the sun, at its perihelion, at 47,000 times that we received from that great luminary; a heat sufficient to convert almost any substance upon earth into vapor, or at least intensely ignite it. The comet was visible for 40 days: the nebulosity of

---

[1] Mädler, *Astr. Briefe*, p. 290.

its head was about 36,000 miles in diameter, and the length of its tail, when most fully developed, 108,000,000 of miles!

"The following circumstances invest the comet of 1843 with peculiar interest: 1st, Its small perihelion distance; being as small as that of any comet whose orbit has been computed, and nearly as small as is physically possible. 2d, The length of its tail; being equal to that of any comet hitherto observed."—Loomis, *Recent Progress of Astronomy*, p. 131.

Beila's comet, discovered in 1826, and having a period of over 6½ years, was seen on its return in 1846, to be divided into two parts, constituting, as it were, two comets sweeping along side by side. This strange phenomenon greatly attracted the attention of astronomers. A subsequent appearance in 1852 has shown that the body is permanently divided. The two nuclei, when last seen, were more than 1½ millions of miles distant from each other—much further than in 1846. It is supposed the comet may, perhaps, have been divided by a repulsive force emanating from the sun.

That comets are composed of ponderable matter, however light and diffused it may be, is proved by the circumstance that they are affected in their movements by the attraction of the planets. It at the same time becomes evident that the density of these bodies is incalculably small, since no slightest effect of theirs can be detected on the planets.—Tr.]

§ 5. *Origin and Stability of the Solar System.*

We might at the close of this glance at the constitution of the solar system, inquire whether Astronomy is capable of furnishing us with any results, bearing the stamp of reliability or of probability, as to the *origin* of this system. But it is at once perceived that to give such information is not the mission of that science, nor is it competent to the task of supplying it. Astronomical *speculation* may, indeed, as we readily admit, with a full acknowledgment of its rights, advance more or less plau sible theories founded upon the condition of the actual present, to explain how it originated[1] from a hypothetical past. But all its claims to vouch for the correctness and reliability of these theories, are in all cases and under all circumstances equally futile and unwarrantable.

[1] The most plausible hypothesis, and the one best accounting for the facts in the case, in regard to the origin of the solar system, is that one brought forward by the celebrated mathematician, *Laplace*, in his *Exposition du Système du Monde* (compare Müdler, *Astr. Briefe*, p. 335 seq.). *Laplace* assumes that our system originated from an inconceivably rare and immensely extended mass of matter without definite form and possessed of a rotary movement. The gradual cooling of this mass caused a contraction or diminution of its volume, which must have produced, according to the law of Kepler, an acceleration of its rotary motion. Hence it would gradually become more and more flattened at the poles, assuming a somewhat lenticular form as its matter gathered towards the equator. The more the latter tendency exhibited itself, and the rotary motion increased from a diminution of the volume as the process of cooling progressed, so much the more must there have been developed a tendency of the parts about the equator to separate from the general mass, as the centrifugal force would be constantly on the increase. A separation at length actually took place, as soon as the centrifugal force was sufficiently developed. In the simplest case, a circular zone must have been separated from the circumference of the mass, which would continue to rotate as a *ring*, and in which the process of contraction would still go on. Were this ring of equal density in all its parts, it might retain the annular form, but so soon as a dynamic preponderance exhibited itself in any *one* point, it must begin to gather itself together into a globular mass. Were there several, or more than one, such points of preponderance, the ring must break, and its several parts assume the globular form. Thus the planets were produced: the outer first. But contraction and rotation, with all their consequences, would still continue in the separated masses, and produce at length, from the substance of

Connected with the question as to the origin of the solar system is another concerning the *stability* and duration of the present order of that system. Here science may speak more decisively. Supported

---

these rapidly-whirling spheres, new rings, for the formation of secondary bodies. In the case of the earth, the ring cast off was collected into a *single* sphere—the moon: in the cases of Jupiter, Saturn, and Uranus, it was respectively broken into a certain number of *fragments*, constituting their satellites; whilst in the case of Saturn, it happened that one of the rings *retained* its original form. The inner planetary masses being last cast off from the great central sphere, when its density was greatly enhanced, did not admit of the production of secondary bodies under the circumstances of their rotation. By far the greater part of the original mass remained undivided, constituting the huge body of the sun.—Most of the phenomena within our solar system are explained by this hypothesis in a satisfactory manner. But it does not account for them all, and there are some contradictory facts: for instance, the body of the sun is not as dense, to say nothing of being denser, than his nearest planet—his density equals only that of Jupiter, one of the most distant of the planetary masses. *Laplace* did not make any account of the comets, or attempt to explain them. *Mädler* supplies this defect as follows: All parts of the original mass were not capable of so great condensation, and hence, as soon as the planetary masses permitted this, these refractory portions separated themselves in their original state; and as such separation took place not only in the equatorial but in all regions of the great mass, we can easily account for the inclinations and eccentricities observable in the orbits of the comets. *Schubert's Anschauung von der Bildung unseres Sonnensystems nach seinem gegenwärtigen Bestande*, comp. chap. 6, § 6 obs. 5.

[We may in this connection note the bold and ingenious application of the nebular hypothesis, to some of the vast forms in the heavens, by our own countryman, *Prof. S. Alexander*, and on a scale so grand that any ideas of world-production we may gather from the application of the principle involved to a mass of matter

by the experience and observation of thousands of years, it may boldly maintain that in spite of all antagonistic forces which are at work, in spite of a wonderfully involved whirl of movements, yea, in spite of all perturbations and disturbances which may here and there occur (themselves controlled by unchangeable laws), the present order of our solar system bears the character of a stability the most unshaken and abiding. Ever since all fear that the world might be destroyed by coming in contact with some roving comet has been got rid of, through a knowledge of the light physical properties of these bodies, no agency or discoverable accident within the whole compass of our system has been known to Astronomy, by which the order of this system might be destroyed or even materially changed.

---

equal to that comprehended by our solar system, can scarcely serve as stepping-stones to an adequate conception of what must have taken place in the production of the thickly-strewn worlds of the fixed stars.

Prof. Alexander says: "The material of which some of the clusters and resolvable nebulæ were formed, may have been—1st. A fluid spheroid of great ellipticity, the gradual cooling of which might increase its velocity, and produce a rupture and dispersion which would respectively give rise to the present forms of the spiral nebulæ observed by Lord Rosse. The Milky Way may have this form.

"2d. A ring may have been the primary form, or a spheroid may have been transformed into a ring, the subsequent rupture of which might give rise to other recognized forms.

"3d. The simultaneous rupture of a ring might give rise to the annular nebula in Lyra and others.

"4th. The simultaneous rupture of a spheroid might give rise to the 'Dumb-bell' nebula and others.

"5th. Globular nebulæ also show traces of similar action."—Tr.]

§ 6. *Parallaxes of the Fixed Stars.*

But it is time for us to mount up into higher spheres. Leaving Neptune and the comets, we hasten towards Sirius, burning in the depths of space, surrounded by his countless thousands of brother stars, who all, as friendly messengers of higher and holier regions, greet us with their sparkling, glowing light. Urging our way deeper into the vault of heaven, we behold through the telescope the milky-way, which to the naked eye appears as a faint zone of whitish lustre, resolved into millions of worlds, radiant as those we have left behind; yea, piercing still further into the unfathomable depths before us, our wondering eyes rest on thousands of nebulous clouds, floating at a distance such as mocks the scrutinizing glance of the best instruments of our day.

Vision and thought, indeed, can travel over this immeasurable distance in an inconceivably short space of time; but the brain which begets the thought, and the eye which casts the glance, cannot follow their swift-footed offspring, nor measure the distance gone over according to any ordinary rules of measurement. "To learn the distance of a *single* star," said a recent astronomer,[1] "is the abiding hope, the most ardent desire of the astronomer. There are some things to be despaired of in every sphere of knowledge. Astronomy is not a privileged science—it too must sometimes despair."

The calculation of the parallaxes of the fixed

---

[1] Pfaff, *der Mensch und die Erde*, Nürnburg, 1834, p. 41.

stars[1] had heretofore been so irregular and arbitrary, that all hope of ever arriving at any sure information on this point seemed about to depart, when the observations of *Struve* and *Bessel* (in the year 1836) brought about results so reliable and happy beyond all expectation. These two noted astronomers succeeded in solving this great problem, by a most careful observation of optically doubted stars (i. e. stars which to the eye appear very close to each other, but which in reality are not related, but separated by immeasurable distances). The principle upon which they proceeded was this: stars which appear when first observed to be in an almost perfect range, must after the lapse of six months' time, when the earth has arrived at a point some 200 millions of miles distant from the place of first observation, appear somewhat altered in their relative positions. The parallaxes of the nearest fixed stars, it was soon discovered, might be determined by the application of this principle. *Struve* chose for observation the briliant star *α*, or Vega, in the Lyre, near which, at the distance of 43 seconds, lies a faint star of the eleventh

[1] By the *parallaxes of the fixed stars* is understood the apparent displacement of those bodies, arising from their being viewed from different or opposite points in the orbit of the earth, or, what is the same thing, the apparent magnitude or diameter of the orbit of the earth as viewed from the fixed stars respectively. The diameter of the earth's orbit being about 200 millions of miles, the points from which a star is viewed at times separated by an interval of six months, must consequently be distant from each other also 200 millions of miles. If any displacement capable of measurement is detected in the position of a star when viewed from points so widely separated, it is said to have a *sensible parallax.*

magnitude. As the former was from its brightness regarded as one of the nearest stars, and both might safely be considered devoid of all connection or mutual dependence of movement, they seemed specially fitted for his object. As the result of ninety-six observations, this astronomer obtained a parallax of 0″·2613 for the star Vega. According to this parallax the latter star must lie at the distance of 789,400 times the semi-diameter of the earth's orbit, or about 75 billions of miles, a distance through which light could not travel in less than 12 years and one month.—*Bessel*, on the other hand, examined the star 61 of the Swan, which indeed is of much less brilliancy than Vega, but which on account of its peculiarly large *proper* motion — the most considerable known—excited a more reasonable supposition that it must be one of the nearest of the fixed stars. He compared this star with two faint ones at the respective lateral distances of 460″ and 705″, and obtained as the result of 402 careful observations, a parallax of 0″·3483, which would make the distance of the star to which it belongs 592,200 times that of the sun, or about 56 billions of miles, requiring 9¼ years for the passage of light. The parallax of the Polar Star was found by *Peters* (who subsequently made 33 other measurements of a similar nature) to be 0″·067. These results he obtained from very numerous and close observations. According to his conclusions the Polar Star must be distant from us three million times the distance from the earth to the sun. Its light cannot reach us in less than 43 years. Further, *Maclear* and *Henderson* having quite recently, at the Cape of Good Hope, examined several southern stars with reference to their parallaxes, found one of them, *a* of the Centaur, to have a parallax of 0″·9213. This result was obtained from several hundred observations. According to it, this star, *a* Centauri, which is one of the brightest in the heavens, possessing also a very large proper motion, and being encircled by the orbit of a star of the fourth magnitude, must be *the nearest to our earth of all the fixed stars:*—distant 223,000 times as far as the sun, or about 20 billions of miles, and requiring three years and a half for its light to reach our earth. *Rümker* fixed the parallax of Arcturus at 0″·34. Hence, nine and a half years would be required for its light to pass to our planet.

Thus we see that the distances of several of the nearest stars have been determined with reasonable accuracy. As respects those lying in the more remote regions of space, we shall perhaps for ever be left in doubt, or to calculations which can never rise beyond probable correctness. Of the latter kind we may mention the calculations of the astronomer *Mädler*, who computes that it would require 2934 years for light to pass from the nearest point in the milky-way to our earth, and from its most distant point 3836 years.

[The parallaxes of some five additional stars seem now to have been determined.

Henderson ascribes to Sirius a parallax of 0″,230, with a distance of 896,780 times the radius of the earth's orbit, requiring over 14 years for the passage of light. *Peters* finds, for the star 1830 of Goombridge's Catalogue, a parallax of 0″,148; hence, a distance which light would not traverse in less than about 22 years. He also finds a parallax of 0″·046 for Capella, placing that star at 4,484,000 times the distance of the sun, 73 years being required for its light to reach our eyes. *M. O. Struve* announces that he has recently found the parallax of *a* Cassiopeæ to be 0″·34, and the parallax of *a* Aurigæ to be 0″·30. He also thinks the parallax ascribed by M. Struve to Vega to be too large, his results giving a parallax of 0″·15 for that star. *M. Peters* has attempted to determine the absolute distances of the stars of various magnitudes, and arrives at the result that the time required for the passage of light from these bodies to us, ranges from 15 to 120 years, on the average, for stars from the first to the sixth magnitude.—Tr.]

§ 7. *Solar Nature of the Fixed Stars.*

The sceptre of the sun, so potent within the solar system, does not extend to the regions of the fixed stars. Unaffected by the influence of the mighty king of day, which to so amazing a distance enchains all things to himself, these worlds pursue their silent, majestic courses adown the ages—light or heat from the sun they ask not. They scorn to be regarded as vassals, and claim a position as brother stars to the sun, sprung from the same source of light, and led by the same almighty hand in a circling dance through the immensity of space, to the praise of Him by whom they were created.

Suns they certainly are—those brilliant points in the firmament, which still remain but points when viewed through the best instruments—if an inherent power of giving light be the ensign of their high position, the distinguishing characteristic of a sun. For the fact that their light is inherent, and not borrowed as in the case of the planets, is indicated by the circumstance of their extreme remoteness, as well as the intensity of their light while they present to the eye a mere point instead of an appreciable diameter. We have, however, a direct means of setting this question to rest. The light of the fixed stars, just as the light of the sun, presents no marks of polarization; while reflected light everywhere reveals itself *as such* by the fact of its polarization. That the proper light of the fixed stars is in general of essentially the same nature, follows the same laws of dispersion, and is possessed of the same velocity, notwithstanding the various colors it assumes, is taught by observation, for the *constant* of abberation is the same in the cases of all the fixed stars."[1]

Again, the multiplicity of *colors* exhibited by these brilliant points in the vault of heaven, is very remarkable. The double stars particularly, but many also of the single stars, display to our admiration, colors bright and variegated, accompanied with the most profuse and delicate shadings. Here glows a star with a red or ruddy light, there sparkles another with a blue or greenish tint, while others cast a more or less intense ray of yellow, or reveal themselves in purest white.

The *intensity* of light with which the fixed stars shine, is, also, strikingly different in the different stars. It is conditioned both by the magnitude and distance of these bodies, as well as the amount of light developed upon them. The *strength* of their ray is capable of measurement, and the *distance* of at least some of the nearest stars has been deter-

[1] Mädler, *pop. Astr.*, p. 391.

mined. From these data may be found the real light-giving power of any star in comparison with our sun. But it still remains undetermined in this calculation, how much of this light is to be referred to the greater or lesser intensity of the star's light on the one hand, or to the greater or lesser magnitude of the star on the other; for astronomy has never yet passed the means of arriving at the magnitude of a single fixed star.

"The light of Sirius is, according to the closest measurements, 20,000 million times weaker than the light of our sun. Hence we may learn from calculation, that were the sun removed from us 141,400 times the distance it now is, it would appear to us as a star of only the brightness and apparent magnitude of Sirius. But those fixed stars which present to us any appreciable parallax, and must for this reason be considered those nearest our system, are removed from us from 200,000 to 800,000 times the distance of our sun. But Sirius is not to be classed among the nearest of the stars, and no well-founded objection can be made to the assertion of the great English philosopher (*Wollaston*), that Sirius shines with such a brilliancy as would scarcely be produced by 14 suns like ours (or a luminous body 14 times the size of our sun) at such a distance. The star Vega, in the constellation of the Lyre, is, in all probability, much nearer us than Sirius, yet the strength of its light is only one-ninth that of the dog-star. The star 61 Cygni belongs, as shown by its parallax, to the nearest of the stars, yet its light is still much weaker." (Schubert, *Naturlehre*, p. 78).

Thus, then, the fixed stars are suns like *our* sun, shining from their own unborrowed light, and some of them far surpassing it in brilliancy, be this on account of a vast excess in size, or from a greater clearness and intensity of light.

However unhesitatingly we may from this point of view accord to the fixed stars the name of suns, the application of the term is of doubtful propriety, as we may here take occasion to remark somewhat in anticipation, if we look upon the position and physical constitution of the sun in other respects, its dark planetary body, and its relation to planets, moons and comets, as essential characteristics of a heavenly body that would be called a sun: for in these respects the analogy cannot at all be established, and indeed there is much, as we shall see hereafter, in direct conflict with such a view in connection with most of the fixed stars.

§ 8. *The Milky Way.*

The unaided eye, when cast aloft at night, beholds a whitish glimmer or band of light traversing the whole vault of heaven, or encircling it like a girdle. What was matter of conjecture from the earliest times—that this band of light was composed of the united rays of countless distant stars, incapable of being separately distinguished on account of their extreme remoteness—was proven by Herschel's telescope to be an indubitable fact.

W. Herschel regarded the Milky Way and the visible stars as all belonging to *one* vast system, " but he did not in accordance with former assumptions ascribe to this system a *spherical* form. He imagined the system to be projected in space in the form of a flat, *lenticular* plane, near the midst of which our solar system was situated, the whole presenting in general the form of one vast circle or plane of stars. Subsequently, however, he assumed the view that the stars constituted a vast *ring*. The most recent investigations have established this view, but with this modification, that the Milky Way consists not of one, but a system of several or at least two concentric rings of stars, encompassing all the *single* visible stars.

The Milky Way does not strictly form one vast circle through the midst of the vault of the heavens, but divides it into two unequal parts, the superficies of which are to each other as 8 to 9. Besides, it is separated through $\frac{2}{5}$ of its course into two arms, which again merge into each other. These two facts find their explanation in the discovery that the Milky Way consists of *two* concentric rings, and that the position of our sun is excentric in the concourse of stars encircled by these rings. For if we held a central position with respect to these rings, the Milky Way would divide the vault of heaven into two equal halves, and the inner ring would so completely cover the outer, that no longitudinal division could be perceived in the whole course of the Milky Way. Our solar system does not therefore lie in the plane of the Milky Way, but outside of it, in that part of the heavens divided by the latter which appears largest to the eye (*i. e.*, in the direction of the autumnal equinoctial point). But we must conclude from the fact that the Milky Way appears to us throughout $\frac{3}{5}$ of its course as a simple band of light, while it is divided for the other $\frac{2}{5}$ of its length into two arms, that our position in space is considerably nearer to that part of the Milky Way where the two rings appear separated, than to the opposite region, where the outer ring is completely covered by the inner. "The middle of the divided portion of the Milky Way lies in the Scorpion, and we must consequently look for the nearest point in the former in the direction of the same constellation."

But there are inequalities and irregularities observed in these two rings which cannot be explained as optical illusions. "At some points they are broader than at others, have an increased brilliancy, and are subject to irregular bends, divisions, &c." The bridge-like arms which pass between and connect the rings in several places are worthy of special remark.

Whether there still exist beyond these rings, other vast belts of stars, enclosing the fixed stars of our system, cannot be satisfactorily determined. "The position of our sun is such," according to the view of Mädler (page 417), "that perspectively no other than the division above mentioned is possible." That still further astral rings do really exist, is by no means impossible. At least, the fact that the most powerful instruments still reveal faint clouds of light incapable of being resolved into stars, perhaps from their extreme remoteness alone, would seem to favor this view. "But even assuming that another series of rings concentrically arranged, were concealed by the outer of the two rings known to us, we cannot certainly regard the series as without end. For, as has been shown by *Olbers*, were the plane of these stellar rings extended without bound, it would be discovered by the naked eye, in the form of a bright line in the centre of the Milky Way, and traversing it longitudinally. And this is so, just as in general it is true, according to the well-grounded remark of different astronomers, that were the starry heavens everywhere extended to infinity, every point in the nocturnal heavens would shine with the light of the sun and the brightness of day, so that there would really be no longer any distinction to the eye between night and day." (Schubert, *Weltgeb*, p. 24). It has been repeatedly affirmed by the younger Herschel, that in many points of the Milky Way the cloud-like, glimmering light can be so completely resolved, that we behold the dark back-ground of the deep, starless heavens, lying beyond. (Compare Humboldt, *Kosm.* III. 188, 213).

It is not improbable that the concourse of stars, also, within the rings which compose the Milky Way, is made up of layers of stars, having an annular disposition in a common plane, and separated by starless spaces traversed here and there by arms uniting the various rings. But the difficulty of determining the number and form of these supposed rings must be almost insuperable, on account of their proximity to us. But this much *Mädler* holds (p. 417) as established beyond dispute, that whether the inner regions of the fixed stars are disposed in layers or rings, or whether it be otherwise, the space filled by these stars is certainly not of a spherical form. The outer parts of these inner regions do indeed evince a somewhat annular form, for here the stars of from the 7th to the 11th magnitude are unusually numerous, both upon the Milky Way and in the neighborhood of its boundaries. This stellar ring when viewed with the naked eye, must, consequently, pretty nearly coincide with the Milky Way; but a telescope of only moderate powers, and one that fails of resolving the Milky Way proper, is fully capable of revealing to us the individual stars that compose it.

§ 9. *The Central Sun.*[1]

Ever since *Bradley's* time (the middle of the last century) the conviction has been growing in the minds of astronomers, that the so-called *fixed* stars, together with our sun, are by no means really *stationary* stars, but that they all have a proper and real motion of their own.

The bold and wide-spread creations of poetic genius in regard to a vast and all-controlling central sun, which enchained the millions of other suns to itself, and caused them to revolve around it in unswerving obedience, through the might of its preponderating gravity,[2] seemed in this grand discovery

---

[1] Compare Mädler, *die Centralsonne*, Dorpat, 1846; *Untersuchungen der Fixsternsysteme*, Mitau, 1847; *Pop. Astr.* 4th ed., p. 404 seqq.

[2] The occasion of this fanciful supposition was furnished by the attempt to transfer the relations and arrangements of our solar system bodily into the regions of the fixed stars, of which we shall say more hereafter. Since moons here revolved round

to have attained a scientific basis. But there was needed only a closer investigation of the proper motion of the fixed stars, to show how inadmissible this view was, however confidently it had obtruded itself upon the world.

Of all the fixed stars, none seemed to have such just claims to this high and sovereign prerogative in the universe, as Sirius, a sun surpassing all others in brilliancy. "But *Argelander* has well remarked that Sirius *cannot* be the central sphere, since it itself has a *proper* and very observable motion through space. . . . If there exists anywhere, visible or invisible, one grand central body, out-balancing and controlling all others through a preponderance of gravity, the most rapid general movement must take place in the region of that body. And as we behold fixed stars in all directions, it is clear that in some one point the whirl of movements must be most conspicuous, and from there the rate of motion suffer a constant decrease. But nowhere in the heavens is such a point to be found — no one of the stars of the first magnitude fulfils the condition here imposed." (Mädler, *Centralsonne*, p. 4, 5.)

These and similar considerations led Mädler to the final result, "that *no such single preponderating central mass is to be looked for in the starry heavens, as there is none such in existence.*"

In this state of affairs astronomers inclined to the view, "that the movements noticed in connection

---

planets, and planets with these about the sun, it was thought that all suns in like manner must be moving round a vast central body of equally preponderating gravity.

with special stars, were occasioned *merely* or chiefly by the *mutual influences* of the stars in *closest proximity* to each other." But still this view could not be made to account for the data supplied by observation and calculation.

It was left to the deep sagacity and untiring diligence of *Mädler*, after six years' uninterrupted investigation and thorough-going examination and comparison of all previous data as to the progressive movement of the fixed stars in the heavens, to arrive at a result no less simple than surprising, which promises at length to explain to us the mysterious movements in the heavens of the fixed stars, and the wonderful harmony in the construction of the universe, or at least to point out the way to such a result.[1]

---

[1] It is indeed true that many of the leading astronomers have thus far refrained from direct assent to the hypothesis of *Mädler;* and a few have been free to express their great doubts as to its correctness (as *Peters* in the *Astron. Nachrichten,* 1849, p. 661, and *J. Herschel's Outlines of Astronomy,* 3d ed., p. 589). *Lamont,* however, expresses himself in favor of it. *Alex. von Humboldt* (Kosm. 3, p. 283), withholds his opinion. But *G. H. von Schubert,* on the other hand, has eagerly laid hold of Mädler's idea, and incorporated it in his ingenious work, *Das Weltgebäude* (p. 27 seqq.). It is true, as we readily admit, that Mädler's grounds are still defective, and his view far from being incontestably established as yet. In order to arrive at a result in all respects conclusive, there is demanded the continued observation of centuries, and in connection with a much larger number of stars than has heretofore been the case. But the care and the close scrutiny with which he received the observations of his predecessors, as well as increased them by his own efforts, and also the harmonious result obtained by a combination of the two, seem to lend to the conclusions of the sagacious and untiring astronomer the character of great pro-

If the assumed centre of the world of the fixed stars, to which all their movements are to be referred, cannot be a body controlling all others by the might of its preponderating gravity, it by no means follows that no common centre exists, around which stars and systems of Milky Ways revolve. Though it be not the amazing gravitating force of one huge central body that induces the movements of all the stars, it may doubtless be the gravitating influence of one star upon another, and of all upon all, which causes the whole to revolve about a common central point; and this centre may just as well be assumed to be an *empty* space, as one filled by a body, which body, too, might be one of the smallest dimensions. For as each body of the system of our world is attracted by all the others belonging to the same system, it is not conceivable that the whole should move with respect to any particular member of the system, but rather, that it should take a course which would satisfy all alike. Thus there would necessarily arise a common movement of all about a common centre (be that an empty space, or filled with a body), and the position of that centre would depend upon the original disposition and arrangement of the stellar worlds.

If it be true that the countless stars of our system suspended in space, affect each other in inverse pro-

bability, and warrant the hope that they will derive new support from future observations. At all events, he has the merit of having given astronomical investigation a new and powerful impulse, and of having marked out a path for it, the following of which, even though opposite results should be obtained, will signally advance the problem of the heavens towards its final solution.

portion to the square of their distance, according to the common and general law of gravity; if, further, these countless attractive forces of all upon all, resolve themselves into a harmonious movement about a common centre, just as a thousand different tones unite to form one grand and swelling accord, — then is the case just the reverse of that which takes place in the movements of our solar system. Here we behold a huge central sphere, out-balancing 700 times the united weight of all the other bodies of the system, and excluding the possibility of a general and harmonious movement about a common empty space or centre; here we behold the several bodies composing the system led like vassals round the all-controlling sun, these carried along with a more rapid movement as they approach their lord, those at a distance moving more or less deliberately according to the increase or decrease of solar attraction. But there, on the other hand, the case must be reversed; with an increase of distance from the empty central space there must be an increase of movement also, so that the time of revolution must in all the fixed stars be about the same. If we suppose, for example, a certain number of concentric rings to be formed by the substance of the earth, from the equator to the earth's centre, it is plain that the atoms composing the rings nearest the centre must have a slower, and those of the more distant rings a more rapid movement about the common centre.

If now these be indeed the laws according to which the movements of our stellar worlds come to pass, it is clear that stars diametrically opposed to each other must have opposite movements. As in a rotating wheel the spokes of one side have a motion from right to left, and those of the other a motion from left to right, so also in the great wheel of the fixed stars whose circumference is represented by the Milky Way, the stars of one side must proceed from north to west, and those of the other from south to east;—and of all known means *this* law, next to the one above-mentioned that refers the more rapid motion to the *greater* distance from the grand centre (and the reverse), is best calculated to point out to us the central point for which we are seeking, if there be any such in existence, to which the movements of the stars are to be referred. Further investigation may have something to go upon, if it can be determined with reasonable accuracy in what direction the supposed centre lies, since the dynamic centre of the system of the fixed stars cannot in all probability vary much from the mathematical centre of the same. Just at this point the two-fold excentric position of our sun comes to our aid. We have already learned (§ 8) that a point lying nearer the constellation of the Scorpion than any other part of the Milky Way, and on the side of the autumnal equinox, marks the position of our sun in relation to the central point. "Consequently, in order from the position we hold to arrive at this central point, the eye must be directed to the opposite side of the heavens, and in the direction of a line leading from the region of the vernal equinoctial point to the Milky-Way about the constellation Taurus." (Mädler, *pop. Astr.* p. 402.)

Mädler at length, after the most careful and thorough measurements, comparisons and calculations, with the use of all the data furnished by previous investigators,[1] arrived at this result, which fully harmonizes both with these data and the laws above mentioned: that the long sought for point lies in the beautiful and brilliant constellation of the *Pleiades* (or seven stars), and probably, too, near by or in the brightest star of this group, Alcyone.

"*I hence regard*," he says at the close of his investigations (*Centralsonne*, p. 44), "*the Pleiades as the central group of the whole system of the fixed stars, even to its outer limits marked by the Milky Way, and Alcyone that star of all those composing the group, which is favored by most of the probabilities as being the true central sun.*" But at the same time he remarks, that in consequence of a change in the constellations in the course of ages, the centre of gravity belonging to the system of the fixed stars may pass from Alcyone for awhile, and perhaps to some neighboring star.[2]

---

[1] A careful catalogue of 3222 star-positions was left behind by Bradley. Renewed measurements of the same stars — after the interval of almost a whole century — must go far towards determining their motion. Mädler applied this in the cases of more than 800 stars which seemed specially to serve his object. Also Bessel's manifold and highly-careful observations in regard to 73 stars of the Pleiades, 11 of which had been before closely scrutinized by Bradley, were very opportune and serviceable.

[2] *Schubert* says of the Seven Stars (*Weltgeb.* p. 27): "A group of stars alone in their kind is to be observed in the heavens, not far distant from the vernal equinoctial point — a group which from the earliest times has specially attracted the attention of man. This is the cluster called the Pleiades. Alcyone, a star of comparatively large magnitude, stands there, surrounded by

It is clear from the foregoing, that neither the group of the Pleiades nor the star Alcyone holds such a conspicuous position in the system of worlds, from the possession of a higher essential dignity than the other stars, — that the ground of this their distinguished position does not lie in themselves, in their nature and individuality, but merely in their accidental situation, if the expression may be allowed. And as the question here is not in regard to a *body*, but to a *place* in the universe, whether that place be occupied by a body or not, the fond application of the term *central sun* to Alcyone, by the discoverer, is not exactly a fitting one, and is much exposed to misapprehension by the uninformed.

---

five others, which may be fairly distinguished by the naked eye. In regard to these six stars, *John Michel*, of England, has shown that they must constitute a physically connected whole, the probabilities against their close juxtaposition arising from accident or optical illusion being in the ratio of 500,000 to 1. The peculiar lustre of this group does not, however, depend merely upon the six stars visible to the naked eye; but also arises from a whole cluster of stars which are brought into view by means of the telescope. As in the case of the double and multiple stars a common centre of gravity must exist, so also in this cluster there must be a common point about which they move; and if this be not in Alcyone, very probably it is not far distant from that star. But it is only from the closely-crowded relation of all the members of this group, that its point of gravity can derive significance as the grand centre of the whole astral system. According to the computations of Mädler, all these bodies are collected and compressed into a space not amounting in diameter to four times the distance from our sun to the nearest fixed star. It is not the single stars of the group, however, but rather the collected might of the whole, which lends to this cluster the character of a connecting bond or foundation-stone for the whole structure of the heavens.

Mädler also made an attempt to determine the parallax of Alcyone, from a sagacious application of facts founded upon the known parallax of the star 61 Cygni. (*Pop, Astr.* p. 427). The result attained was a parallax of 0″,006533, according to which Alcyone is removed from us 31½ million times the distance of the sun, a distance requiring 498 years for light to traverse.— Our sun in its course about Alcyone, moves at the rate of 8 geographical miles in a second, and requires 18½ millions of years to complete one revolution.

Notwithstanding the amazing distance to which our sun is removed from the true centre of the system to which it belongs, "we still hold a position," as *Schubert* says, "deep within and proportionably near the centre of the vast circle bounded by the rings of the Milky Way as walls of light."

We shall close this discussion by giving Mädler's view of the arrangement of the whole stellar system, as deduced from these his observations and discoveries. He says:[1] The starry girdle of the Milky Way probably consists of two broad concentric rings, which at their most distant point from us perspectively coincide, and in most part cover each other; but at their nearest point, on the other hand, form such an angle with each other as to leave an open space where they appear separated. Since now the inner and pretty well defined limits of the Milky Way, indicate that it is separated from the hosts of fixed stars it encloses, though that separation be not a complete one, and since, on the other hand, in the

[1] *Centralsonne*, p. 46 seq.; *Pop. Astr.* 415 seqq.

neighborhood of the Pleiades *particularly*, a quite observable starless space exists, we may imagine the whole constitution of the system of the fixed stars to be as follows: The centre of this system is marked by a group very rich in stars closely crowded together, and contains single masses of considerable size. Around this there extends a vast zone proportionably devoid of stars, having a diameter somewhat over six times that of the central system. This is succeeded by a broad annular stratum, teeming with stars, which is again followed by an interspace containing but few stars, and so on for an indefinite series of starry strata and partially empty zones, until we at last arrive at the two outer rings composing the Milky Way. These vast rings are not equally well developed in all their parts, but exhibit here and there a tendency to resolve themselves into groups and clusters, though they are chiefly made up of isolated and double fixed stars. They are connected at various points by starry formations which traverse the empty interspaces and bind the rings together.

§ 10. *Variability of the Fixed Stars.*

We have ever been accustomed to connect with the heavens of the fixed stars the ideas of immutability and sameness. But modern astronomy has revealed to us an exceedingly rich variety of cosmical formations, groupings and movements in those same heavens, as well as the circumstance that changes and transformations take place in connection with many of their stars, to which the facts of our solar system furnish no analogy.

If it be true that we can observe changes in bodies which notwithstanding their overwhelming magnitude appear as mere points of light through the best of telescopes, those changes must certainly be so mighty, grand, significant and influential, so comprehensive and complete, that none of the changes or revolutions with which we are conversant in our domain of life are worthy to be compared with them.

Of all the changes experienced by the fixed stars, we can detect only those which affect their light. All else that there takes place must for ever remain hidden from mortal eye. Their light alone, which traverses with the rapidity of thought the immeasurable spaces of the universe, reaches our eye only after years of travel, and its changes alone can make known to us the cosmical changes which are there experienced.

These are revealed in part by a variation in the *color* of the light, but chiefly in an increased or decreased *intensity* of their light, which in the same star swells now to the brightness of Sirius, and then fades away, down to the light of a star of the lowest magnitude, or dies away entirely.

In regard to a change in color, that is mostly observed in connection with the double stars (§ 11), where it generally occurs periodically. But it may take place in single stars oftener indeed than has yet been detected by observation. The ancients describe the color of Sirius as red, while this star at present shines with the purest white light.

As to the physical cause of this variation in color, science has not thus far been able even to conjecture.

Much more significant, however, is the change in *strength* of light, which has been observed in not a few stars (which hence are called VARIABLE STARS), and which is peculiarly fitted to give us an intimation of the great variety and peculiarities of the laws of life and movement which obtain in the celestial regions. There has been observed with respect to more than thirty stars, a more or less strongly marked increase and decrease of brilliancy (or apparent magnitude), which mostly recurs periodically. The two most remarkable stars in this respect, are Algol, in the head of Medusa, and Mira (so-called on account of this singular characteristic), in the Whale. Mira attains its state of greatest brilliancy 12 times in 11 years, while the period of Algol is only 2 days, 20 hours and 49 minutes. "With respect to most of these stars, however, the variation observed in them is itself subject to variation. The process of increase and decrease in brightness, the period itself, and the maximum and minimum brilliancy, are all subject to change. Specially worthy of note is the fact that in most cases the increase in brilliancy is more rapid than the decrease, and that all these stars, with the exception of Algol, remain a longer time at their minimum, or near that condition, than they do at their maximum." *Mädler*, p. 440.

Various methods have been tried to account for this strange phenomenon. One of the first was the assumption of a dark, invisible body, revolving about the bright body of the star in the given period, and covering its disc in part, just as in an eclipse of the sun by our moon. But, however well this seemed to account for the striking phenomenon, it was open to many objections. Says Schubert:[1] "A dark planetary body, which in passing over the disc of our sun should so darken it, that the obscuration would be as obvious in the regions of the fixed stars, at a distance of billions of miles, as the variations in the light of Algol are to us, must be of such enormous bulk and so close to the sun, that, according to the mean proportion existing between the bodies composing our system, its revolution must be completed once in less than 14 hours. But the period of revolution in the supposed dark body must be about five times that length, which would lead us to infer a density in the region of Algol 25 times less than exists in our system." Besides, the fact that the increase of brilliancy is much more rapid than the decrease, could not well be harmonized with this hypothesis.

Another attempted explanation would find the cause of the periodic variation of light, in the star's rotation upon its axis, so that (in a similar manner, but in an incomparably greater degree than occurs in connection with the solar spots and facules of our central sphere,) at one time its brightest side must be turned towards us, and this be then succeeded by the side less intensely lighted. But this theory also meets with many difficulties, such as the one before mentioned, that the increase of light in almost all the variable stars is more rapid than its decrease, and that the amount of increase and decrease is not the same in every recurring period.

---

[1] *Naturlehr*, p. 99.

"A third hypothesis supposes the form of the star in question to be flat or lenticular, so that in rotating the edge and broad side of the star are in turn directed towards us." But such an unusual rotation would contradict the laws of gravitation which every where obtain, and still not account for the irregularity in the periodicity of the star's variation.

This remarkable phenomenon appears most satisfactorily accounted for, upon the assumption that the stars to which it belongs *are subject to a periodic, but varying increase and decrease in their development of light, grounded in the constitution of the stars themselves.* "The increase and decrease of their brilliancy," says *Schubert* (*Weltgeb.* p. 64), "reminds us of the recurrence of days and seasons with us; but with this distinction, that in our system these changes are caused by the influence of the sun upon the planets, while with respect to the stars the cause probably lies in their own constitution. The changes we here experience from the higher to the lower grades of warmth, from morning to noon, to night, and to mid-night, or from winter to spring, summer and autumn, are, in the variable stars, the changes from the lowest stage of brightness to the medium, and then to the highest, with a return again to the medium, and finally to the lowest. Spring frequently comes earlier, the summer is hotter and longer, and the winter milder than in other years, where the opposite of this takes place. So in most of the variable stars, the changes are more or less marked, and the different steps of their progress of varying length."

In addition to those stars which at their lowest

stage of brightness still remain visible, though it be only with the aid of a telescope, there are others (as those in Sagittarius, Cygnus and Leo,) which have for ages appeared visible at periodic intervals of many years, becoming again *totally* invisible after each appearance. Perhaps to this catalogue belong the NEW STARS, which have been repeatedly observed, appearing *suddenly* with the *greatest* brightness, and which after shining awhile, have gradually died away until they have utterly vanished, becoming LOST STARS. Humboldt (*Kosm.* p. 220,) mentions 21 such stars, and Mädler has added to this catalogue one which was visible for a short time in January 1850, in the constellation Orion.

Hipparchus, so early as the year 125, B. C., observed such a phenomenon. In A. D. 389, a new star broke forth near the star Altair of the Eagle, so bright, that for three weeks it equalled Venus in brilliancy; but in a short time it vanished completely. In like manner, a large new star appeared repeatedly in the years 945, 1264, and 1572, on the borders of Cassiopeia. Tycho de Brahe closely observed the last appearance. In the course of a few minutes the star kindled up to the brightness of Sirius, in a month's time began to fade, and at the end of a year and a half had totally vanished. A similar star appeared five times (in the year 134, B. C., and in A. D. 393, 827, 1203 and 1584,) in the Scorpion.

Are we to suppose that we have, in these instances, examples of stars really newly formed, and soon thereupon vanishing into nothing from whence they seemed to come? Such an assumption lacks probability, and is assuredly not in accordance with the analogy of the heavens. As the years 945, 1264, and 1572, are separated by about equal intervals of time, it has been conjectured that the star observed by Tycho may be one recurring periodically, about every 300 years, from a sudden kindling up of its surface, dependent upon some inner unknown cause; the star after that, gradually so diminishing in brightness as to remain hidden from view for 300 years again. The close of the present century will prove whether this conjecture be well-grounded or not. The same reasons may lead us to believe the appearance of new stars in the Scorpion to be periodic in its occurrence. That the sudden kindling up and subsequent dying out of these stars, is dependent neither upon the manner of their rotation, nor upon the intervention of a dark body between them and us, is obvious enough to every mind. The conjecture possessing most probability, is that one which supposes the stars in question to be *dark* bodies in themselves, which periodically or at irregular intervals, and through a native, independent action, or from excitement from without—perhaps through the medium of a magneto-electric process—are brought into such a glow or intense excitement, that they for awhile shine like stars proper, from an innate light.

§ 11. *Double and Multiple Stars.*

A better acquaintance with these stars, whose connection with the characteristics of the heavens is of such special importance, is marking quite an epoch in the history of astronomy. Frequently two or more stars, mostly of different magnitudes, lie so close together, that to the naked eye or a glass of moderate powers, they appear as a *single* star. This effect is in many cases dependent on *optical* illusion; but in many others, continued observation has clearly proven that the stars in question have a *physical* connection, and circle about a common central point[1] in such manner, that, if they be of equal size and weight, the

[1] We call this central point a common one, and the motion of the stars about it a reciprocal one, although frequently the smaller moves round the larger body. But the occurrence of the latter case does not destroy the fact that the motion is reciprocal, and it takes place only when from a vast disparity in size or weight of the bodies concerned, the reciprocal effect is so unequal that the centre of gravity lies very near the surface or within the larger body. Our planetary system supplies a good example in this connection. The motion here is properly a reciprocal one. Not only does the sun attract the earth, but also the earth the sun; and they both move round a common centre of gravity. But in our system the case is such that the attraction of the earth, yea, the attraction of all the planets and their satellites, upon the sun, affect this huge body comparatively very little. The mass of the sun is, for instance, 345,936 times that of the earth, so that the common centre of gravity lies 345,936 times closer to the centre of the sun than to that of the earth: or, as the semi-diameter of the earth's orbit amounts to about 95,000,000 miles, not more than 275 miles from the centre of the immense body of the sun, which is over 870,000 miles in diameter. Were all the planets and satellites belonging to the system to take their positions on one side of the sun, and there expend their united powers of attraction, still the centre of gravity would lie but little without the body of this great all-controlling sphere. The earth and the moon furnish another case in point. As the mass of the moon is 68½ times less than that of the earth, and its distance from us only 60 semi-diameters of the earth, the centre of gravity still falls within the body of the earth. "The *true* central point in a system is

paths of their orbits will coincide; but if they differ somewhat in these respects, as is frequently the case, their orbits will be marked by concentric circles. Our more intimate acquaintance with this so long neglected sphere of celestial life and movement, is connected especially with the names *Herschel* and *Struve*, names as conspicuous in the history of astronomy, as are the stars they investigated in the heavens. It is to the astonishing activity and persistent diligence of *W. Struve*, before all others, that we owe, in connection with greatly improved instruments in his possession, the most of what has been accomplished in this sphere of research. He described, in the year 1827, amid the almost 120,000 stars of from the first to the tenth magnitude visible in the heavens at Dorpat, which he reviewed in 2½ years with his gigantic refractor, 3112 double stars, of which only 340 had been noticed by Herschel the elder. Ten years later appeared his greatest work, under the title: "*Mensuræ micrometricæ stellarum duplicium*," which gives the results of repeated micrometrical measurements of 2710 double stars—some hundreds of those previously catalogued being excluded from further examination on account of the faintness of the accompanying star. Of this great work, by the way, *Mädler* says: "It may be regarded as the true basis of all present or future researches of a similar kind, and is a work to which, in the sphere of physical astronomy, there is none other worthy to be com-

that point of gravity about which all the connected bodies forming a system move in sustained equilibrium — an ideal point, not necessarily occupied by any body." — Mädler, *Astr. Briefe*, p. 86.

pared, either in regard to magnitude of labor or perfection of details." Subsequently, through the continued labors of the younger Herschel, Struve, Mädler and other astronomers, the catalogue of double stars has been gradually increased to almost 6000.

In the above review of the heavens by *Struve*, it was found that every 38th or 39th star, on an average, was a double one. He found, in addition to the systems composed of but two stars, 113 triple, 9 quadruple, and 2 quintuple stars. The quadruple stars are in most cases composed of two pairs of double stars united. It is worthy of remark that in some of the triple stars, not the larger or chief star is the double one, but the smaller or accompanying one; somewhat as with us the moon revolves about our planet, and with the latter around the sun, except that there, not only the planets, but the moons also, are of a solar nature.

The systems become still more complicated as we ascend into higher regious. In the constellation Cepheus we find one composed of 4 pairs of stars, and in Orion, one of 3 pairs bordered so closely by one of 4 double stars, that we are led to conclude that a union subsists between the systems of these two orders. "Such a union of astral systems of a lower to a higher, and this perhaps to a still higher order, may, possibly, form the transition to those assembled hosts of celestial worlds revealed to us by the telescope under the name of *clusters of stars.* Hundreds, and sometimes thousands of stars, as easily separately distinguished through the telescope as those composing the inner ring of our astral system, are in some of these stellar clusters and by the bonds of mutual attraction, assembled around a visible central star. By far the greater number of these clusters, as well as most of the double and multiple stars, lie in the Milky Way, or on its borders. They are very frequently separated from the crowded stratum of the astral ring, by a dark and almost starless space, as though they had drawn themselves together from the surrounding bed of stars, leaving a dark zone between."

"A space not greater than that which lies between our sun and the nearest fixed star, there contains, frequently, hundreds of thousands, yea, perhaps millions of suns; so that one sun cannot in proportion be further separated from another, than in our system a planet is from its nearest neighbor. For, if we assume in our computations, that the moderately bright stars of these crowded clusters are further removed from us, we must also at the same time greatly increase their supposed individual diameters, so that the inexplicable fact of their close connection still remains the same as when, with a less bulk, we suppose them to be nearer at hand."

Highly significant in connection with a knowledge of the mutual relations existing between the double and multiple stars, are the alternating contrasts to be observed in the strength of their light, and in the quality and beauty of their colors. "Continued observation has clearly proven that many of them experience a change of brightness, which clearly betrays a reciprocal relation and influence subsisting between them—such a relation, that now one is caused to shine with a stronger light, and then the other. . . . . . . The careful eye of *W. Struve* succeeded in detecting 71 such stars, in which a periodic variation showed itself as always very probable, and generally very decided." No less remarkable are the strong contrasts in color which they exhibit. While one appears of an emerald green, the color of the other is ruby red; while one casts a deep yellow ray, the other shines in clearest blue, and the like. "The accompanying star generally receives the blue or violet tint, while the chief star appears white, yellow, or red, less frequently, of a greenish tint." That this phenomenon does not arise from optical illusion, as in the case of the so-called complementary colors, has been proven by the careful observations of *W. Struve.* He repeatedly examined these stars by excluding one of them at a time from the field of the telescope, so that had the special color been merely complementary, it should have disappeared, which however was not the case. "The degree of the tinting sometimes greatly increases in both stars of a pair at the same time, and in evident reciprocal relation, as in number 163 of Struve's great catalogue, which at one time, in the year 1831, exhibited its chief star of a coppery red, and the accompanying one of a bluish cast, and soon thereafter, the former of a rosy red, and the latter of a sapphire blue color."

The most significant and important result gained by a careful and laborious examination of these double and multiple stars—a result the correctness of which is now, as it would appear, fully established—is the thence derived fact, that amid the spaces of the fixed stars on high, the same laws of movement obtain as with us. The investigations of *Struve* and *Mädler*, for example, have clearly proven that the orbits of those distant spheres, just as the orbits of bodies belonging to our system, partake of an elliptic form, reminding us more, however, from their great eccentricity, of the orbit of a comet than that of a planet. "The human mind experiences a peculiar satisfaction and secret delight, in learning that thus the *first* of Kepler's laws, so significant and comprehensive, silently receives acknowledgment so far amid the depths of space. But, also, the other laws of cosmical motion discovered by *Kepler*, as well as the great law of *Newton*, bear uncontrolled sway over those remote worlds; though it by no means necessarily follows that attraction of mass *alone* effects the movement, since magneto-electric attractions, just as all attractions of the higher order, obey the same law."[1]

With respect to the two other laws of Kepler, which demand "an accelerated orbital motion with a diminution of distance," it may be remarked that their sway has been detected in several of the astral systems composed of double stars. The Newtonian law also "holds good in connection with the orbital motions of the double stars, so far as these have been learned; for it is generally only in stars of the first (more approximate) order, that an orbital motion can be distinctly observed, which, in other orders, exhibits

[1] Schubert, *Urwelt*, p. 88.

only feeble traces of its presence; and at the same time, it is in the stars closest to each other, that, according to the rule, the greatest velocity of movement should be observed."

["In the great work which M. Struve has lately published, containing the record of his labors on double stars at Dorpat, he gives, as the result of his careful examination and comparison of the whole body of facts in stellar astronomy, some conclusions of a novel character respecting the number and constitution of the double, or multiple stars. He examines, especially, the brighter stars—those comprised between the first and fourth magnitudes—and arrives at the conclusion that *every fourth star* of such stars in the heavens is physically double. He even ventures to assert that when we have acquired a more complete knowledge of double stars, it will be found that *every third* bright star is physically double. Applying these considerations to the stars of inferior orders of magnitude, he finally arrives at the following conclusion, which he admits to be of an unexpected character — that the number of isolated stars is indeed greater than the number of compound systems; but only three times, perhaps, only twice as great."—*Annual of Scientific Discovery*, 1856, p. 379. — Tr.]

§ 12. *Dark Bodies in the Heavens of the Fixed Stars.*[1]

Beyond the limits of our solar system we behold none but *self-luminous* bodies. No *Frauenhofer's* refractor, no gigantic telescope of a *Herschel* or a *Rosse*, shall ever be able to discover for us, whether or not there exist amid the countless hosts of the starry heavens, dark celestial bodies, in addition to those resplendent suns. Were such bodies indeed in existence, and exposed to the rays of a blazing Sirius, or

---

[1] Compare Humboldt, *Kosmos*, III., p. 267 seqq.; Mädler, *Nachträge*, p. 16 seqq.

the concentrated light which streams from a double or multiple stellar system, or were they situated in the midst of some thickly-crowded starry cluster, buried in a very sea of light emanating from thousands of suns, still the immeasurable distance intervening between them and us, must ever prevent their borrowed rays from reaching our telescopes.

But that which the bodily eye of man can never reach, with all the mighty helps his invention has contrived, may, perhaps, still in time be disclosed, through the untiring efforts of the human *mind*, by means of observation, combination and analysis.

Though we may not be able to discover those hypothetical dark bodies by means of the influence exerted upon *them* by luminous orbs to which they belong, there is a plain contrary possibility of our being able to ascertain their presence by means of the influence *they themselves* exert upon *those shining orbs.* This may result from either a partial or complete obscuration of those bodies at each periodic revolution of an invisible sphere about them, or a discoverable perturbation in their orbits caused by the gravity of existing dark bodies. In the two cases, the detection of their influence must be different, and the proportion in size between sun and planet must there be wholly different from what it is here; for an observer stationed upon Sirius or some other fixed star, would not, with the closest observation, and under the most favorable circumstances, be able to discover the least trace of an eclipse of our sun, or of the disturbances in its movements caused by the planets of our system. The effect of gravitation in this case could not be observed in mere disturbances, but only when it was raised to such a pitch that our sun should, from a preponderance of *mass* in the planetary body or bodies, be forced to assume an orbital movement; and a solar eclipse could only be visible when a dark body of full the sun's *size*, or even larger, fairly intervened between it and the distant beholder.

However strange and contrary to our notions it may be, to conceive of a sun being controlled by the superior weight of a dark body, and forced to revolve about the latter, the most recent discoveries seem to make the real occurrence of such a fact probable.

A paper was produced in the year 1844 by *Bessel*,[1] a famed hero of astronomical science, in which it was shown that two of the most brilliant fixed stars, *Procyon* and *Sirius*, were, besides the general motion to which all the stars are subject, participants of another, which, instead of leading them in wide orbits, caused them to describe very small, contracted circles; and that hence the motion of these stars can be accounted for, only on the supposition that they revolve about some point of gravity near at hand, and in all probability—as these are not double stars in the accustomed sense—some central body which, however mighty in mass or bulk, is invisible to us, and consequently must be a totally dark[2] or but feebly lighted body.

---

[1] *Astronom. Nachrichten*, 514–516.

[2] [This supposition of Bessel's is acquiring additional support with the progress of time, and from the more careful observations of astronomers. Captain Jacob, of the Madras Observatory, a

Bessel was fully convinced of the legitimacy of his conclusions, and so remained. Other astronomers, as *Struve*, for example, doubted, and were inclined to refer the phenomenon of peculiar motion to mistaken observations. Still others, as *Airy* and *Pond*, regarded it as proceeding from some variation in the proper motion of the stars in question. In the meantime *Mädler* took a decided stand with Bessel. In the case of a beautiful double star in the Twins, his observations would not agree with the results furnished by calculation. *Mädler* then assumed the existence of a triple system, in which one of the bodies was invisible, and immediately arrived at a satisfactory result.

Finally, in the years 1850 and 1851, there were published about the same time, in the astronomical journals of Europe (so says Mädler, *Nachträgen* p. 17), four different investigations, those of *Schubert*, *Pierce*, *Peters*, and *himself*, in regard to the stars Spica, Sirius, and Procyon. In regard to Sirius, the second of these stars, *Schubert* and *Peters*, independently of each other, and with striking agreement,

year or two since made communications respecting the binary star, 70 Ophiuchi, the exact orbit of which is yet in doubt, although nearly a whole revolution has been completed since Sir W. Herschel first discovered the character and motion of the star, in 1779. There must be some perturbing cause, as all the orbits thus far computed fail at certain points in representing the observed positions. The facts are best accounted for by supposing the existence of a third and dark body perturbing the other two. The same observer, also, in a letter to Prof. Smith of Edinburgh, dated Jan. 26th, 1856, conjectures the existence of a dark body in the vicinity of α Centauri, as otherwise unaccountable perturbations are observed in connection with that star. — Tr.]

ascribe to it an orbit of from 49 to 50 years about a point 2½″ distant from that star. If the parallax of Sirius as calculated by *Henderson* be assumed as correct, this point of gravity must be occupied by a mass of at least ⅗ the weight of the sun.

Mädler's investigations in regard to the orbit of Procyon have not yet been brought to a close, but he estimates the period of that body to be from 50 to 60 years, and the distance of the assumed point of gravity to be 2½″. *Peters*, who more recently has turned his attention also to Procyon, fixes the period of that star at 50,096 years, and the mean distance of the point about which it revolves at 2″,56.

Mädler closes his remarks thus: "So far as regards myself, I have not the least doubt, as the matter at present stands, that *Bessel* was entirely correct in his suppositions, and that we really owe the greatest and most important of all that immortal man's discoveries, to the last evening hours of his life, when he was hopelessly confined to a bed of disease which he was never more to leave."

Thus, then, if Bessel's interpretation of these phenomena be the correct one, all possible variations in the relations of the celestial bodies to each other in the system of our Milky-Way, are exhausted. We see dark bodies revolving about dark bodies (moons about planets); further, dark bodies moving around shining bodies; again, suns about suns; and finally, suns also about dark bodies.

The two last mentioned forms belong exclusively to the world of the fixed stars — no analogy to them is to be found in our solar system. Whether, on the other hand, the two other forms belong just as exclusively to our partial system, or whether the arrangements which here obtain are extended into the regions of the fixed stars also, is a problem which cannot be satisfactorily determined by astronomy at present, and probably never will be.

But this much, at least, astronomy can assert with full assurance, that the view so fondly advanced by *Fontenelle*, that all the fixed stars were suns like our sun, with solid bodies similar to it, and like it, encircled by planets, moons, and comets; in short, that all in the universe was "tout comme chez nous," is a view not to be entertained for a moment. Modern science has, since the great discoveries of *Herschel*, caught glimpses enough of the infinite variety of formations and physical relations existing in the universe, to force it to take a decided stand against such a tedious and ever-recurring monotony, and to turn with aversion from such a contracted and miserable theory of the world.

True, astronomy will not bear us out in opposition to the view that arrangements similar to those in our solar system, *may*, indeed, be found without the bounds of this system, though not a *single fact* can be brought forward *in proof of it.* Those stars which lie scattered singly in the vault of heaven, and which may be supposed the nearest stars to our sun, both from the intensity of their light and a discoverable proper motion, as also from a measurable parallax — those stars lie so far apart, that if we look only to distance, there is room sufficient, as we are free to confess, for dark bodies, massive as the planets and moons of our system, to revolve about them.

But on the other hand, when we pass from the single to the double or multiple stars, the idea of transferring the arrangements which prevail with us to those stellar systems, at least so "nude crude," as often happens, seems so out of place, that the likelihood of any such transference actually taking place scarcely retains a shade of probability in our minds.

Herschel the younger went so far as to attempt to explain "the wondrous influence green or red, blue or yellow light, streaming from a double or multiple star, would exert upon the inhabitants of the planets belonging to such systems." *Schubert* also, though not at all charmed by a theory embracing so much monotony, followed out this idea further. "If it be true that they are suns," says he (*Naturlehre*, p. 106), "which give forth light and heat in the same manner as our sun, and which are carried by the potent force of mutual attraction through one revolution in a time exceeding but little one of Saturn's years, if it be true that instead of two, three, or four, perhaps still more such solar bodies are in close connection, then is it not possible that either night or winter should even occur in the *planetary* spheres, lost as it were in such a whirl of suns, and no more could any mortal eye possibly endure such a dazzling sea of light."

But though it were even proven not directly impossible that the orbits of planets may be entwined with the orbits of double and multiple stars, still the favorite "tout comme chez vous" will not by any means apply in connection with them. For the double stars in part compose systems so compact, that we cannot without difficulty ascribe to them a mass of the bulk, density and weight of our sun, and still less could a host of dark bodies of the weight of our planets and moons, thread their ways between the suns of these systems, without endangering the harmony of the movement. Further, let us imagine all those thickly crowded hosts of worlds exhibited to us by the stellar clusters, in all respects like our sun, *i. e.*, with solid, massive central bodies, and each of them surrounded by numerous solid, planetary spheres. Were this really the case, those worlds could never so peacefully and undisturbedly pursue their silent and majestic courses through all time; nay rather, overcome by the domineering force of gravity, the supposed planets, moons and suns, would be hurled against each other with appalling and destructive power.

But apart from the foregoing, there are many other facts which seem to conflict with the supposition, that the arrangements of our solar system are extended into the regions of the double and multiple stars. We may mention as of special note in this connection, in addition to the marked progression in the formations of the heavens, observable from the centre of the astral system to the outer ring of the Milky-Way, the magnificence of color displayed by so many of the single stars, but particularly, and almost universally, by the double and multiple stars. In color, light and darkness are united, their antithesis being resolved and removed. A complete and permanent union of light and darkness so as to form colors, something that is altogether foreign to our system, seems to be what is common in the stellar worlds. And this very fact it is which justifies the conclusion, that as light and darkness do not there exist as such, in separation from each other, but are harmoniously united, so also the solar and planetary principles, as media of light and of darkness, are there united and combined in the same vital and harmonious manner; both the solar and planetary principles being indeed present in the stars, though not apart and in juxtaposition to each other, but intimately united. Not in polar opposition and complete divorcement, but in vital union, concrete fulness, and eternal harmony; not in mechanical connection (as in a sense our sun and its luminous atmosphere are), but mutually pervading each other in the most thorough union.

§ 13. *The Nebulæ.*

To the inquiry, as to the number of the solar or fixed stars contained within the system bounded by our Milky-Way, it must be answered that an approximate computation, less reliable as the distance increases, has to take the place of any attempt at enumeration; and that computation itself hopelessly fails of arriving at any reliable results, in the most distant regions of the Milky-May, where our best instruments are wholly incompetent to reach and separately distinguish the myriads of thickly-strewn worlds.

The most acute human eye is capable of distinguishing fairly, under the most favorable circumstances and without a glass, stars of the seventh magnitude, whilst a common eye can plainly distinguish stars up to the fifth, or at the highest, the sixth magnitude only. Of the stars of the 1st magnitude there are but 18, equally distributed between the two hemispheres of the heavens. Of the second magnitude there are 55, of the third about 200, of the fourth some 460, of the fifth 1160 or more, and of the sixth and seventh the rapidly increasing number of over 20,000.

The number of stars visible through the best of telescopes in the Milky-Way of the northern hemisphere, is, however, according to the ingenious and laborious computations of *W. Herschel*, about 18 millions. If we allow a number somewhat less for the southern hemisphere, we have the astonishing sum of about 30 *millions of suns within the bounds of our Milky-Way.*

Let us endeavor to grasp all that is conveyed by the words, *thirty millions of suns!* to bring within the compass of our minds all that belongs to a *single* sun; let us strive to picture in imagination the infinitude of details comprehended in the short expression, "a million!"[1]

But is this all? Have we now reached the limits of the universe? and will not telescopes superior to those of the present day, reveal still thousands and millions of suns in the outermost regions of the system of our Milky-Way, of which no glimpse can be caught by the best modern instrument? Is it true that the system of our Milky-Way is the only con-

---

[1] At the rate of one hundred each minute, it would require uninterrupted counting from morning till evening for fourteen days to number a single million.

tinent in the ocean of immensity? May it not rather be a *mere island* itself, one of thousands like it scattered over this shoreless ocean?

The reply to this inquiry is still a problem of science for the satisfactory solution of which observations heretofore made are wholly incompetent.

A nebulous ground still remains unresolved in the Milky-Way, under the power of even the best telescopes; and similar clouds of light are to be seen in other parts of the heavens. These are called *Nebulæ. Mädler*, (p. 447,) describes them as follows: "A good naked eye beholds in many places in the heavens, a faint glimmer of light, which lessens the darkness of the background; and also stars, which instead of presenting sharp, well-defined points, like most stars, seem to be, as it were, grown together. But this spectacle gives one scarcely the most distant intimation of the scene presented through a large glass. . . . . We have there laid out before us, tracts of all shapes and sizes, interrupting the deep darkness of the ground of the heavens with a cloud-like light, similar to that of the Milky-Way. The very best glasses frequently succeed in resolving what through less powerful ones appears as a nebulous cloud, similar to the tract of the Milky-Way, wholly or in part into separate astral bodies, presenting to view a thickly-crowded *cluster of stars*. In other cases, the resolution is not so complete that the individual stars can be separately distinguished, but still of such a degree that the mind cannot escape the conviction, that the whole nebula is composed of myriads of stars, just as in the case of a heap of grain or sand, the individual grains cannot be clearly distinguished at a certain distance, though seen with sufficient distinctness to assure the mind that the heap is made up of such grains." "Where the resolution succeeds, the starry structure presents to the eye an indescribably magnificent spectacle. A nebula in the constellation Hercules presents to the eye when resolved, from 6000 to 10,000 simultaneously visible stars, which are so compacted together at its centre as to form a sphere or ball of light." "But very many nebulæ are still to be found in which not the least approach to a resolution can be detected."

W. Herschel, who, as his epitaph says, broke through the barriers of the heavens ("cœlorum claustra perrupit"), directed his gigantic instrument towards 2500 of these remarkable formations of the heavens, but he succeeded in resolving but 197 of them into stellar clusters similar to those of the resolved nebulous tract of the Milky-Way. *His* investigations first brought us into a more intimate acquaintance with this exceedingly important and interesting part of astronomy. Europe listened with astonishment to the accounts of his grand discoveries, and to the interpretations which were put upon them by the great discoverer himself. Astronomers and natural philosophers made them the foundation of their hypotheses, and joined in fierce conflict one with the other. But that which alone could lead to the desired end, a careful and uninterrupted continuation of such investigations as Herschel's, was neglected for almost a whole generation. It was *John Herschel*, the son, who, inheriting both his father's name and his great fame, first took up again these investigations, and in a few years advanced them in so astonishing a degree. It was subsequent to the year 1825 that he turned his chief attention to this subject, and, in order to bring the nebular structure of the southern heavens within the sphere of his observation, sailed to South Africa. He there instituted a most comprehensive series of observations, in the years 1834–1838, the results of which were published in the year 1847. Since that time it has been *Lamont*, of München, who has in particular applied himself to the task of investigating the nebular formations of the heavens. But that which neither these nor any other astronomers could attain to, seems left to be accomplished, slowly but surely, by *Lord Rosse's* gigantic telescope, the most powerful instrument upon earth.

Some parts of the heavens are exceeding rich in nebulæ and stellar clusters; in others they appear to be entirely wanting. According to Herschel the younger, they are most accumulated in the northern hemisphere of the heavens. In the southern, on the other hand, while their number is much less, they are much more equally scattered over its surface. The individual forms of these structures exhibit an endless variety. We condense from *Humboldt* (III. 329 seqq.) in regard to this matter. The form of the nebular structure is at one time regular, spherical, more or less elliptical, annular, planetary, or like the photosphere surrounding a star; at another time, irregular, and no less difficult to classify than the clouds of our atmosphere. The elliptic may be mentioned as the normal form of the regular nebulæ, with a great variety of transitions from a round to a long elliptic and awl-shaped form. The more the form approaches to the spherical, the more readily is it resolved into a cluster of stars. It is only among the round and oval formations that *double nebulæ* are to be found. *Annular* nebulæ are some of the most rare occurrences. But seven such are known in the northern hemisphere as seen by *Lord Rosse's* telescope. The space bounded by the ring is sometimes of a deep black color, sometimes faintly lighted. They are probably stellar clusters disposed in annular form. The *planetary* nebula are much more numerous than the annular. They have the most striking resemblance to planetary discs. They vary much in size and strength of light, and several of them shine with a bluish light. To the regular nebulæ belong, besides, the so-called *nebulous stars*, *i. e.*, true stars, surrounded by a milk-white veil or nebula, which in all probability is related to and depends upon the central star.

Very different from all these are the numerous large nebulous masses of *irregular* form. No two of the latter are alike; but what may be observed in connection with them all, and what gives them all their peculiar character, is this, that they are always found in or very near the borders of the Milky-Way, of which, indeed, they may be regarded as offshoots or extensions. The regularly shaped and well-defined small nebulæ, are, on the other hand, in part scattered over the whole heavens, and in part crowded together in a region of their own, far distant from the Milky-Way. Modern observation *has not* established the once wide-spread theory of a Milky-Way of nebulæ, crossing and cutting the Milky-Way of the stars, almost at right angles. The most remarkable of all the irregular nebular formations are the Magellanic Clouds, in the neighborhood of the south pole. These striking objects enchain the astonished gaze of the wandering mariner, both from their great size, and their brilliancy to the naked eye, which is equal to that of the Milky-Way at its brightest points, as well as from their completely isolated position with regard to all the other astral and nebular formations of the heavens. There are two of these *clouds*, the larger containing about 42, and the smaller 10 square degrees of surface. We are indebted to *Sir Jno. Herschel's* residence on the Cape, for a closer analysis of these wonderful structures. They are composed of an assemblage of the most diverse elements. *Herschel* discovered a large number of single stars (of from the 7th to the 10th magnitude) scattered through their substance, also, groups and globular clusters of stars; oval, regular and irregular, closely-crowded nebulæ. These *clouds* are not connected with each other, nor yet with the Milky-Way. Opposite to them, but somewhat more distant, there circle about the south pole, *dark spots*, called by the old mariners *Coal-sacks*, which are very prominent in contrast with the showy splendor of the *clouds*. They are not, indeed, wholly devoid of stars, but contain comparately few (only *one* star of the 6th or 7th, though many telescopic stars of from the 11th to the 13th magnitude), and it is the marked contrast between them and the adjacent splendor of the Magellanic clouds which accounts for their remarkable blackness.

Two inquiries of high import press upon the thoughtful observer as he regards these mysterious luminous structures: is the distinction between resolvable and irresolvable nebulæ to be referred merely to the imperfection of our instruments, so that as the latter improve the class of resolvable nebulæ will constantly increase, until at length, when we have reached a supposed perfection, all nebulæ heretofore discovered or yet to be discovered, shall have resolved themselves to our astonished eye, into millions or billions of single stars? Or is this distinction founded in nature, so that there are really nebulæ in the heavens absolutely irresolvable?

Then follows the second question: Are these nebulæ members of our own astral system, forming with it a complete whole, and bound to it by the strong bonds of an intimate and essential relation? or are we to regard them as wholly separate and independent systems of worlds, so that each one of the thousands of nebulæ is in itself a distinct system, similar to and of equal importance with that one to which our sun and our Milky-Way belong?

Ever since the more general use of the telescope has revealed the nebular formations of the heavens with more distinctness, and in greater numbers, the most remarkable difference of opinion has exhibited itself in connection with this question. *Galileo*, *Cassini*, *J. Michell*, and others, regarded all nebulæ as distant clusters of stars; *Tycho de Brahe*, *Kepler*,

Halley, *Derham*, *Lacaille*, *Kant*, and *Lambert*, on the other hand, maintained the existence of starless nebulous masses. *W. Herschel* was at first attached to the view that all irresolvable nebulæ are extremely remote systems of Milky-Ways. "As he, however, towards the close of his life, again examined some of those nebulæ supposed to be immeasurably distant, he observed in them a progression towards a certain adjacent star, quite manifest even in the short course of his own life. These observations forced upon his mind the probability, that those were not so much inconceivably distant starry formations, as luminous masses without form, situated inside the boundaries of the heavens visible to the naked eye, not by any means at so great a distance. *Schröter*, another no less careful observer, remarked variations (for example, a sudden extension or contraction of its boundaries) in the nebula of Orion, which took place so instantaneously, and over such a vast extent of celestial space, that they reminded him much more of the electro-meteoric phenomena of our atmosphere, than anything else. Similar changes are maintained to have been observed in the cloud-like formations of the heavens by other observers.

W. Herschel was but strengthened in his new opinion, from his examination of the so-called nebulous stars, and led back to a view formerly held by Tycho de Brahe and Kepler, if it were in a different connection and under a different apprehension. It was this — that the unresolved part of the nebula consisted not so much of thickly-crowded stars, as of star-material, cosmical matter; so that the universe is a constant scene of world-production, a place where worlds are being formed. Also, that there was a time when nothing existed but floating, unbounded nebulous matter, and that what now appears as a nebulous mass incapable of resolution, will yet in the future glitter as a cluster of stars. Perhaps, too, many of these worlds have been completed for thousands of years, whose rays, sent out since that epoch, have not yet reached us, but are still under way—it being reserved for our remote descendants to behold them in their perfected state. All grades of progress in this process of forming worlds are still to be observed in the heavens, from the unlimited dispersion of nebulous matter without shape, to the gathering together and compacting of the same into well-defined, regular forms; from the first beginnings of nuclear condensations, to the full completion of suns and solar systems. "Just as in a forest" (thus *Humboldt* illustrates this view, *Kosm.* I., 87), "the same species of trees are seen coexisting in all stages of growth, from whence we derive the impression of progressive development of life, so also in the great nursery of worlds we behold the greatest variety of gradually-progressing cosmical formations."

This view has also been adopted, at least in part, by *G. H. von Schubert*, and independently developed and carried out by him, particularly in his ingenious work, "*Die Urwelt und die Fixsterne.*" "The eminent *Herschel*," says he, p. 60, "has in the most convincing manner established the fact of the origin and formation of the fixed stars from such luminous nebulous matter, and has indicated many points in the heavens, where may be seen, as it were, those great golden birds coming forth from the egg, or still covered with parts of the shell — the remains of the unconsumed nebulous matter." And again, p. 145: "There in the heavens we behold the element from whence the stellar suns derive their form and being, a uniform, ethereal, mildly-shining matter, which is dissipated through the wide spaces of the universe like a phosphorescent vapor, everywhere transparent, and possessed of the greatest mobility, assuming now one form, and then with the rapidity of light seizing to itself new boundaries; but with all these high qualities, still deprived of the proper and higher form, which is founded upon polarization alone. The difference between a well-defined fixed star, shining with a bright ray—like the thousand-fold concentrated light of the electrical flame—and governed by forces of a higher order, and the irresolvable glimmering nebula, is no other than that which exists between the crude shapeless bodies of our earth and the crystalline. Pierced by a ray of creative power the coal becomes a diamond, the nebula a star."

Schubert further attempts to show that all the nebulæ and clusters of stars, both within and without the Milky-Way, form with the latter a well-arranged, closely-connected system, mutually conditioning and completing each other as its constituent parts. He sees a faint image of the grand whole comprehended in the organism of the universe, in the luminous atmosphere of our sun, where there are discovered by the telescope, darker and lighter portions adjoining each other, caused by the gathering together of the sun's light-ether into solar faculæ, and a simultaneous formation of solar spots in the places vacated by the ether, where, as through a rent vail, may be seen the dark body of the sun. These alternations of lighter and darker regions are most striking in the region of the sun's equator. And just at this place it is that the vast body of the sun is surrounded by a still more extensive and immeasurably wide-spread nebulous light, the so-called *zodiacal light*,[a] which extends beyond the orbit of Mars, or which is (according to the latest views[1]) a free and mobile ring of light, revolving about the sun, between the orbits of the earth and Mars. This ring is seen only previous to sunrise and after sunset, being overpowered by the blinding light of the denser solar atmosphere by day, and presents to the eye a faint luminous appearance, similar to the luminous tract of the Milky-Way, having a pyramidal form, with its base resting upon the horizon. "Were the eye of an observer in the region of the sun's equator to be cast into the depths of space, it would behold everywhere in the direction of the zodiacal light, a girdle or tract of nebulous light traversing the whole vault of the heavens, and extending outwards to an immeasurable

[1] "The *zodiacal light*," says Humboldt (*Kosm.* I., p. 89), "which presents itself in pyramidal form — from its mild lustre, the constant glory of the tropical night — is either a large rotating nebulous ring, between the orbits of the earth and Mars, or, less probably, the outer stratum of the solar atmosphere itself." Compare further, *Kosmos* I., p .142–148. The plane of the zodiacal light deviates from the plane of the earth's orbit but 7½ degrees, for which reason it cannot be seen in circular but only in perpendicular form, and was called by the ancients, *trabs*, δοκός.

distance. . . . . What is represented on a small scale by the solar atmosphere is repeated in the heavens of the fixed stars, on a scale inconceivably more grand and extensive. The eye of an observer directed from our planet to the lofty, brilliant, and far-reaching expanse, which takes in the whole of our cosmical structure, beholds in almost every direction a nebulous light — like an atmosphere — which, being here and there interrupted by dark open spaces, is generally upon their immediate borders, as in the formation of solar faculæ, intensified and gathered together into brighter, denser nebulous clouds, clusters of stars, and clear, brilliant single stars. The unfailing mutual accompaniment of a dark space and a cluster of stars, which so clearly favors the view that all the visible realms of light belong to one connected whole, being the offspring, as it were, of *one flood* of light, is so obvious to every careful eye, that Herschel in his age frequently called attention to it."[1] It would appear, further, that the Milky-Way of the heavens possesses the same significance as the girdle or ring of the zodiacal light, which resembles the former not only in its lenticular or annular form, but also in the circumstance that just here those dark solar spots appear with most constancy and in the most striking manner, just as the brightest nebulæ and most dense strata of the Milky-Way are bordered by dark and almost starless spaces.

"If we compare more closely" (continues Schubert, p. 117), "the later observations of Herschel, we can scarcely escape the conviction that all the ne-

[1] *Urwelt*, p. 114 seq.

bulæ and Milky-Ways belong with our own, to one and the same closely-connected and on the whole pretty equidistant system, whose luminous masses and generally globular clouds of light, have, just like the nebulous belts which surround Jupiter, crowded themselves together in a special manner only in certain direction, leaving the other portions of the heavens wholly or comparatively bare. For by far the greater part of nebulæ discovered up to this day, lie not as it were accidentally scattered in every region of the heavens, but form pretty regular zones and strata, of which one at least traverses the whole circle of the heavens. . . . The Milky-Way, according to Herschel's observations for many years, does not by any means consist in general of equally-distributed stars; for in the half of this interesting formation known to himself, 225 clearly separated clusters of stars were to be pointed out, all belonging to one closely-connected whole. But why except from such a close connection with the great whole of our Milky-Way, those hundreds of additional nebulæ discovered by Herschel in or upon the very borders of this starry zone, and which because they formed more closely-crowded and generally small globular systems, or nebulous structures not capable of resolution, he held to be inconceivably more remote than the Milky-Way proper?"

The oft-mentioned alternation of portions of the heavens bright and rich in stars, and portions dark and containing but few stars, "justifies (*Schubert*, page 128) the conclusion, that all those nebulous formations and cosmical masses in the heavens, the distant as well as the near, have proceeded from the same once uniform and wide-spread, unbroken nebulous cloud, which not until it felt the energy of the Divine command that it should bring forth worlds, resolved itself into those single and separate luminous nebulæ and glowing spheres." This productive and vivifying luminous ether which fills the whole starry heavens, the true fountain and store-house of all created light which animates and supplies the visible realms of immensity, he calls the atmosphere of atmospheres, which is to all the concourse of the worlds of the fixed stars, what our atmosphere is to the earth and its inhabitants, and regards it as the medium whereby all the thousands of stars and stellar systems are bound together into one vast and closely-related community of the created.

The view of the elder *Herschel*, in regard to a still ever-progressing formation of worlds out of luminous cosmical nebulæ, has in the meantime, however, continuously lost credit. The very astronomers most conspicuous in this sphere of investigation, *Herschel* the younger, and *Lamont*, have declared themselves *against* it, and *in favor of* the notion of the permanent stability of the starry heavens as the result of a long-since completed formative process. In regard to this question, *Herschel* the younger speaks thus:[1] "All cosmological arguments founded upon the observation of such a transition, lie open to the objection, that however unequivocally the existence of a gradually upward tending series amid a large number of contemporaneously existing individuals

[1] Mädler, *Pop. Astr.*, p. 455.

may be established, still we have no reason for the belief that each individual has passed or could pass through all observed stages, or that it is indeed in a state of gradual advancement.—The grades of animal life, from man down to the lowest orders, are almost infinite, and certain naturalists would fain introduce a system of development, which, commencing with the simpler forms, rises to the more complex, and finally to the highest of all; but so long as the real presence of such a development is not perceived — so long as every animal through all generations inherits the defects of its progenitors, we can at best but assume the *possible original* existence and fruitful manifestation of a tendency towards improvement or advancement, all progress in the *now existing* state of nature having long since reached its end."—In like manner *Lamont:* "If we examine the oldest records and sources of information relative to the state of the heavens, all is found to agree with what is still at present observed. . . . When I take into careful consideration all the circumstances concerned, this appears to me with great probability the legitimate result, that the whole cosmical structure, after the close of some sort of a formative period, long since passed over into a state of sustained equilibrium or repose, of permanent and all-preserving order."

Schubert also, now, in his latest work (Weltgeb. p. 105), refuses his assent to the theory of the elder Herschel. His words are these: "In contradiction to the theory of development in connection with nebulous matter, heretofore mentioned by us, we must now, from the stand-point of present knowledge, decide in favor of the view that all the various forms of the stellar systems and nebulæ of the universe, are parts of one vast organic co-ordained whole, which, just as the earth and its atmosphere, just as the gelatinous medusa or tremella, and the more highly organized animal or the cedar, may have originated and may henceforth subsist together, with each other and side by side." But he still yet, however, stands by his previous view of the unity or close connection and relation of all the stars and nebular groups in the firmament. He also, according to the above-mentioned work (p. 94), now holds that "those nebular formations are not situated in regions of space indefinitely remote, but within our own astral system, nearer to us perhaps than the inner ring of our Milky-Way."

Mädler, however, has shown (*pop. Astr.* p. 450, seqq.) this view which would refer all the phenomena of the heavens, without exception, to a space bounded by the system of *our* Milky-Way, to be inadmissible; and most modern astronomers seem inclined to side with him, even though they may hold the matter to be one which must long remain incapable of a definite or satisfactory solution.

In regard to the more regularly-formed, though not as sharply defined as the planetary, nebular masses, *Mädler* has declared the supposition admissible, "that they do not consist of stars, but of light, rare, luminous matter, star-material as it were, holding somewhat the same relation to the more dense bodies of stars proper, that comets do to planets;" and he also allows the possibility, that "they may belong to our astral system, and take part in the complex web of its attractions." The planetary nebulæ proper seem also best explained by the same supposition, since the exact rounding off of a cluster composed of myriads of very distant fixed stars, could be at best but a rare accident, which amid the countless number of *equally* possible forms, could not occur more than 78 times out of 2500. (?) With regard to the nebulæ which have already been resolved into distinct, separate stars, there exists little doubt of their all belonging to the stellar system whose outer boundaries are marked by our Milky-Way. There are, however, a few of these clusters so closely crowded with stars, and of such apparent insignificant diameter, that besides the above explanation, the other also might be adopted, according to which they are regarded as lying beyond our Milky-Way, and as not belonging to it. *These* may, indeed, *Mädler* hesitates not to say, be astral systems similar to the system of our Milky-Way. But that which may be allowed as possible, in the case of these resolvable nebulæ, can scarcely be avoided in the case of the nebulæ which are of irregular form, and wholly incapable of resolution. It seems not possible, according to the laws of gravity, that they should be composed of rare, luminous nebulous matter. Were this their constitution, they could not for centuries have retained the *same* form, but would have long since been drawn together into more or less globular form, through the influence of a mutual attraction of their constituent parts. Hence it must be assumed that by far the greater number of the nebulæ are, absolutely considered, capable of resolution—that they are true clusters of stars.

The whole question here depends, first of all, upon whether the absolute invariability of form for centuries, supposed by Mädler, has been duly ascertained and established. But we have already seen that *W. Herschel*, *Schröter*, and others, have remarked changes of form in connection with nebulæ, such as sudden expansions, contractions, &c. From the comparative rarity of such experiences, however, we are left to conjecture that some illusion of vision may have taken place in the cases mentioned. From the universal attention now directed to these mysterious formations, we may look forward to no remote day, when a much more correct and satisfactory judgment may be pronounced in regard to this matter than is possible at present. At all events, an important and decisive point in regard to the whole question of the nature of the irresolvable nebulæ, lies in the determination of the correctness or falsity of the above observations of Herschel and others.

But astronomy in the mean time seems to approach nearer the desired object in another way. The number of the resolvable nebulæ is increasing with every year. *Mr. Bond*, of Cambridge, has discovered 1500 small stars in the borders of the nebula of Andromeda, which has hitherto been regarded as absolutely incapable of resolution; and although the centre of the nebula still remains unresolved, *Humboldt* hesitates not to set it down (III. 316) among the clusters of stars. Of signal and unwonted service in this domain of astronomical science is the colossal telescope of *Lord Rosse*. "The 40 nebulæ chosen by Rosse for examination, have, by the aid of observation, been divided into three classes: uniform circular surfaces, round nebulæ with one or more marked nuclei, and finally, such as are lengthened in certain directions, or in general deviate materially from the circular form. The first, 10 in number, may be completely resolved into distinct stars, under the moderate magnifying power of 360. In the second class it was observed that the brighter star, frequently indicated by previous observers as a single central star, resolved itself into a cluster of closely-crowded brighter stars, which again were surrounded by more scattered and fainter stars. The nebulæ of the third class were scarcely resolvable on account of too great optical condensation of their inner parts. The stars of most of the nebulæ are comparatively very small." (Mädler, *Nachträge*, pp. 24, 25). In regard to the celebrated nebula in the sword of Orion, heretofore considered incapable of resolution, Rosse expresses himself thus: "I may safely say, that there can be little if any doubt of the resolvability of this nebula. We were unable on account of the state of the atmosphere, to use more than half the magnifying power the speculum bears; still we could see that all about the trapezium is a mass of stars. The rest of the nebula equally abounds with stars, and exhibits the characteristics of resolvability strongly marked." At a subsequent period (1848) *Lord Rosse* had not announced that his fondly cherished expectation of being able to resolve completely the nebula of Orion, had, as yet, been fulfilled, though he still hoped it would be.

Robinson, an American astronomer, expresses it as

his decided conviction, "that there does not exist in the heavens a single nebula, in a physical sense;" and *John Herschel* writes: "Although there are nebulæ to be found, which still appear only as nebulæ, through the colossal instrument of Rosse, without betraying any signs of resolution, we may nevertheless argue from analogy, that there is in reality no distinction between nebulæ and clusters of stars."

If we hence take it for granted that all the nebulæ without exception, will eventually turn out to be thickly-crowded myriads of stars, what will become of the unity of the astral system, so ingeniously set forth by *Schubert?*

It does not seem to us, that, even were all the nebulæ shown to be capable of resolution, we should be forced to assume the existence of several, perhaps, indeed, thousands of separate and independent astral systems like our own. The very fact of their being resolvable would favor the view that they cannot be more distant than the outer ring of the Milky-Way, and the more strongly so as the resolution were more easily accomplished. And most assuredly if our best instruments have revealed a nebulous stratum in the Milky-Way which they cannot resolve, but which must notwithstanding be regarded as belonging to the same system as this Milky-Way, we may place the still unresolved nebulæ in the same category as to distance, and regard them as belonging to the same all-comprehending system.

We should not without regret (and we frankly acknowledge it) relinquish the view, which has grown far into our favor, that all the astral and nebular formations of the heavens belong together in one unique and comprehensive system; but at the same time our preference for this view does not go so far, that we should for its sake ignore or lightly regard the results of science. Whenever the fruits of scientific investigation shall have shown the correctness of the opposite opinion, we shall just as readily take our stand by it. Certainty does not belong to either aspect of the question, as the matter at present stands, nor can the abettors of either the one or the other view compel submission. And just here lies the reason why we should ever hold in proper regard the view opposite to our own, and leave room for it in any general theory we may adopt.

But were the unity of all astral and nebular formations rejected on good grounds, we should *then* have, in addition to our own astral system, with its perhaps hundreds of millions of suns, some thousands of completely isolated systems of worlds, similar to our own and entitled to the same distinction. Indeed, just in the *same* ratio that those remote systems were laid open to view by the increasing perfection of our instruments, and separated for the first time into millions of suns, new nebulæ might be espied, which are now hidden from our best glasses, but which, nevertheless, might be resolved into stars by greatly improved instruments. *Mädler* (*pop. Astr.* p. 454), from a calculation of probabilities in regard to the distance of these systems, arrives at the conclusion, that the nearest astral system lies at such a distance from us, that its light must be under way for about 30 millions of years before reaching our telescopes—

and yet light travels with the inconceivable velocity of 192,000 miles in a single second! What an amazing distance! But think! the distance of the *most remote* astral system! ——

---

*[The foregoing representations, involving the zodiacal light, may require some modification from the following.

One of the most important recent additions to our knowledge in regard to the solar system, pertains to an object belonging to our own planet—the *Zodiacal Light.* This phenomenon, which has been the subject of such different opinions, and referred to such different places and connections in our system, seems at last to have been located. It is now regarded as a nebulous ring belonging physically to the earth, and revolving round it at the distance of some thousands of miles. It is the nearest known body in the firmament. It suggests an analogy between the earth and the planet Saturn, and, like the inner ring of the latter body, it is transparent. As this new discovery is of such importance, and concerns a body so close at hand, yet so little known until within a year or two, we shall quote at some length from a paper by the Rev. George Jones, U. S. N., announcing his fruitful observations in regard to this object.

Rev. Mr. Jones says: "Only some vague notices of the *zodiacal light* occur in ancient authors, before it is distinctly and briefly mentioned by Chauldry, in 1661. It was first carefully observed by Cassini, an Italian by birth, at the Observatory of Paris. He thought it an emanation of the sun. His associate, Tacio, thought it a ring around the sun. Miran, in 1731, thought it an atmosphere connected with the sun. In all subsequent speculations, no new observations, after Cassini's, were used, till 1832. In 1844, Biot observed that the nodes of the zodiacal light did coincide with those of the earth, and suggested that it might be more local than had been supposed. He found that it gave more heat than the tail of a comet did. . . . . . The zodiacal light appears best when it is on the ecliptic. When at the summer solstice and on the tropic of Capricorn, the zodiacal light was visible from 11 to 1 in both horizons at once, with their apices approaching each other. In the centre of the light is a condensed part, with a boundary of its own. On this voyage [when Mr. Jones made his observations], Jan. 31, 1854, at Loo Choo, he first noticed the pulsations of intensity in the space of a minute or two that it exhibits on some occasions. He made 331 sets of observations. He found that, if by the revolutions of the earth he receded from the ecliptic, the zodiacal light moved a little in the same direction, and *vice versâ.*"

Mr. Jones then stated that the following facts he had noticed could be explained by one supposition, viz., that of a nebulous ring surrounding the earth. The following are the results of his observations:

"1. This light cannot be from any body involving us in its matter, else we could not get boundaries to it, any more than we could to a mass of fog or a column of smoke in which we were involved. We must be apart from it in order to get bounds.

2. It cannot be from a planetary nebulous body revolving around the sun, but must be from a nebulous ring; for it is to be seen every morning and evening in the year, when the moon or clouds do not interfere, which could not be the case were it anything else than an unbroken ring.

3. If a ring, with the sun for its centre, it cannot be within the orbit of our globe; for then it could not be seen simultaneously over the eastern and western horizon at midnight, the spectator's horizon then extending far above it, on either side; nor could its vertex ever be in the spectator's zenith, or indeed any great distance above his horizon, which is contrary to the facts.

4. Is it a solar ring extending beyond the earth?

On this subject I must refer to the data afforded by these observations, for it is only from facts that we are able to argue in the case. Any one examining these data will see, I think, that the *lateral* changes from hour to hour in the boundaries of the zodiacal light, especially towards the horizon, could not have taken place in a ring so distant as a solar ring would have been at the point where reached by the horizon. . . . If, in the course of four or five hours, the earth's rotation carried me from the southern to the northern side of the ecliptic, or the opposite, the zodiacal light changed with me, its lateral boundaries shaping themselves according to my change of place. This was not always the case, but it was the general fact. When I was far in southern latitudes, the greater mass of the zodiacal light, instead of being on the ecliptic as here, had shifted over to the south; and as we came from Rio to New York, as rapidly as steam could carry us, the mass of light came with us to the north once more; still, however, in its varying positions, having a reference to my position with regard to the ecliptic. I ask, supposing that the zodiacal light is from a solar ring, which would make the base of its light at its first and last appearance, nearly or quite 180,000,000 miles off, would that light at its base show such changes as it actually does in half an hour or an hour, when the spectator's place on the earth has been so slightly changed? I have taken a few of my observations, and have submitted them to calculation, not making much of a selection, for almost every observation in the book would give similar results."

Mr. Jones then goes on to detail some observations and calculations too lengthy to be introduced here, and arrives at the conclusion that a solar ring will not meet the data of the case. But he says, "an earth-ring will do; that is, a nebulous ring around the earth will readily allow such lateral changes to be produced by such a change of the spectator's place."

"But there is another view of this subject which may be considered still more conclusive against such a solar ring. Take cases which very often occur when the ecliptic is somewhat toward a right angle to the horizon, and circumstances therefore favorable for a good display of the zodiacal light. Say it is morning, an hour and a half before sunrise. The base of this light will be exceedingly brilliant—as much so almost as if the sun were just going to rise—while the vertex of the light overhead will be so dim as to be scarcely made out. Yet on the supposition of a solar ring reaching beyond the earth, the base of that light must be 180,000,000 of miles from us, and the vertex comparatively only a very short distance, while also the whole circuit of the ring is equally illuminated by the sun, and those portions near our zenith, as far as I can judge, also more favorably situated for reflecting his light than those portions at his base. We can scarcely imagine such a state of things.

"Believing that this query, as to the data of the case being met by the supposition of a solar ring, must be answered in the negative, I am driven to the only alternative, of *a nebulous ring around the earth.* The moon's zodiacal light seems also to show that matter lies within the orbit of the moon. . . . . We may well query—if the zodiacal light comes from a nebulous ring around the earth and within the orbit of the moon, may not the shooting stars, and even the aerolites have their origin there. Observations, I think, show that there is a constant commotion within the ring itself; may not the nebulous matter, half agglomerated here and there, be shot by these commotions beyond its sphere, and, caught by the attraction of the earth, be drawn down, till, striking our atmosphere, they glance in any casual direction, and taking fire become consumed, thus giving us the shooting stars? And may not this nebulous matter, still further solidified and with a same fate, afford us the aerolites? For if such matter could have once afforded us our moon, it may easily afford bodies such as aerolites are found to be. What is nebulous matter? My observations throw no light upon the subject. It is very transparent, for I had no difficulty in seeing stars of the sixth magnitude through its most effulgent, and therefore densest portions. But transparency does not argue tenacity as a matter of course; for rock crystal and the diamond are the most transparent, while they are densest and hardest of all bodies. But of whatever composed, I do not suppose the ring of the zodiacal light to be composite, for its internal disturbances are opposed to this. But with our present knowledge, such reasonings cannot satisfy us: they only beckon us to be searchers and further collectors of facts." —*Annl. Scient. Dis.*, 1856, p. 374, seqq.

The theory of Mr. Jones is favorably received by the scientific men of the country. Prof. Peirce thinks that if his theory had been first proposed, no second would ever have been entertained. "His only objection had been that one satellite could never maintain a ring, and he is still of that belief, but is convinced that it has many other satellites too small to be seen, and that these satellites furnish the meteors which fall to the earth."

Prof. S. Alexander arrives at the conclusion that the earth's ring revolves in about 12 hours, at a distance of nearly 17,000 miles from the centre of the earth. He deduces the distance in connection with the fact, that the apices of the zodiacal light are 35° apart when simultaneously visible.—Tr.]

§ 14. *Retrospect.*

Nowhere in the known universe is absolute rest to be found. All celestial bodies are involved in a whirl of a varied and complicated movement, and none of them can escape its magical influences.

Rotation about an axis is the simplest form of movement. This species of motion is a constant and invariable law in our solar system, to which the king of day himself, no less than the smallest of his subjects, is compelled to conform. The same law in all probability prevails even to the utmost limits of the worlds of the fixed stars.

A second form of movement, which may either coincide with the former (as in the case of the moon), or be different from it (as may be seen in the planets), is that in which one body revolves about another, or rather, the two revolve around a point of gravity common to them both. This movement is also conditioned by laws, the general prevalence of which throughout the whole universe is scarcely any more a matter of doubt.

Our moon revolves round the earth, forming with it a partial system; the earth with the moon circles about the sun, and the sun itself, with all its planets, moons and comets, is involved in that majestic circling dance of the spheres, in which millions of suns move harmoniously around a common centre. And if the thousands of nebulæ capable of being reached but not resolved by our telescopes, are systems of Milky-Ways similar to our own, we may certainly with great probability suppose that all these island-worlds scattered throughout the ocean of immensity, are related to each other, forming a higher, yea, the very highest system of created life and movement.

Our solar system *does not* occupy the very centre of the great city of worlds surrounded by the rings of the Milky-Way as magnificent walls of light. But still our position is comparatively quite near this central spot—"somewhere upon the great central square," as it were, of the city of worlds.

If we compare our solar system with the rest of the worlds belonging to the system of the Milky-Way, we are at once struck with the effectual contrast that reveals itself, and which would represent our solar system as *one* in fashion, constitution and arrangement (and perhaps the *only one* of the kind in the universe)—which, while it is supported and animated by the same *fundamental* laws of life and movement as all the rest, still, in all other respects, maintains its independent and peculiar character.

But it is at the same time to be observed, that when our solar system, with its independent and peculiar character, a structure perhaps without a counterpart in the whole cosmical system, is compared as a whole to the great whole comprehended by the system of the Milky-Way, it represents the latter in many respects, as a diminutive model of this system of the fixed stars. For just as our solar system from its outer limits to its centre, gradually assumes a different nature, does, mutatis mutandis, a repetition of the same take place in the system of the fixed stars. We have previously remarked the fact, that, in our solar system, as bodies increase in distance from the sun, the common centre of the system, they also increase in size and in the number of their accompanying bodies, with which, also, they form partial system; but that they in a similar ratio decrease in density. The analogous occurs in the astral system of the heavens.

Our solar system may be compared in the effectual contrast it everywhere presents to the astral system, to an island in the wide ocean;[1] in loosing from the island we forsake the firm land, it recedes from our view, the waters gather more around us, until, the traces of that which is firm and stable dying out, we are left in the midst of a very different element. As in our solar system there is a decrease of density with an increase of distance from the centre, so also in the system of the fixed stars as distance increases density grows less. The stars nearest to us seem still, in a measure, to remind us of the solidity of the bodies of our system, from their density and fixed outline; whilst in the more remote celestial regions the similarity gradually completely disappears. The same analogy holds good with respect to size. And as with us the planets nearest the sun pursue their courses alone, and groups of closely-related bodies multiply and grow larger in the number they contain, in proportion to their distance from the centre of the system; so also amid the worlds of the fixed stars, but on an infinitely grander scale. "If we," says

[1] According to *John Herschel*, there is scarcely a region to be pointed out in the whole known universe where the fixed stars lie so distant from each other as does our sun and the nearest fixed stars. — Schubert, *Weltgeb.*, p. 29.

Schubert, "consider the stellar clusters as double or multiple stars of a higher grade, it seems worthy of note that the clustering together of stars of the latter kind, should always take place in the regions of the universe supposed to be most distant from the centre of the astral system; while the association of a few small stars into a system is frequently observed in that part of the heavens thought to be much nearer to us, until at last, comparatively quite close at hand, such collections of a few stars even become more rare, and the state of isolation more general." We may here also remind the reader how analogies are furnished throughout the starry heavens to corresponding conditions and movements in the solar faculæ and solar spots, referring to our remarks about the variation of brightness in the stars, the expansions, contractions and condensation of nebulæ, &c. We may further refer to the correspondence to be observed between the zodiacal light and the Milky-Way, and to the fact that Saturn, with its remarkable system of rings, is best fitted of all known cosmical formations, to illustrate to us the arrangement of the rings of the Milky-Way; and finally, we may note how remarkably our countless comets remind us of the nebulæ of the firmament, which resemble the former, not only from their being in part infinitely light, mobile, with a nuclear, star-like condensation, and composed of a similar luminous vapor—though of a brilliancy vastly superior—but also present other phenomena which remind us much more closely of those mysterious erratic bodies. The close and crowded position of the remote celestial bodies, so in contrast with the wide distances of our system, finds a distant analogy in the extremely close approach of the comets to the sun. But still much more striking analogies are offered by the appearance and configuration of nebulæ and comets respectively. In 15 nebulous stars, for example, there is to be seen, in addition to the more dense nucleus, a pencil-like or fan-like tail, while others have a hooked shape, and the like. On the other hand, comets have been observed, the nuclei of which—similar to the nuclei of nebulæ — are composed of numerous quite small single stars. *Schubert* is of opinion (*Urwelt*, p. 56), that were the comets self-luminous in the same intense degree, "the millions of these erratic bodies which belong to our planetary system, would then appear to a distant observer, as so many suns moving in opposition to each other in a narrow and closely-crowded space."

CHAPTER SIXTH.

CONFLICT AND HARMONY BETWEEN THE BIBLE AND ASTRONOMY.

§ 1. *Design of this Chapter.*

THE results of astronomy have in the present day become the common heritage of all well-informed and cultivated minds, and are signally deserving of our high consideration. For no *human* science is so well fitted as astronomy, to loose the fetters which bind the human mind to the narrow sphere of earth; none so able to widen the contracted horizon of earthly views and efforts, or to waken up and sustain within the breast of man a consciousness of his high calling to make everything in time and space subservient to the wants of his mind.

But the high significance of this science in the culture of the mind, can only be asserted, when the results of astronomy are not merely laid up in the memory, but also brought into vital and fruitful commerce with the other possessions and efforts of the mind. It may very readily happen in this process of fullest appropriation, that new ideas which we just now acquire, may stand in conflict with previous knowledge or information, gathered from other quarters. This is signally the case in the sphere of religion, or rather theology, its scientific mode of apprehension; for both this science and astronomy extend their spheres of knowledge into the higher, supramundane regions, and hence frequently come in contact or extend into each other.

Natural religion, which the human mind, enriched by the contemplation of nature and history, has acquired from its own depths, and by its own intellectual efforts, is ever ready without much difficulty to yield to the really or apparently superior claims of new knowledge, if it contradict the previously acquired possessions of the mind, and to permit itself to be reconstructed from the new material. But the case is different with revealed religion. This, as objective, divine truth, demands unconditional submission and acknowledgment on the part of all contradictory or irreconcilable knowledge—a demand which cannot but be conceded to so long as we recognize it as *revealed* religion. But were the truthfulness and superiority of the contradictory astronomical results real and incontestable, and the contradiction itself an absolutely irreconcilable one, our faith could then no longer retain its position as a *revealed* religion.

Such is found to be the sad case of the Christian religion at the present day, according to charges made from various quarters, with great assurance. It is maintained that astronomical results of undoubted correctness have been obtained, which irresistibly force us to a theory of the world which is wholly irreconcilable with the Christian theory, and completely overthrows it.

Three grand points in the Biblical theory of the world, with which at best the Bible and its consideration as a record of Divine revelations, stand or fall, are here called in question. First, the Biblical teachings in regard to the *creation of the world*, both with respect to its general features and also to the special detailed facts of the process; next, the doctrine of the *redemption of the world* through the incarnation of the Son of God, with its preliminaries and consequences; and finally, the Biblical teachings in regard to the *end of the world* and *the judgment*, as the close of all historical developments in the world.

But we design looking the pretended antagonist somewhat more fairly in the face. We design to see whether our faith in the Divine origin and authority of the Scriptures, which have heretofore so abundantly verified their power in life and in death, which have transformed and renewed the world, must yield to the light of this human science, must really blush to be taught its own credulity. And finally, we design to see whether there may not be effected a reconciliation and a union, whether the forbidding antagonist may not, after all, on a better understanding, become our willing friend and ally.

§ 2. *The Doctrine and History of the Creation.*

Infidelity has ever felt itself specially called upon to contest the Biblical *doctrine* and the Biblical *history of the creation*. Deism and Pantheism, partly in alliance and partly at dissent, have been equally prompt in entering the lists against these hated opponents. Pantheism has been specially bitter against the *doctrine* of the creation; while Deism, on the other hand, has chiefly contested the *history* of the creation.

As *Deism* found its views to harmonize with the Biblical theory of a creation out of nothing, at the bidding of the Almighty, it had nothing to object to the *doctrine itself*, but merely to the assertion of its *Divine origin*, which in its denial of the possibility of revelation or inspiration from above, it cannot allow. But, in order to give this denial a specious coloring, it has been over-zealous in contesting the *history* of the creation; endeavoring to show that the latter is full of contradictions with itself and with the results of physical research, full of childish notions, laughable blunders, and absurd suppositions, so that even were revelation possible, such a book could not pass as Divine.

The sympathies of *Pantheism* were enlisted in a wholly different direction. The transcendence of God, his exaltation over time and space, with a creation proceeding from the will of this transcendent Being, the very point in which Deism agreed with the Bible, was to it an exceedingly bitter draught. Its hostility was hence chiefly directed against the Biblical *doctrine* of the creation, against the scriptural views of a creation in time and from nothing, through the will of a personal God, distinct from the world, and infinitely exalted above it. Its hostility towards the *history* of the creation was founded altogether on secondary considerations. This history was obnoxious to it, merely because that hated doctrine lay at its foundation, and had assumed in it a concrete form. Hence it blushed not to make common cause with Deism in combating the history of the creation, and to appropriate to itself the trivial objections and absurd charges of the former. It hesitated not a moment to stoop to the disgrace of so treacherous and dishonorable an alliance, with an antagonist formerly despised from its very heart, and held in proud derision, and one which urged the contest against the historical account of the creation, for the direct purpose of getting rid of those views of that account which alone seemed of any consequence to Pantheism itself, if indeed even they were more than half reasonable—its teachings respecting the immanence of Deity, the observation of succession in creation, &c.

It is not our object here to combat Deism and Pantheism, *as such;* but merely to show the groundlessness of their mutual appeal to astronomy—to show that astronomy in its established results does not join with them in opposition to the Bible, but that it agrees with the latter in combating the positions of both Deism and Pantheism.

From this stand-point we have not the least fear as to the issue of the assault upon the Biblical *doctrine* of the creation. But as the *aggression* is made, not with the weapons of astronomy, but of speculation, the charge may be *repelled* with *similar* weapons, if any one consider the matter to merit so much consideration. No astronomer has ever pretended to maintain that the results of his *empirical* research have forced him to deny the possibility of a creation out of nothing. Wherever astronomy has departed from its legitimate object of experimental investigation, and has built up hypotheses touching the *probable* origin of the celestial bodies, upon the basis of results obtained, it has ever come to a boundary where it was said, "Hitherto—but no further." We might, perhaps, not be mistaken in supposing that with the aid of this science, the origin and progressive advancement of the celestial bodies to the state in which they are now found, might, from the analogy of beginnings and developments which are still matter of observation, be made in a measure intelligible to the human mind. But astronomers have never seriously thought of attempting to decide whether the primeval matter and forces concerned in the production of these bodies, existed from eternity, or were created in time; whether the coöperation of this matter and these forces in the formation of cosmical bodies, was merely the result of accident, or whether it was produced and directed by a higher, personal, superintending will.

It hence merely remains for us to adjust the pretended contradiction between astronomy and the Biblical *history* of the creation.

§ 3. *The Creation of the World in Six Days.*

It is objected first of all, and from different quarters, that the Bible limits the process of the creation to *six days.*

Minds have in time past been stumbled at the circumstance, that God, of whom it is said: "He spake and it was done; he commanded and it stood fast,"[1] should have spent six days in the creation of the world, and not have accomplished it in a single moment. But more lately, since our theory of the

[1] Ps. 38 : 9.

world has been modified by Herschel's ideas of a still progressing astrogenesis, and the hypotheses of geology concerning the formation of the earth's crust, it has been deemed by many inconceivable, yea, even absurd, to suppose that the heavens and the earth should have originated and attained their present structure and perfection in the space of *merely* six days. At least thousands or myriads of years, it is said, if not millions or billions, must necessarily have been spent in such a creation as we behold.

It is not our intention to combat the astronomical or the geological suppositions upon which this argument is founded, or to cast suspicion on them, although those are in reality at best mere hypotheses which cannot lay claim to complete certitude, but are founded merely on a greater or less degree of probability. We shall pass altogether by such often used and often misused arguments; for one reason, because we can do without them. For another reason, because, say what we will against the reliability of the hypotheses touching the formation of the earth and the stars, there still remains an impression of which we are wholly unable to rid ourselves, that the process of the formation of the whole world, from its first beginning to its last finishing touch, must have required a much longer time than merely six times twenty-four hours.

For the same reasons we shall pass by the theological argument, "that one day is with the Lord as a thousand years, and a thousand years as one day," or, that the question is not one of length of time, but of the measure of the Divine influence exerted, in the more rapid or slower process by which the worlds were completed, and the like, although we will by no means admit that they are without significance and value.

We shall pass by all these, as we have said, since we do not stand in need of them, but at the same time we shall not deny that they possess both truth and weight. A proper understanding of Genesis 1, such as we have arrived at, Chap. 4, § 8 and 17, is amply sufficient to prove the futility of this and all similar objections.

The *six days' work* had nothing to do with the first creation of the *earth*, to say nothing of the creation of the *universe.* The heavens and the earth were already in existence (v. 1): they were both created and indidualized before this work began. But the earth, at least, was still devoid of light and life: it was "tohu va bohu." Both these it received during the six days' work, in continual progress from their lower to the higher grades. It was during this time that the earth received its *present* form, its *present* physical forces, its *present* inhabitants, and its *present* relations to the rest of the heavenly bodies. These are all points in which neither astronomy or geology has the least right to pass an opinion as to the length of the process. Astronomy may have a right to maintain that the heavens of the fixed stars must have been in existence for hundreds of thousands of years, but it has no right to say that the sun, moon, and stars, may have regulated and ruled our earthly night and day, *prior* to the fourth day. There was necessary, in addition to a power of exciting light, which may have been possessed by the stars since their first origin, a susceptibility to light on the part of the earth, in order that their agency might affect the latter; and no dogma of astronomy can disturb our clear conviction that this commerce of influences was opened at the time represented by the *Bible.* Equally ready are we to concede to *geology*, that vast periods of development, and demolition of previous forms of matter, may have preceded the present form of the earth. These periods occurred before the *tohu va bohu,* and there is not a word in the Bible antagonistic to such a view. But never can geology convince us that the last preparation of the surface of the earth for the residence of man, must have required a time of either more or less than six days. We have already stated why God did not choose to give the earth its present form in a single moment, rather than extend and distribute his creative influence over six days, should any one here still feel disposed to object. Besides, the objection has been sufficiently answered heretofore, by the reference made to Gen. 2 : 3, by the Church Fathers. The form, distribution, and duration of God's creative agency, were determined with respect to man, just as the earth itself was prepared *for* him. God's employment on the earth was to be a pattern and type of man's future earthly activity.

A second objection against the representations of the Hexæmeron, arising out of the later astronomical and geological results, rests upon the unequal distribution of creative activity between the six days. The fourth day is particularly conspicuous in this connection. Whilst five whole days were spent in the completion of our earth, which is at best but a mere point in the whole universe, it would be maintained, it is objected, that all the rest of the universe, with its millions and perhaps billions of suns and worlds, was finished in a single day. But it is clear that the same misapprehension lies at the foundation of this objection, which gives rise to the controversy about a six days' work in general. If we confine the work of the fourth day to the establishment of a permanent relation between the earth and the celestial bodies, which is not merely justified, but even required by the record, all difficulty immediately vanishes, and the distribution of creative activity, as mentioned in the Hexæmeron, exhibits the fairest and most equal proportions.

§ 4. *The Creation of Light before the Sun.*

A host of severe and clamorous complaints besides, which prefer the charge not merely of contradictions with modern astronomical results, but also of puerile narrowness of mind, absurdities and self-contradictions, is heaped up against the account of the fourth day's work.

The remark is often heard, how laughable, absurd, and insufferably puerile it is, that the record should represent the sun to have originated on the fourth day, while light, which as every child knows, can proceed alone from the influence of the sun, should have been already created on the first day.

One scarcely knows whether to be provoked at the inconsiderateness shown in such an argument, to laugh at its utter shallowness, or to commiserate the pitiable mental condition of those who make use of it.

For assuredly only the most unbecoming and inexcusable want of consideration, or the narrowest mental capacity alone, can account for the circumstance that the author of the record of the creation should be imagined so stupid and shallow, that, had his communications been but the offspring of his own speculations and fancies, he should not have known that it was the sun which now gave to the earth its light and shade, its morning and its evening. No less unaccountable would it be, otherwise, that any one could imagine he should (as he subsequently, verse 16 seqq., expressly mentions the office of the sun to be to give light upon the earth,) have forgotten this fact, and speculated in the face of it. And he a man, too, profoundly wise and acute, just in the measure that his communications are confined to the wisdom of his own mind!

The difficulty here is not that the author does not appear to know what is plain to every child of two years, but that being *doubtless fully aware* of the facts (v. 16–18), he should not hesitate to teach that a light which illuminated the earth, was created before the sun,—and not once to have thought of this difficulty, is ground enough upon which to convict his accusers of culpable want of consideration, or the most lamentable shallowness of mind.

But what shall be said to the fact that a mere glance at the page of a modern text-book in physics or astronomy, is sufficient to show us that the earth and probably the rest of the planets also, still possess, since a permanent relation has been established between themselves and the sun, countless inherent sources of producing light; and that, just as the Bible says, the sun is not a light, but a bearer of light, a body which excites and developes light, and the like? And should we not rather, instead of seeking excuse for the stupidity of the author, who pretends to be a divinely-illumined Prophet, try to understand how it has happened that he should have unconsciously and undesignedly, obtained profound views into the nature and modes of light, such as have for thousands of years escaped the acute and untiring investigations of the physicist; that he in immediate prophetic contemplation should have anticipated the profound and happy investigations of modern days, in regard to the nature of light?

We may here add for further consideration, in addition to what we have already (Chapter 5, § 1.) gathered concerning the nature of light, and its development through the influence of the sun upon the planets, as well as through an independent agency belonging to the planets themselves, a passage from Humboldt's Cosmos (I. 207), where he speaks of the *northern light:* "This phenomenon derives most of its importance from the fact, *that the earth becomes self-luminous,* and that in the capacity of a planet, besides the light which it receives from the central body, the sun, it shows itself capable, *in itself, of developing light.* The intensity of the *terrestrial light* exceeds somewhat, in cases of the brightest colored radiation toward the zenith, the light of the moon in its first quarter. Occasionally printed characters have been read by this light, without difficulty. This almost uninterrupted *terrestrial* development of light in the polar regions of the earth, leads us to the interesting phenomenon presented by Venus. The portion of this planet which is not illumined by the sun, often shines with a phosporescent light of its own. It is not improbable that the moon, Jupiter, and the comets, shine with a light of their own, in addition to reflected solar light, noticeable as such through the polariscope. Without speaking of the problematical but very common species of cloud-lightning, in which a heavy, lowering cloud may be seen to shine with an uninterrupted flickering light, for many minutes together, we still meet with other instances of *terrestrial development of light* in our atmosphere." *A. Wagner* adds: "The northern light being an intermitting phenomenon, and exhibiting to us a change from light to darkness independent of the sun, we may find in it an analogy to a similar change occurring upon the earth before the creation of the sun." And *Schubert* says (*Weltgeb.* p. 218): "May not that polar-light, which is called an aurora of the north, be the last glimmering light of a departed age of the world, in which the whole earth was enclosed in an expanse of ærial fluid, from which, through the agency of the electro-magnetic forces, streamed forth an incomparably greater degree of light, accompanied at the same time with animating warmth, almost in a similar mode to what still occurs in the luminous atmosphere of our sun?"

Let us not, however, be understood from the foregoing, to assert that that light which, according to the Mosaic account, was created before the sun was formed to serve the earth in the capacity it now does, was a northern light, or even merely a phenomenon related to the northern light. No: we desire only to show that even yet, since the establishment of the relation which now exists between the sun and the earth, the latter still possesses in itself a capacity of developing light; and that there is nothing to prevent us from ascribing to it *prior* to that point of time, the same capacity in a degree much greater and vastly more magnificent and effective.

It is not pretended that those inherent powers of producing light which manifest themselves in the earth, are either numerous enough or strong enough to develop a light wholly equal to that of the three first days, which appears to have been strong enough for the origination of the vegetable kingdom of the third day. Consequently, it must be assumed that the first and provisional production of light, was essentially the same as that which is now brought about through the influence of the sun upon the earth. So long as the present existing relation between sun and planet was not yet ordained and established, the powers of exciting light, which ever since have belonged to the sun, may have dwelt in the planetary bodies themselves also, producing very much their proper and corresponding effects. Not until the fourth epoch of development, when the bodies of our system had progressed so far in their individual development, that a positive and permanent relation could be established between them, was it possible that the polar opposition between sun and planet should discover itself, in which the sun, perhaps on account of the preponderance of its mass and gravity, summoned to itself and retained the *powers of developing light.*

The facts of astronomy appear very well to harmonize with this view, as the body of the sun itself is found to be dark, and of a planetary nature, and its light-producing power, to dwell in the photosphere which surrounds it. Not the creation and fashioning of the *body* of the sun, but the formation of this photosphere, or the concentration about the planetary sphere, of the previously-created but heretofore diffused agency for the production of light, probably marked the point of progress attained on the fourth day.

§ 5. *The Creation of the Fixed Stars before the Earth.*

A fresh objection is founded upon the alleged representation of the Hexæmeron, that all the starry worlds should have been first created on the fourth day, subsequently to the complete formation of the earth. It is in itself unreasonable, it is said, that we should ascribe priority in time to the earth, which is but a subordinate member of the solar system, in preference to the sun which rules both the earth and all its brother and sister planets. But this Biblical representation in regard to the fixed stars amounts to an absurdity, when viewed in the light of modern astronomy, especially when it is taken in connection with the chronology of the Bible. The earth to have been created before the fixed stars! and still it not in existence some six thousand years ago! But does not astronomy teach that the stars nearest our system could not be seen within less than from eight to twelve years after their creation? and stars of the twelfth magnitude not within less than 4000 years! Consequently, then, the starry masses of the Milky-Way, scarcely resolvable or wholly incapable of resolution by the best telescopes, and the nebulæ, must have been created thousand upon thousands, yea, perhaps millions of years before their light could have reached the regions of space traversed by our earth, (comp. chap. 5, § 6, 13.) But, instead of their light having just now become visible upon the earth, has it not even as far back as human recollection extends, shone in just the same measure it does now?

We shall not open our defence of the Biblical cosmology or chronology by attempting to combat these dicta of astronomy, although it is by no means so certain that a ray of light, which in the ether of our planetary system is limited to a motion of merely 192,000 miles "in a whole long second," is everywhere in the universe confined to the same "snail's pace." For, even could we make up our mind to undergo the disagreeable necessity of having to claim a ten, a hundred, or perhaps even a thousand-fold greater velocity for light in the supra-planetary regions of space, still the idea that the earth was really created before the fixed stars, would meet with many other not less formidable difficulties.

Let us rather take the assertions of astronomy without any hesitation, leaving it to that science itself, whose office it is, to establish or combat all possible doubts which may arise as to the correctness of these assertions. Let us here also see if we may rather seek the cause of the pretended contradiction, in an erroneous apprehension of the Biblical account, than in the errors of astronomical science. We have already found (chap. 4, § 8), as the result of an exegesis made without prepossession, that the Mosaic account of the creation confines itself exclusively to the earth and its appurtenances; that it is only from *this* point of view and from *this* motive that the sun, the moon, and the stars, are included in the representation of Sacred Writ; and that the latter does not treat of their creation as such, but merely of the creative influence by which they became what they were to be *in their relation to the earth.* But whether these two points, which in itself is a thing not at all impossible, were identical in point of time, or diverse, was a question which had to remain undetermined in the interpretation of the record.

But that which is left undetermined in the Mosaic history of the creation, is placed beyond all doubt at a later stage of revelation. The Book of Job, for example, as we have already seen (chap. 5, § 17), puts the assertion in the mouth of God himself, that the stars were present as admiring witnesses and jubilant spectators, when the foundations of the earth were laid. If, therefore, the Biblical view touching this point be asked for, it must be maintained that priority in time in the stars over the earth, is a veritable and clearly expressed point in the Sacred record, and that the Bible and Astronomy here at least strikingly coincide. The Bible refers us clearly enough, both in the passage mentioned, and also by many other hints and indications, to which we have already in the fourth chapter given sufficient attention, to a two-fold creation, in which the creation or rather the new-creation of the earth, takes the second place in point of time, against which astronomy will assuredly find nothing to object.[1]

[1] A point distantly related to our object in connection with the above question, may here be presented in the words of a celebrated French investigator in the domain of nature. *Marcel de Serres* says, in his work *de la création de la terre* et des corps célestes, Paris, 1843, p. 17: "S'il n'y avait eu qu'une seule création, on devrait voir chaque année, presque chaque jour, apparaître de nouvelles nébuleuses au milieu de la voie lactée. L'observation est loin de confirmer cette continuelle apparition, et qui prouve, que cette dernière supposition est tout à fait gratuite. Le nombre de ces nébuleuses ne s'accroit que par la puissance des télescopes ou des lunettes, que les astronomes emploient pour les découvrir au milieu de l'immensité de l'espace. Du reste, si cette hypothese, tout à fait contraire au système d'une création primitive et d'une organisation posterieure des corps célestes qui en aurait été l'objet, était exacte, le spectacle que le ciel aurait présenté aux premiers âges du monde, à Adam et à ses descendants, aurait été aussi extraordinaire que singulier. Le premier homme n'aurait pas vu, lors de sa venue sur la terre, une seule étoile au ciel; le soleil, la lune et les planètes auraient été les seules astres, qu'il y auraient apperçus et dont il aurait joui pendant les premières six années. Au delà de cette epoque, les étoiles auraient commencé à apparaître successivement et dans un ordre inverse de leur distance à la terre. La voie lactée n'aurait donc présenté l'aspect, qu'elle offre actuellement qu'au delà d'un certain nombres de siècles. Enfin aujourd'hui encore des étoiles et des nébuleuses devraient se montrer pour la première fois dans le ciel. Il faut l'avouer, de pareilles conséquences sont tout à fait inadmissibles; dès lors on est en droit de rejeter la supposition qui y a donné lieu. La création des étoiles et des nébuleuses a donc precédé la création de l'homme actuel d'un

As it has further been objected, that it is taking a very narrow view, and one unworthy of Divine revelation, for the Mosaic record to represent the stars as having been created *merely* for the purpose of casting their scanty, flickering light amid the thick darkness of the earth, it may be well for us to observe that the complaint is chargeable to an interpolated "*merely*," of which the record is wholly innocent. It is doing unpardonable violence to the author's meaning to maintain, in the face of the unequivocal design of the record to mention only that which was of significance to the earth, that he really did believe all the starry worlds were created for no other purpose than to give light to the earth and adorn its nights. But if any one seriously maintain that such a purpose would be too insignificant and unimportant to claim notice in the Biblical geogony, we would merely inquire of him, if the thought has never yet suggested itself to him, when gazing on the splendor of the nocturnal heavens, how much of comfort and delight we poor inhabitants of the earth owe to the mere outward manifestation of these celestial bodies.

§ 6. *The Creation of the Planetary System.*

It is further said, that the connection of all the planets of our solar system, as well as the similarity

---

grand nombres de siècles. On est ainsi amené, comme forcément, à admettre deux époques bien distinctes dans la création : la première ou la plus ancienne est celle aù l'ensemble des corps célestes est sorti du néant à la voix du Créateur ; la seconde, bien posterieure, serait celle où le soleil, les planètes et particulièrement la terre ont reçu leur organisation définitive et sont parvenus à leur état actuel."

of their physical constitution and their reference to the sun, points unmistakably to the fact that their origin was a common one, both with respect to the matter out of which they were formed, and also to the time when they were individualized and finished. This we readily admit. But when it is further argued, that this supposition is neither acknowledged nor allowed by the Mosaic history of the creation; and that the latter here speaks of the formation of the earth, and there of the formation of the sun, moon, and stars, as though wholly independent of each other, the earth being completed and furnished with its mountains and valleys, continents and oceans, before the others were made—when all this is done, we must enter our most decided protest. The object of the record, Gen. 1., is to convey to our minds merely an account of the process through which the *earth* arrived at its present state, to tell us how it was prepared as a place of abode and activity for man. The sun, moon, and stars, are first mentioned in it, where they begin to play a part in the history of the gradually improving earth; and claim attention merely in so far as they do this. The record was not designed to tell, neither could it or should it have told, whether the earth, the sun, the rest of the planets, and the satellites, were formed out of the same original matter, nor any more, whether the individualization of these various bodies was simultaneous, their individual completion being subsequently effected side by side. That such was the case seems at least probable, in the light of astronomy.

With respect to the view to be drawn from astronomical investigations and reasonings, touching the probable mode of origin of the planetary system, and the bearing of this mode on the Biblical account of the creation, we shall recur to what has already been said on this subject, in chapter 5, § 5, for further elucidation of the question. It was there discovered, that astronomy itself is wholly unable to say anything on the subject; and that all theories which speculation has built up on the grounds of astronomical observation, or may yet build up, lack any solid or satisfactory foundation.

Still, however, we have regarded it as not out of place to devote a note to the theory of *Laplace*, the most plausible and interesting of these theories; and a word or two may be added in regard to the question whether, granting the correctness of this theory, something neither yet proven nor capable of proof, it may be harmonized with the 1st chapter of Genesis.

It will, however, be immediately discovered, on a mere hasty comparison of the two, that the Bible leaves room enough for this, as for all similar theories, and conversely, that no such theories can in a single point contradict the teachings of the Biblical record; for the Bible never enters upon the question whether the bodies of our system were formed out of the same common original matter, and if so, how their formation was effected; but rather, at first mention[1] represents them as proceeding forth from the hand of their Creator as worlds already individualized.[2]

---

[1] Genesis 1 : 1.

[2] *G. H. v. Schubert,* proceeding to carry out the view deduced from Scripture, that the domain of the world to which our earth

§ 7. *The Celestial Worlds in general Inhabited.*

Intimately connected with the above-mentioned pretext, that the Bible, in *opposition to* all sound human reason, should teach that the sun, the moon, and the stars have no significance in any other direc-

belongs was the scene of a history of the most comprehensive relations and important consequences, prior to the creation of man (comp. chap. 4, § 20 25), concludes (*Weltgeb.* 559–565) that probably in this, the first great period of its existence, this domain of the universe may have been represented by a single and unique astral formation, which only since the catastrophe which brought the history of this period to an end — or rather, upon that second creation which prepared it for a new and no less important and influential phase of history — was separated into different individual bodies, but connected and harmoniously adjusted into one complete system. He imagines that it was similar in the first period of its existence, to the planetary nebulæ with a dense nucleus, the photospheres of which extend themselves to the circumference of millions, yea, billions of miles (chap. 5, § 13). "Such an astral photosphere may have contained a fulness of elements sufficient for the formation of altogether other worlds than our small earth; for had it the size of even the smallest planetary nebulæ revealed to us as such by the telescope, it must fill a much greater space than does our present solar system, including the orbits of all its planets and comets. . . . . It is to be supposed that the primeval photosphere of the earth was also the special abode, not only of the forces of the electro-magnetic species, but of the higher primary forces of life, with corresponding forms and movements. . . . . It gives light and warmth to the nucleus beneath it; it is the more essential part of the star. The star itself, as the inner solid mass of a planet, does indeed form the supporting centre, binding the lighter substance of the envelope or photosphere to itself, by the force of its gravity: but this envelope is related to the nucleus, as the surface of the planet, upon which alone organic life dwells and flourishes, is to the inorganic basis upon which it rests. . . . . The sacred record speaks primarily

tion, and no office, but to give light to the earth, there is to be found another objection. Such a view, it is said, would do away with the supposition that the rest of the celestial bodies are inhabited by reasonable, spiritual beings, endowed with existence for their own sakes. The Biblical theory of the world, it is maintaned, is so narrow and inadequate as to

---

of the days' works in that new creation of things, in which man appears as the last and highest creation, on the eve of the Sabbath. The measure of time first begins with him and his history; the succession of years is first introduced by the formation of the sun and a heaven of planets from the primeval photosphere belonging to our domain of the universe. As to the history of the principality and powers of the preceding period, and their influence upon those works which were preparatory to the decree of the further future, that shall never be taught in time, nor understood in time." — We have nothing to object against this view, yet think we are justified in pointing to other formations of the astral heavens which perhaps may with equally good reason be regarded as analogies to the original condition of our domain of the universe. We refer to the closely-connected and related families of the double and multiple stars (chap. 5, § 11), or even to the presence of dark, extinguished bodies (chap. 5, § 12) involved in the orbits of resplendent suns, as seems so probable from the indications of the latest astronomical observations. Perhaps this region of the world was originally represented by such a closely-related family of different individuals, whose primeval harmony and glory were destroyed by some great catastrophe, and restored again in another and peculiar manner through the new creation of the six days; or perhaps it was occupied by a double star, one member of which was destroyed and broken up by that catastrophe, thus furnishing the substance for the formation of the planets and comets of our system, the relation of which to the sun was again established in a peculiar manner on the fourth day of the Hexæmeron. — The Scriptures are silent here, leaving the widest room to the play of conjecture.

represent the earth only as inhabited; life and activity, history and development, here only to manifest themselves; while assuredly the simplest common sense forces upon us the irresistible conviction, that the countless celestial worlds, some of which, as may be shown, possess a like nature and a like cosmical position and importance with the globe we inhabit, but by far the most of which infinitely surpass our poor earth in outward extent, as well as in inner significance, in glory and dignity, must also in the same degree be the theatre of like and infinitely higher manifestations of created life and activity.

But this objection is completely overthrown by an acknowledgment of the fact, that the Bible, while it does indeed regard the stars as dispensers of light to the earth, does by no means exclude the idea that they may, in themselves, be objects of infinitely higher significance and design than the earth.

We find our minds somehow deeply possessed of the idea, that wherever there are worlds there is to be found a proper place for the life and activity of spiritual beings; and neither faith nor philosophy, when not led astray by narrow or false views of Scripture, or blinded by a Pantheistic deification of the human mind, will find itself able to imagine the countless hosts of the celestial worlds as wholly uninhabited. It is to sound human *reason* we owe this lesson—*not* to astronomy, which with the greatest extension and closeness of its observations, will in all probability *never* be able to discern the evidences of created life in the moon even, the nearest of the celestial bodies; and hence cannot claim the right to say a word as to whether the stars are inhabited or not. We may seek to prop up so meagre a theory of the world as the one we oppose, with any number of analogies; we may take for example, if you choose, the favorite one of a royal saloon, with its thousand brilliant lights and profusion of costly articles, not of immediate service, but merely used to set off the glory and majesty of the king; still this as all other pseudo-analogies will have no effect upon the mind uncultivated perhaps, but possessed of common sense and unprejudiced, upon the creed that grows out of the depths of faith and reason.

It is *one* and the same God who sits enthroned in the heavens and displays His omnipotence and omnipresence here upon the earth; *a* God who upholds all the systems of worlds, and sustains the mote in a sunbeam; a God of *life*, who everywhere that the tread of his foot is seen, or his breath felt, calls forth an abundance of life. If it be true that our poor earth is inhabited, from man who walks with countenance erect, to the veriest worm of the dust; that a drop of water, a grain of sand, or a leaf of the forest, contains a whole world of living beings—if it be true that the whole restless sea of living organisms which manifests itself in millions of varied forms upon the earth, finds its unique completion and end only in *that* being endowed with reason and the capacity to know and praise his Creator, in man alone, a mediator between it and Him for whose glory it was created—how can it be that yonder starry choirs should be devoid of life, that we should not there expect to find fresh domains of life and spiritual movement, self-conscious and free creatures, endowed with capacities to know, to praise, and to adore their Creator?

It is untrue that the Bible excludes the supposition, that the stars, too, are inhabited by corresponding personal beings,—yea, even more than that, it even contains, as we have already seen (chap. 4, § 23), determinate and almost unmistakable references to the fact, at least positive intimations, that the celestial spheres[1] are really inhabited. The Scriptures regard the heavens, and of course, all the single worlds which help to make up the heavens, as the abode of countless hosts of spiritual creatures, called in general terms, *angels*, and represented as the messengers and servants of God, as those who execute his will and join in anthems of praise and adoration to the glory and majesty of Deity. And in *one* passage at least,[2] these holy and blessed spirits of the celestial regions are placed in such close relation not to the heavens in general, but to the concrete, individual celestial worlds, that our view of the matter as heretofore expressed, that the angels inhabit these worlds, seems fully justified.

§ 8. *The Angels as the Inhabitants of the Fixed Stars.*

Astronomy does not and cannot teach us any thing concerning the nature and destiny of the spiritual

[1] We speak primarily of the heavens of the fixed stars only. Much more difficult and doubtful is the question, hereafter to be considered, whether we are to imagine the rest of the bodies of *our* solar system inhabited, and if so, by what species of beings? (Comp. § 10).

[2] Job 38.

inhabitants of the stars, as predicated upon the general grounds of philosophico-religious reasonings; while it opens to us a few glances into the physical constitution of these bodies, however incomplete and meagre these may be. The *Bible*, on the other hand, was designed only and is able only to teach us religious truth; but nothing touching the nature and constitution of the *stars*. But still it does contain intimations which lead us fairly to the supposition, that these very stars of the firmament are the abodes of the angels. Harmony or conflict between the Bible and astronomy, can therefore, on this point, depend alone upon the agreement or non-agreement of the physical constitution of the *stars*, as taught us, or rather (vaguely enough) conjectured by astronomy, with the nature of the *angels* as represented in the Scriptures—upon the fitness or unfitness of the material abode for the spirit which is to inhabit it.

Since astronomy, enriched by the magnificent investigations and views of *Herschel*, has entered upon a new and healthful path of development, it has left behind those old and narrow hypotheses of a monotonous repetition in all regions, of the order and arrangement which obtain amidst the bodies composing our solar system, and of the reproduction everywhere of a physical constitution similar to that impressed upon this system.

The constitution of nature is wholly different in the celestial worlds, and the latter bear different and higher relations toward each other—hence the beings which inhabit these worlds must be of a different species, and altogether differently constituted, and must possess a different calling and destiny, different capacities and duties from ours.

Modern astronomical results have not indeed shown it to be strictly impossible, but still, improbable, that the glowing worlds of the fixed stars should be suns precisely like the sun of our system, having like it dark, solid, planetary central bodies, and being accompanied by secondary bodies dependent upon them for light and heat. They too, indeed, have—at least some of them—their faithful companions; but the association there involved is not conditioned by the despotic sway of mere physical force, but through the bonds of close affinity and mutual sympathy; not through subordination, but rather through co-ordination; for there we behold as it were, suns circling about suns, one glorious sphere about another, its equal in kind and prerogative, however different in brilliancy or extent. To all appearance there does not there exist that physically opposite, or sexual character, as it might be called, of the world's organism, which here manifests itself in the contrast between the solar and the planetary principle, as that which on the one hand excites and imparts, and on the other is excited and receives. We there no longer find that marked characteristic of mass and of gravity which here rules and bears sway: nor do we there observe that alternation of light and darkness we here experience; there is then no night to break in upon the life and activity of those spheres, nor frost nor winter to benumb their energies.

But, although those celestial worlds are not possessed of those coarse, material characteristics which everywhere meet our eyes here below, it by no means follows that they are *immaterial:* although the contest and the change which are here carried on between light and darkness, do not extend into those regions, it does not hence follow that light there fails of a corresponding substance to which it may attach itself, and thus arrive at a fixed character and full intensity. There materiality is merely not limited to the characteristics of passive, dead matter only; and light and darkness are not there hostile to each other, but rather, like the body and the soul, pervade each other in a true and complete union. We may appeal to the fact that even in the single, but especially in the double stars, a province of colors displays itself, "rivaling in beauty and variety the flowers of spring, or the wings of the butterfly." Color is light manifesting itself through darkness, and thus attaining a determinate quality and intensity of brilliancy: it is a vital union of the two. "While in our planetary system," says a profound thinker,[1] "sun and planet — light and darkness — as such, are distinct and separate, abstracted from each other, forming a totality in a mere outward respect, they are in the celestial regions intimately united and pervaded by each other. . . . Thus does each part become the whole, and yet remain on the whole." Here harmonious unity resolves itself into conflicting contrasts: night contends with day, light with darkness, heat with cold, death with life, and the body with the soul. But there all contrasts are reconciled:

[1] C. Fr. Göschel, *Unterhaltungen zur Schilderung Gœthescher Denk- und Dichtweise,* vol. 3, p. 192, Schlesingen, 1838.

light and shade, day and night, are intimately united; the one shining through the other, the soul animating the body. There we find no alternation of light and darkness: a million suns at the same time shed forth the radiant light of an eternal day, yet so mildly as to avoid excess of heat no less than destructive cold. The dark material structure is pervaded and animated by a higher breath of life, and the latter through a most real and intimate union with the former attains a concrete manifestation, a vital existence, a harmonic fulness and entireness. For everything living and real is "unity amid diversity, union of soul and body. Light becomes color through the medium of darkness only, so also the soul manifests its active presence through the medium of the body only. The fruit of like and like is a dead production: where like and unlike resolve into one, there a sweet sound is produced."[1]

If, therefore, the worlds there, instead of carrying heavily through their orbits, such a coarse and crude materiality as belongs to the bodies of our system, possess a physical structure infinitely refined, light, and glorious, and hence pursue their silent and majestic courses with signal freedom, lightness and ease, "the restless and ceaseless workings and counter-workings of our powers of attraction and repulsion, which cause such painful swayings to and fro between friendship and hostility amid the ponderous bodies of the solar system, must certainly find no counterpart among those celestial worlds." Here, in the cosmical domain we inhabit, the laws of grav-

[1] Göschel, p. 192.

ity bear iron rule: the force of gravitation is an outward, despotic power, and the bodies of the system are held together by it alone: without it, they would fall to pieces and become utterly demolished. The same *law*, indeed, obtains among the celestial worlds; but *love*, which also in this respect may be regarded as the *fulfilling of the law*, shuts out *slavish fear*. The effect is the same, but the cause is different. The categoric imperative of physical force exacts not a slavish obedience, but a higher will, in which freedom and necessity have become one, calls forth similar effects, in nobler form and higher potency. But still, other forces also may there manifest themselves, as not impossibly the magic forces of the electro-magnetic species, which with the rapidity of thought traverse the entire earth; but in an inconceivably more imposing manner and greater degree, and with consequences vastly more magnificent and glorious. Hence "we there behold one sun fraternally linked to another, and hosts of glorious worlds peacefully pursuing their courses, held together by the bonds of a higher relationship than those which here impel one ponderous rock upon another, with crushing force."[1] Mysterious bonds of sympathy and secret affinity must indeed bind those worlds together, where "gravity is no longer the tendency of each individual to seek in some other material structure the central point wanting in itself, but the free impulse which centralizes all single bodies, all single central points with each other in the highest centre."[2]

---

[1] Schubert, *Weltgeb.*, p. 85. [2] Göschel, p. 192.

There move on in familiar intercourse and association, thousands, yea, millions of celestial worlds; here are immense, immeasurable wastes, empty celestial spaces, insusceptible of the illuminating, heating, or animating influence of light, and filled only with the blackest night. There the spaces between the different worlds are filled up with intervening nebulæ, the channels and highways, as it were, of a communication to be carried on with the speed of a celestial electricity; here are impassable and all-devouring gulphs; there familiar and gladsome intercourse; here separation, distance, and non-intercourse. What a plenitude of life, and what an energy of its allotted functions, must there unfold itself, where worlds are so wondrously crowded together, hold such lively intercourse, and exert such varied reciprocal influences; while at the same time the particular agencies of any individual world, are excited and strengthened by the constant influence of countless kindred worlds! And what a variety of formations, what a rich abundance of forms, transformations and renewals are indicated by the inconceivable expansions and contractions of those luminous cosmical masses, their volitalization as well as their condensation; by the variation in the light and brilliancy of the stars, as well as the reciprocal relations of their colors: truly, a mobility and freshness of life of which we, with the ideas of slowness of motion, weight, and vis inertiæ we ever attach to material bodies, can form but a very inadequate conception! And whilst most of the changes and revolutions which occur here upon the earth, are disastrous, and followed by suffering, tears, and distress, the revolutions there experienced must take place so gently and peaceably, that "to the countless myriads of eyes which near at hand behold them, and carry on their functions in the midst of them, they must be despoiled of all those terrors they would have in this terrestrial sphere, and cost no tears of sorrow, but rather, if weeping be possible there, tears of joy."[1]

But what, then, are we to imagine the inhabitants of such worlds to be? If the principle be well-founded, that there exists everywhere throughout the sphere of the created, the same connection between abode and inhabitant, as between body and soul, then can astronomy throw much light on this question. The physical world we inhabit speaks everywhere, as well in small as in great things, of blessing and cursing, of love and hatred, of sorrow and joy, of longings and hopings; and a deep response to its tones is wakened within our own breasts, as we feel that nature around us is adapted to us and we to it. But in those worlds we seek in vain for the ominous shadows of sin and death; there we behold light without its antagonistic darkness, life without death, harmony without strife and discord, day without night, and waking without sleeping. Hence they must be the abodes of such beings as are altogether unacquained with sin and death in themselves, and having natures not requiring the alternation of light and darkness, of day and night, nor visited by the

[1] Schubert, *Urwelt*, p. 108.

cruel changes of heat and cold we here experience. Life, which here displays the opposite poles of generation and corruption, birth and death, is there unity and fulness. There the sexual contrast, the opposition of the solar and planetary principles, the contrast of that which excites and that which is excited, is done away with; and hence it is there we may expect to find the exalted sphere, where they neither marry nor are given in marriage.[1] And as in those worlds the planetary "flesh-and-blood structure" of the dark and crude earth, is quickened and illuminated, rendered clear and glorious, as instead of physical inertia and constancy of form, we behold the greatest aptitude for rapid movement and the assumption of new forms; so also must we deny the inhabitants of those worlds bodies of an opaque, flesh-and-blood character, such as that by which our corporeal frames are hopelessly confined to the surface of the earth, and the soaring flight of our thoughts so oppressed and crippled, and ascribe to them a refined, ethereal, and infinitely mobile corporeity, capable of renewal and rejuvenation, ever the willing servant of the indwelling spirit, and adequate to all the wants and exigencies of spiritual life.

But such holy inhabitants of light are revealed to us in the Scriptures, under the name of *angels*, and are placed by them in manifold relation to the celestial worlds, so that here science significantly coincides with faith.

---

[1] Matt. 22 : 30.

§ 9. *Continuation.*

We shall yet notice a few objections against identifying the fixed stars with the abodes of the angels, which might possibly be urged, as well from the Bible as from the astronomical stand-point.

And first of all, it might appear that the almost infinite distance, according to our rules of measurement, of the stars from the earth—a distance which a ray of light requires ten, one hundred, or a thousand years to traverse (chap. 5, § 6)—would be but little in harmony with such a supposition. For this immense distance seems to accord unsatisfactorily with the Biblical teachings in regard to the frequent influential presence of the angels upon the earth, since they are represented as appearing not *only* at the grand crisis of the development of the kingdom of grace, but also at all times necessary for the aid and protection of the children of the kingdom.

It is clear, however, that such an objection is of any weight, *only* so long as we attribute the limited conditions of our terrestrial sphere of life, to the life and activity of the angels, something we are by no means justified in doing. There are here, even in the sphere of sublunar physical agencies, and within the compass of our own knowledge and experience, velocities with which the velocity of light itself can bear no comparison. In the electric telegraph we see an all-pervading physical agent employed as the messenger of the mind, with a velocity of movement which is not capable of measurement in our longest distances. And the rapidity with which the influence of gravity passes and repasses from one celestial sphere to another, must, "according to very fair and reasonable deductions, be at least 10 million times greater than the velocity of light."[1] All these velocities, however, are still mocked by the flight of thought, of the mind. True, our *corporeal forms* cannot begin to keep pace with that; but shall not the body, in those spheres where nature is refined and rendered glorious, and in those holy beings which are by way of pre-eminence termed *spirits*, obey and carry out the will of the mind?[2] And shall not those privileged beings be able, with the rapidity of thought and without dispensing with their bodies, to transport themselves whithersoever duty may call?

Again, it may be objected, that the diversity of formations amid the starry worlds of the firmament, of which modern astronomy has enabled us to catch some significant glimpses, cannot be regarded as in harmony with the generic unity of the nature, the existence, and the calling of the angels, as represented in the scriptural doctrine of those beings. Two things, however, must be overlooked in offering such an objection. First, that the Scriptures beyond doubt comprehend and point out within this generic unity, great specific differences between the several classes of angels, the existence of many grades of dignity, might, and the like,—and then, that where the angels are designated as a common and like whole, this expresses merely the general contrast and distinction between them and of man. But the contrast between

---

[1] Schubert, *Urwelt*, p. 18.

[2] Comp. J. P. Lange, *Leben Jesu*, II., 1, pp. 58, 59.

the character of fixed stars and the planetary nature of the earth is great enough and thorough enough to accord with the contrast between angels and men.

We are well aware of the fact, that we have derived the foregoing proof of the correlation between angels and stars, in part from astronomical views which have by no means yet been proven undoubtedly correct or fully admissible, and which may perhaps ever remain problematical. We may make special mention in this connection, of the view that the antagonistic relation of the solar and planetary principles, which here below so determines all cosmical conditions, and imprints upon the system to which our earth belongs its peculiar character, should not have any existence in those distant regions of the universe. But such a difficulty can scarcely be surmounted, from the unavoidably problematical character of all perceptions and revelations coming from those extremely remote regions of creation.

But supposing our views be not strictly and thoroughly in accordance with the reality, supposing also that those glowing spheres be surrounded by planets, which, just like our earth, are dependent upon solar influences for light and heat, indispensable to their inhabitants, still the Bible contains not a word in opposition to the view that such planets may be inhabited by angels. Much rather, there still remain many peculiarities of physical structure which correspond to the characteristic distinction which obtains between angelic beings and the inhabitants of the earth. In the systems of the double and multiple stars, at least, and the thickly-crowded clusters, where thousands upon thousands of stars join in forming one system, such hypothetical planets would have to be, were they not to be hurled against each other or against their suns with destructive force, composed of a material so refined and so light, that even in them, as opposed to the earth, should be mirrored the contrast presented to the flesh-and-blood character of the human body, by the light, ethereal structure of angelic bodies. No less would such planets be the theatre of a constant and inexhaustible manifestation of light, from the simultaneous influence of hundreds, yea, thousands of suns, and in this respect also might be the fitting abode of glorious angelic natures, the inhabitants of light.

But what shall we say in the face of the discovery of *Bessel* (chap. 5, § 12), that, conversely, in the regions of the fixed stars, suns, and those the largest and most brilliant, revolve around bodies which are in all probability dark? What shall we do with this discovery, in case the observations upon which it is founded be proved correct beyond a doubt? where shall we find a proper place for *it* in our Biblical theory of the world? We frankly admit we know not where; but without yielding the principle that what is still a mystery from mere want of knowledge, is by no means proper evidence of the inadmissibility of what we do know. Meantime, we may console ourselves with the thought, that the friends of the purely astronomical theory of the world will find themselves in no less a dilemma in connection with this problem, should the physical fact be established. Besides, it may be remarked, that with all the reliance to be placed on *Bessel's* observations, and with all the corroboration they seem to have received from renewed observation in the same direction, the inferences drawn from them must ever retain a highly doubtful character. Just in the same measure that this doubtful phenomenon may be explained through a reversal of the existing solar and planetary relations, in the face of all previous views and knowledge, it may assuredly be referred to agencies of life and movement in the celestial regions, of which we have here not the most distant knowledge.

§ 10. *Inhabitants of the extra-mundane Bodies of our Solar System.*

But what shall we say of the rest of the planets of our solar system, and of the sun itself? Are we to suppose them also inhabited, and what might be the nature of such supposed inhabitants?

The Scriptures mention but the two species of self-conscious, personal, free, and spiritual creatures: *angels* and *men.*—We have already seen that there are physical grounds upon which to oppose the idea, that these bodies should be inhabited by *human* beings, with flesh and blood like *ours;* and further, that that the fundamental point in the Biblical theory of the world — the unity of the whole human race, and its derivation from *one* original pair — would be still less reconcilable with such a view. Are they, then, inhabited by some kind of angelic beings of a different nature, of different orders and grades from those which dwell upon the other starry worlds? We cannot adopt such a view, since thereby neither the marked contrast between angels and men, nor the general oneness of calling and nature belonging to the angels, both matters expressly taught in Scripture, would seem to be sufficiently regarded or acknowledged. For, notwithstanding the diversity in details presented to the earth, by the separate bodies belonging to our system, the similarity of their physical constitution and arrangements in general to those of our earth, is too great, on the one hand, and the contrast they in common with the earth present to the fixed stars, too strongly marked, on the other, that we should concede them to be inhabited by any sort of angelic beings.

Or are we to imagine, as has been frequently intimated, that the souls of the departed dwell there? The righteous upon the pleasant Mars, upon the bright and fair Venus, and the glorious Sun, perhaps; and the souls of the unblest amid the dreary and stormy wastes of Jupiter, and in the dismal craters of the moon?[1] — Against any such opinion we are also bound to protest, as it seems to us, from the Biblical stand-point. It seems to us next thing to a mere groundless and fantastic chimera, to suppose that these bodies were intended to subserve no other purpose than to supply prison-houses for the dead of the earth, and should have been created for that distinct purpose, even before death was introduced and had become general through the sin of Adam. Such a view, at all events, is not in the least justified by the

[1] Comp. J. P. Lange, *Land der Herrlichkeit*, pp. 8, 9, and his *Verm. Schr.* II., p. 270 seq., as also Tholuck's *Stunden der Andacht*, p. 549.

Scriptures. For the Bible never speaks of the locality of Scheol or Hades, as the place of the departed, in any other than the common usage of Scripture; and if we may ascribe to its expressions any agreement with the reality, the locality of Scheol should much rather be sought under the earth than in the heavens.

Or, further, are we to suppose that those apostate spirits, which according to Holy Writ, are consigned to waste places,[1] and the barren regions of the air,[2] dwell in those volcanic wastes and dungeons, or amid the darkness and tempests, the scorching blasts and severe cold of these bodies of our system?—Most probably not—if we follow the expressions and views of the Bible, which place the abodes of these evil beings, rather, in close proximity to the earth; amid the waste places, the darkness and the tempests of this sublunar world. For the expressions: ἐξουσία τοῦ ἀέρος,—ἐν τοῖς ἐπουρανίοις, from the Epistle to the Ephesians, certainly refer primarily to the terrestrial atmosphere, even though a wider meaning may be ascribed to the latter of them.

Or, finally, has that supposition the most probabilities in its favor, which regards those regions of creation, for the time at least, as devoid of reasonable beings—perhaps analogous to the regions of our earth uninhabited by man, if not by living creatures in general: its primeval forests, its uncultivated plains, and its wide seas, which none the less on this account are destined to be made arable and habitable by the hand of man, to serve his convenience, and

[1] Matt. 12 : 43; Luke 11 : 24. [2] Eph. 2 : 2; 6 : 12.

are hereafter to be purified and renewed with the rest of the earth, and brought into a paradisiacal condition? A similar disproportion between the earth's fitness for habitation and its actual occupation, certainly existed for centuries before the first human pair, on the strength of the command, "Be fruitful, and multiply, and replenish the earth," had peopled the whole earth, in ever-widening circles from their original dwelling-place.

When we consider the close association and the physical relationship of all the members of our solar system, their organic connection, their complete and symmetrical unity, much that seems unacceptable in the above conjecture of the non-habitation of the planets, appears to vanish. For this unity in the organization and articulation of the whole solar system, seems to point clearly to unity of relation and destiny, and to condition and presuppose a history in unison, as regards its relations and bearings.

Had man been true to his destiny, and in a godly, and hence, also, divinely-supported development, peopled the planet which was allotted him as an abode, from the spot which served as the cradle of the race to the outmost boundaries of the globe, perhaps, as may at least be conceived, his destiny would have been extended to those kindred worlds, at no great distance, and pertaining so closely to his abode, so that they also might be included in the circle of his activities and conducted to their destined perfection. Perhaps man should have been able, in the ever-increasing energy of his destiny from a godly development, and in the complete subjection of the physical forces to his control, to open up as safe and practicable a passage through the sea of ether which stretches from the shores of one island-world of our system to the other, as he has through the trackless watery wastes which separate land from land upon his own earth. And perhaps he would have succeeded in devising means of overcoming the incongruities and inaptitudes of nature in those neighboring worlds, just as he has succeeded in transforming and rendering habitable the forests and wastes of the earth.

But as the sin of the first man gave the whole development of his race a perverted and godless direction, so that the appointed end can be reached only through the incarnation of the Son of God, who as the second Adam took the place of the first, in order to restore what had been destroyed—the destiny and predestined perfection of those worlds, like much upon the earth itself, remains suspended and incomplete, until Christ the second Adam take them up again and conduct them to the end. And as this perfection with respect to the world of our earth can be brought about only through a mighty final catastrophe, followed by a general renovation of the world, in which all that is godless, the fruit of Satan's fall or the sin of man, shall be separated as dross, so also may the same in fitting measure be the case in those kindred and neighboring worlds. This is the rather to be assumed, as it is not in itself improbable that they were more or less affected by that catastrophe which brought about the *tohu va bohu* (chap. 4, § 6, 25) of the earth.

If from the foregoing it be found so difficult and doubtful an undertaking to assign to the planets of our system a fixed and characteristic position in the Biblico-astronomical theory of the world, we may well be excused from attempting a similar procedure with respect to the comets, and the innumerable asteroids called by the unpoetical name of shooting stars, satisfied that here, where Biblical intimations are wanting as well as characteristic astronomical data, no result can be obtained having much beyond a mere shade of probability.

§ 11. *The Astronomical Theory of the World.*

Copernicus, in so triumphantly combating the deeply-rooted cosmological errors of antiquity, gave illustrious proof that science no less than faith, is possessed of a world-surmounting power; but in a different manner, and with respect to a wholly different sphere. This world-surmounting power of each, is ***truth***, which is derived *from* God, is rooted *in* God, and tends *toward* God. *It* insures them both a final and permanent triumph over all the conflicting powers of self-interest, blindness, ignorance, folly, and error of the world, however wide-spread and inveterate these may seem to be: *it* insures the ***true faith*** against all the assaults of that false ***science***, which would combat, not Divine truth, but its own delusions and errors; and no less does *it* warrant ***genuine science*** the most complete triumph over all ***error*** and ***superstition***.

Copernicus triumphed: truth, with him as its champion, succeeded in casting out from the world those hosts of errors and prejudices which had been deeply rooted in it for thousands of years. It were not only vain, but singularly foolish, to resist and refuse to receive truths which have gloriously withstood all the fiery trials of doubt and direct assault, the severe examinations and tests imposed by friends and foes. But it is a different matter altogether, to protest against the false applications and groundless consequences which have been assigned to the same, through misconception or ignorance.

When at length, after a severe conflict, the truth of the Copernican system became generally acknowledged, the earth was no longer able to maintain *that* cosmological significance which had been previously so willingly accorded it by astronomy. It *could* no longer be regarded as the *physical* centre of the whole cosmical structure. It *had to be placed*, in an astronomical point of view, on a par with the rest of the planets, subordinate to the sun; and the latter, as the centre of the system to which it belonged, assigned a place in the series of countless equally privileged suns, scattered through all space. But the idea was very soon conceived, something which never entered the mind of *Copernicus*, that along with the *physical* significance which had previously without reason been ascribed to the earth, its religious significance also, as represented in the Bible, could or should be set aside. It was held, that since the earth had become such a subordinate and insignificant point in the whole universe, our faith in the Bible which speaks of it as the theatre of the most glorious deeds and revelations of Deity, could no longer stand its ground. And further, that upon a world so small, so poor and unimportant, the infinite and sovereign God, who had created innumerable other infinitely higher and more glorious worlds, and such as were better fitted for and deserving of high honor, could not condescend to manifest his miraculous power in the midst of us, and in due time Himself became man upon the earth, in personal union with a body to all eternity.

Next came *Herschel* with his brilliant researches, by which the thousands of suns having equal, if not higher claims to distinction than our own glorious luminary, were increased to millions, yea, to billions; and in which with the increasing penetration of his telescope into the depths of space, new and countless hosts of worlds were laid bare to the astonished eye, and space beyond the utmost bounds of telescopic vision, as well as the worlds it contained, seemed to the excited and enchained fancy, to be lost in absolute infinity. When the mind became bewildered amid such a wide maze of worlds, not only did the earth seem too *unimportant* a spot for the incarnation as mentioned in the Bible, but also the universe in its imaginary infinity of space and time, *too immense* for longer belief in the Biblical doctrine of the finite nature of all creatures and created things, and the infinity of God *alone*, whom the heaven of heavens cannot contain, who *alone* is *from eternity to eternity*, while all the heavens were created out of nothing by the word of His mouth. Thus did the mind get rid of not only the immanent Redeemer of man and the earth, but also of the transcendent personal Creator of both the heavens and the earth.

But if the investigations of *Herschel* gave to the astronomical theory of the world, a centrifugal and ever outward-tending direction, the modern discovery of *Mädler* (chap. 5, § 9), on the other hand, seems well fitted to call it back from its wild and daring flight in immensity, where it threatens not only to forget its proper home, but utterly to lose itself; and promises to give it again the centripetal direction, which is the necessary complement of the centrifugal, and prevents the latter from wildly straying and becoming lost in an infinite—nothing.

§ 12. *The Infinity of Space.*

It is wholly beyond the power of human reason or understanding to comprehend that there exists, or rather, *how* there should exist, an absolute limit to space and the bodies included in it. But is an absolute infinity of space any the less incomprehensible? It is inconceivable how and where space should cease to be, but is it any less inconceivable that it should be absolutely without limit, and how this should be? The incomprehensibility of the one as well as the other, arises, therefore, not from the nature of the things themselves, but from the inadequacy of the mind which attempts to grasp them. *Reason* here utterly confounded, refers us to *faith.* To faith alone remains the decision, and according to the true or false position of the former will be the issue of the latter.

Theism, which regards the existence of a *personal,*

living, and *eternal* God, who is as infinitely exalted *over* time and space, and distinct *from* them, as he is at the same time everywhere and omnipotently active in time and space—*Theism*, which regards this as in itself the most indubitable of all facts, and not requiring proof, must decide unconditionally for the finite, limited nature of creation with respect to time and space, however far we may desire to extend its boundaries. But *Pantheism*, which cannot imagine a *personal* God *without* and *above* the world, must decide no less promptly for the infinity of both time and space.

Our object here has reference to Biblical *Theism* alone; to settle *it* upon a firm astronomical basis, and defend it with the weapons furnished us by astronomy, is the task we have assigned ourselves.

With regard to the question whether *space* is to be considered as *finite* and limited, or *infinite* and without limit, all primarily depends upon the idea we have of *space*. Space may be understood in two different senses, as formal, or real. *In itself*, it is a *form* merely, which acquires substance, an *idea* which acquires reality, only through the bodies which fill it. *Empty* space is nothing more than the negation or absence of material bodies, but at the same time is a possibility or capacity for the manifestation of such bodies. In *this* sense, which regards space as a mere susceptibility or capacity for the actual existence of bodies, we hesitate not to ascribe to it absolute infinity. For this capacity coincides with the omnipotence of Deity, in which as a potency there dwells the possibility of an unceasing creative activity.

But to *real* space, *i. e.*, space which manifests itself or attains reality through the presence of bodies which occupy it (be these the coarsest or the finest, the most solid spheres or the most impassive fluids of a cosmical ether), we cannot ascribe infinity, without thereby wholly destroying the idea of a personal and transcendent God. Though we imagine creative activity unceasingly progressive, so that potential space is ever becoming raised to actual space, through the creation of new worlds; still this actual space must ever be considered, at any given period, as finite and limited. The potency of creation lies in God, and is hence infinite like himself; but the realization of this potency is a manifestation of it on what is finite, and hence its results also must always be finite. For, in the process of creation, the created, which heretofore dwelt *in* God as a potency, proceeds forth *from* Him: it is distinct from God *immediately* upon its creation; hence, also, finite. It were infinite, only when God had fully *exhausted* the infinite creative potency dwelling in himself, *i. e.*, when in creating, God had absolutely *done away with* Himself, and left an infinite universe in his stead. But it is a contradiction in itself to speak of an *infinite* potency becoming exhausted, and such a thing is not to be conceived.

Hence it is clear that the idea of a *transcendent Creator* is wholly irreconcilable with the idea of the infinity of actual space; so that if the latter overpower the mind, and cannot be got rid of, the idea of a transcendent Creator must be at once abandoned. This is the origin of Pantheism.

Since therefore actual space, *i. e.* space containing bodies, is finite, and must consequently have a limit within which it is comprehended, and without which it would dissolve away and be utterly lost, we can but suppose that God Himself is this limit. And also, as God, being a Spirit, is immaterial, that the boundary which surrounds and keeps together the whole creation, is likewise an immaterial one—of a spiritual nature, or a pure force. But this power can and must have a two-fold character, first, it must act from the periphery toward the centre and all points lying between the two; and then it must act from the centre toward the periphery, in all directions and toward all points. The former we perceive in the *Transcendence* of God, the latter in His *Immanence*.

§ 13. *The Transcendence and Immanence of God in the Mirror of Astronomy.*

Let us now inquire how these results of theistic faith and reason comport with the results of astronomy.

It must be understood, in the first place, that when astronomy speaks of an *infinity of space*, or an *infinity of worlds*, as the fruit of its researches, no *absolute*, but ever a mere *relative* infinity can be meant, i. e. that *its* investigations and observations have not succeeded in carrying themselves to the utmost limits of space or of worlds.

Astronomy may have a right to maintain that those nebulæ scattered profusely over the whole heavens, and refusing to be resolved by the most powerful instruments, are new systems of Milky-Ways—although this right even might be very much called in question, as we have previously seen: it may boldly assert, something for which it has no foundation, that yet more space-penetrating instruments will reveal still more remote systems of Milky-Ways, nebulæ respectively: it may extend this ideal and perhaps altogether fanciful construction of worlds, as far as it will,—still it is far, very far, from being justified in the assumption, that this heaping of systems upon systems of Milky-Ways is absolutely infinite, and knows no bound.

Such a *relative* infinity of worlds as the above taught by astronomy, assuredly cannot appear to any one irreconcilable with the theistic or Biblical doctrine of Deity. Indeed, we cannot see why we should apprehend any danger to our faith, were this idea of the relative infinity of worlds so extended, that not only reason but even imagination should grow giddy; we object not in the least against astronomy increasing and exalting the creative glory of our God to such a height, that thought and reason should utterly fail, and we fall prostrate in the dust to wonder and adore; for even here we might appeal to Scripture, in which the glory, majesty, and omnipotence of Deity, are praised in the most exalted terms.

But astronomy, on the contrary, offers us positive data which serve signally to strengthen us in the results of theistic speculation, as obtained in the previous pages, and to give concrete form to the abstract necessity of such results.

We here refer particularly to *the* magnificent discovery of the present day, which we owe to the profound sagacity and fruitful diligence of *Mädler* (chap. 5, § 9).

The tendency of all cosmical bodies and cosmical systems towards a common centre, which is *ever ideal* and *immaterial* (chap. 5, § 11, note 22), and which represents itself as material, *only* in a case where the masses are so widely disproportioned as in our solar system, points very clearly to that Eternal Centre, which, in the language of Scripture, supports and sustains all things by the word of His power;—it bears clear witness to the *immanence* of God, as the eternal, original power, itself immaterial and uncreated, which with omnipotent energy pervades and upholds all things material and created, the sun no more than the mote floating in his beams; and which itself partaking of unity, places all single phenomena of the world of the created under a single (einheitlichen) point of view. As in *Mädler's* central theory each cosmical body exerts its influence of gravitation upon all the rest, and conversely, is itself acted upon by all the other bodies, we have here a symbol and evidence of the omnipresence and efficient power of God, by which all things in the universe are referred to that unity which is Himself, and all the most varied and manifold relations of things subordinated to one another. Gravitation is the *immanence* of God; the *embodiment* of Deity, if it might be so spoken, in the sphere of the *cosmical.* But there exists a corresponding gravitation in the sphere of the created mind, through which all individual minds are placed in relation to the Centre of all mind, as well as in the most varied relations to each other;—only with *this* distinction, that the gravitation which obtains in the spiritual world is made

40

dependent upon the self-determination and voluntary development of the individual, and therefore subjected to the changes and disturbances conditioned by the *use* or *abuse* of personal freedom; whilst the purely cosmical gravitation was finished by the Creator in the beginning, but not connected either with freedom or personality, and hence not exposed to such an abuse of powers, and to the disturbances which would follow.

As we have recognized in the centripetal force of cosmical bodies, a symbol and evidence of the *immanence* of God, so also their centrifugal force points to the *transcendence* of the Divine Being. The latter evinces the fact that, *besides that* force which attracts all bodies towards a single point, and *opposed* to it—completing it and sustaining the equilibrium with it—there must be *without* and *beyond* the world, a no less powerful force at work upon all the single worlds, attracting them no less powerfully and irresistibly. To this striving of the cosmical bodies toward the circumference there corresponds in the sphere of created personality, a need of the spirit to seek God not only in the created but without it and beyond it.

The centripetal force alone would dash *world* against *world*, and lead the created spirit to deification of self and the world; the centrifugal force alone would rend *world* away from *world*, and deprive the *spirit* of all basis and self-dependence. Regarded singly, the centrifugal force corresponds to Deism, the centripetal to Pantheism, and the living union of the two, their mutual completion and harmonious coöperation, to Christian Theism.

§ 14. *The Incarnation of God.*

We now come to the essential point in the nominal contradiction between the Biblical and the astronomical theory of the world. It has respect to the fundamental and leading doctrine of all Christendom, with which the latter stands or falls—the doctrine of the *Incarnation of God in Christ*—which it is asserted can no longer maintain its ground, in the face of the results of modern astronomy which bespeak an infinity of worlds.

How is it possible, it is inquired, how is it to be conceived that the Lord and Creator of all those countless, immeasurable and glorious worlds, before which our earth shrinks away as a mote in the presence of a mighty globe, or is lost as a drop in the ocean, should have chosen this little point, earth, out of all the rich depths of the universe, in order here to appear and here take upon Himself the sorrows and infirmities of humanity; veiling Himself for man's sake in mortal flesh, in order through agony and death to redeem the children of men, and establish the throne of his glory in the midst of them; in order, as their brother and friend, and the partner of their flesh and blood, to make them partakers of all his majesty and glory? Is there indeed none to be found among all the countless celestial worlds, all of which are infinitely glorious, more worthy and fitting to be the place of the most glorious revelation of Deity, the centre of the universe, the eternal throne of His immediate presence? And have not all those worlds—each singly—the very same, and even still higher claims to such distinction? Or can it be that He, the Unchangeable and the Just, is so arbitrary and partial as to give to one what he denies to another?

It must be allowed that the disproportion here presented is so great and so overwhelming, that our minds may well be startled and filled with hesitation the moment it first strikes them. But is He, who created all these worlds, and among them this little point, earth, any less great and powerful? True, no human mind is able to reconcile the contrasts of great and small as here presented, or to fill up the wide spaces between them; but is the Infinite Mind bound by the fetters of human reason, and the Divine will measured by human penetration? Does it become us to pronounce what is possible and what impossible with the Almighty? Shall we presume to set bounds to his power, and say: Hitherto—but no further? To question his demeanor and decide what is worthy of Himself? To set bounds to his workings, lest prejudice should arise? Shall we instruct Him to measure his free grace by cubic miles, or his love by the magnitude of fixed stars? Shall we presume to say how many square miles a planet must have, that it may be a fitting place for the incarnation of the Eternal? Shall we forbid Him, that He should, in his wisdom and grace, choose "the foolish things of the world to confound the wise; and the weak things of the world to confound the things which are mighty; and base things of the world, and things which are despised, and things that are not, to bring to naught things that are: that no flesh should glory in his presence?"[1] Has He not power to do what He will with his own? Is our eye evil because his is good?[2]

The revelations of the *microscope* have frequently, and not without reason, been opposed as a corrective[3] to the discoveries of the *telescope*, which have filled

[1] 1 Cor. 1 : 27–29. [2] Matt. 20 : 15.

[3] *Chalmers*, in particular, has pursued this course, in his *Astronomical Discourses*. We quote the following from the third discourse of the series: "It was the telescope that, by piercing the obscurity which lies between us and distant worlds, put infidelity in possession of the argument against which we are contending. But, about the time of its invention, another instrument was formed, which laid open a scene no less wonderful, and rewarded the inquisitive spirit of man with a discovery, which serves to neutralize the whole of this argument. This was the microscope. The one led me to see a system in every star — the other leads me to see a world in every atom. The one taught me that this mighty globe, with the whole burden of its people, and of its countries, is but a grain of sand on the high field of immensity — the other teaches me that every grain of sand may harbor within it the tribes and families of a busy population. The one told me of the insignificance of the world I tread upon — the other redeems it from all its insignificance; for it tells me that in the leaves of every forest, and in the flowers of every garden, and in the waters of every rivulet, there are worlds teeming with life, and numberless as are the glories of the firmament. The one has suggested to me that beyond and above all that is visible to man, there lie fields of creation which sweep immeasurably along, and carry the impress of the Almighty's hand to the remotest scenes of the universe. . . . . Every addition to the powers of the one instrument extends the limit of the visible dominions of the great King. The advancing perfection of the other peoples every point of immeasurable space. The bold assertions made by infidelity on the strength of the revelations of astronomy require no other refutation, as we view the question, than that furnished by the little instrument, the microscope," etc.

so many minds with hesitation and doubt; for in that the microscope has shown us that every atom of the earth, as well as every drop of water, contains a world of wonder and life, such progress at least has been made that we have arrived at another and a better gauge of the greatness, wisdom, might, and majesty of God, than that founded on distances of fixed stars. We have arrived at the clear consciousness, that the earth, however small, puny, and insignificant it may be, compared with the whole universe, still contains a like infinite plenitude of richly-varied worlds, to that which exists in the whole universe, in proportion to its magnitude.

Further, it has been shown, and clearly, too, that in this question of puzzling contrasts, two wholly incommensurable spheres have been confounded and compared with each other, as of equal title to consideration — the sphere of nature and of spirit, of materiality and of personality, of space and of will. But assuredly the greatest deeds and most wondrous revelations of spirit may unfold themselves in the smallest space! and it is in this very fact that spirit evinces its greatest glory, that it makes the smallest spot, and indeed the rather as it is small, the theatre of its most grand and comprehensive revelations.

But still, such considerations advance us but little toward the desired end. *One* astonishment is merely counteracted by *another;* but contrast opposed to contrast does not really bring about a reconciliation and remove all difficulty. No sooner has the mind recovered itself from the second astonishment, than it again recurs to the first, with its "*but still,*" as a new protest. Its earnest desire, and that not without reason, is to see the wondrous revelations of the telescope no less than the microscope, each in their own sphere, harmonized with religious views. Let us see if such an accordance may not be effected, *without* having recourse to the desperate measure of attempting to destroy and do away with one inexplicable problem by means of another no less inexplicable.

§ 15. *Continuation.*

What if the earth alone, of all worlds, *stood in need of* such a testimony on the part of God; if it were alone fallen into sin and misery, so that it alone should have *stood in need of* redemption? Would not the idea that it should alone have been *worthy* of redemption, give way before the idea that it alone *stood in need of* it, the former be lost in the latter?

"*What think ye?*" say the lips of eternal Wisdom, "*What think ye? If a man have an hundred sheep, and one of them be gone astray, doth he not leave the ninety and nine, and goeth into the mountains, and seeketh that which is gone astray? And if so be that he find it, verily I say unto you, he rejoiceth more of that sheep, than of the uinety and nine which went not astray.*"[1] And shall not the sovereign, the everlasting Shepherd, who tends his innumerable golden flock within the pavilion of the heavens, leave there those millions to hasten after the member that may have strayed, be it the smallest, the weakest, and most sorely stricken of the whole flock? Is it not most in need of his tender care, without which it

[1] Matt. 18 : 12, 13.

should utterly perish? Shall He not, in infinite love and never-ending compassion, seek it out, and greatly rejoice over it when it has been brought back in safety? Left is not necessarily forsaken: the rest suffer not from special care bestowed on one; but are securely kept and guarded, and whether they be hundreds or millions, can that make any change in the counsels of eternal love?

Were this earth the *only* province in the immeasurable domain of Deity, and it the smallest and most insignificant, too, in which rebellion had broken out, where unhallowed claims were set up, where all hostile rebellious forces are concentrated, should the eternal King care less for it, than would under similar circumstances an earthly king for the smallest and poorest province of his realm? Would not all his powers be enlisted to put down and extinguish the mutiny, and would not the inhabitants, who through infatuation alone could have permitted themselves to be enticed to revolt, and to become so unhappily caught in a rebellion, be chastised, indeed, but the penitent received into favor again, delivered from their unfortunate delusion, and peace and order be restored? "But what," in the language of an illustrious writer,[1] "if this be applicable to beings of a higher nature. . . . . If, on the one hand God be jealous of his honor, and on the other there be proud and exalted spirits, who scowl defiance at Him and at his monarchy; — then let the material prize of victory be insignificant as it may, it is the victory in itself, which upholds the impulse of the keen and

[1] Chalmers, *Astronomical Discourses*, dis. 6th.

stimulated rivalry. If, by the sagacity of one infernal mind, a single planet has been seduced from its allegiance, and been brought under the ascendency of him, who in the Scriptures is called 'the god of this world;' and if the errand on which the Redeemer came, was to destroy the works of the devil[1]—then let this planet have all the littleness which astronomy has assigned to it—call it what it is, one of the smaller islets which float on the ocean of immensity; it has become the theatre of such a competition, as may have all the desires and all the energies of a divided universe embarked upon it. It involves in it other objects than the single recovery of our species. It decides higher questions—it stands linked with the supremacy of God. . . . . To an infidel ear, all this may carry the sound of something wild and visionary along with it; but though only known through the medium of revelation, after it is known, who can fail to recognize its harmony with the great lineaments of human experience? Who does not recognize in these facts much that goes to explain why our planet has taken so conspicuous a position in the foreground of history?"

The foregoing course of reasoning must be recognized as in itself admissible, and astronomy has nothing to say against it. But, more than that, the results of this science appear very well to agree with it. For the thorough difference between the nature of the fixed stars and that of the bodies of our planetary system, and the fact that amid the former all the varied contrasts and conditions which here below

---

[1] 1 John 3 : 8.

symbolize and attest the dominion of sin and death, man's recreancy to his original destiny, and his need of redemption, seem to be in fact wanting, very much favor the conclusion, as we have previously shown (§ 8), that the worlds on high are the abodes of pure and holy beings, who stand in no need of redemption or restoration.

This apprehension of the matter also agrees very well, at least in *one* respect, with the teachings of Holy Writ; for man is there certainly regarded, if not the only fallen, at least the only personal being *capable* of being redeemed, and hence *needing* the provisions of salvation. But just at this point we begin to see how unsatisfactory and one-sided this apology or reply is. For the Scriptures speak of a duplex fall, a fall in the angelic world as well as that one in our human world. True, the theatre of the former as well as the latter was this earth; but this fact in itself throws no new light on the present subject; for the incarnation upon the *earth* has no saving power with respect to the first inhabitants of the earth, the angels, but only to its second inhabitants, fallen men.

But this reply is discovered to be incomplete and unsatisfactory in another respect. It falls short of the objection; so that its success can at best be called only a partial, and hence a doubtful one. The Biblical doctrine of the Incarnation, beyond doubt, comprehends something more and something higher than a mere restoration of the human race to an *equal* level with those happy beings which kept their first estate. In it we behold, since God remains man forever, the means and pledge of man's exaltation *above* all other creatures; and in the same measure are we confidently to expect the earth, which through the incarnation has been appointed as the ever-enduring throne of the most immediate divine presence, to be exalted *above* all other worlds.

§ 16. *Continuation.*

The unsatisfactory nature of the reply furnished in the foregoing, accounts for the procedure of some minds in abandoning altogether the view on which it is founded, and building up in its place a directly antagonistic, but not less one-sided theory; instead of holding fast to truth already obtained, and seeking to supply its deficiencies. The earth, it is said, has been honored as the place of the one all-glorious revelation of Deity, not through its necessities and its lowliness, but on account of its dignity and worth. Its nature and destiny, it is asserted, were from the beginning higher and more glorious than the same attributes of any of the other worlds, and were not made so through the accident that revolt from God should have begun its daring career just upon it. And further, that the earth is not to be advanced to the glorious state of the rest of the celestial worlds, but that, on the other hand, all the other worlds of the universe are now involved in a process of development, which is to conduct *them* to that state of cosmical perfection which even now belongs to the earth, in spite of the catastrophe of the fall.

We may, in this connection, cite the words of one who has penetrated far into the arcana of nature.[1]

[1] H. Steffen's *christl. Religionsphilos.*, vol. I., pp. 204–206. Breslau, 1839.

"The discoveries of modern astronomy touching the double and nebulous stars," he says, "show clearly that the universe, on the whole, begins to assume a historical character. It is daily becoming more probable *that* THESE BODIES *exhibit actual grades, even to the complete development of our planetary system.* It were a point gained for the Christian religion, no less than for *speculation*, to learn that our planetary system, yea, this very earth itself, is the centre of the whole universe. . . . . . We may here venture the assertion, that modern astronomy is fast approaching the time, *when our planetary system shall be recognized as the most highly organized point in the immensity of the universe;* and that then the time will not be far distant, when, in like manner, our earth shall be recognized, not as the apparent, but as the real central point of the planetary system, spiritually considered, as is man, in the whole organism. . . . The sacred spot where the Lord deigned to appear, is destined to be regarded *as the absolute centre of the whole creation.* And no less are wild flights of fancy, by which souls are transported to distant stars, a Sirius fitted up as the future paradise, while other minds scrupulously hold that each of the celestial worlds has a history of its own, similar to the history of our human world, destined to halt in their upward course, and return contented to the earth."[1]

---

[1] *Hegel* has expressed himself in a similar manner (*Encyclop.* 3d, § 270): "The planetary bodies are from their marked *concreteness* the *most perfect* of the cosmical masses. The sun is customarily regarded as preëminent, since the mind prefers the abstract to the concrete; as *likewise, and for the same reason, the fixed stars*

It must be confessed, we cannot adopt such an apprehension and theory of the starry heavens: we say, in all fairness, that whatever has been gathered through the medium of modern astronomy, concerning the nature of the fixed stars, of however incomplete, equivocal, and unsatisfactory a character it may be, our mind has received from it the irresistible impression that we do not behold in the constitution of the remote regions of the heavens, lower and undeveloped, but higher, more noble, and purer grades of cosmical structure; for a satisfactory view of which, we refer to what is said in § 8. But still, we cannot denominate such an apprehension of the matter an absurd one, as has sometimes been done, and least of all, from a purely astronomical[1] point of view. For the results of this science are so equivo-

are assigned a higher claim to regard than the bodies of the solar system.

[1] From the empirical stand-point it must ever be conceded by astronomy, that *possibly* the same disproportion between expectation or fancy on the one hand, and the naked reality on the other, might be discovered in the case of the fixed stars, were they brought as near us as the moon is by our telescopes — the same disproportion as exists between the fancied and real state of the moon's surface. How have our poets and sublunary *sentimentalists* praised the quiet, tranquil moon, with its mild lustre, picturing it in imagination as the most peaceful and blessed abode in the universe, and longing to escape the noise and din of our terrestrial world in leading the pure, angelic life to be experienced in such a happy region! But how miserably barren and waste, and wholly devoid of all that here upon earth begets, possesses, or cherishes life, does the telescope reveal it to be! What a leap from a paradise for sentimental, imaginative and romantic lovers, to an unblest prison-house for the spirits of the lost, for which the moon has latterly been found much better adapted!

cal, that in one aspect they seem to favor the view that our planetary system is the most perfect of all cosmical structures; so that both the abettors and the opponents of this theory may claim to have the support of astronomy.

From the aids of science it has become probable, that the like of our planetary system is nowhere to be found in all the known universe; and we have already learned that the said system presents thorough contrasts to the rest of the stellar heavens. Here we have a complex organism, exhibiting co-ordination and subordination; the poles of the solar and the planetary, of the lunar and the terrestrial, are here separated; while there co-ordination only, and unity of these opposing principles, bear peaceful sway. But this result may be appropriated by either party and made to bear in either direction. The one beholds as an evidence of the greatest perfection, the fact *that* these poles *are* separated, and appeals to the analogies of the organic world, where the most perfect forms are distinguished by the separation of the opposite poles (the sexual, for example), while the most imperfect of all creatures are without sex, or hermaphrodites; and thus, perhaps, sees in the full manifestation of these contrasts, the most energetic potency of life, the most complete development: while the other beholds in this opposition of principles, only antagonism, conflict, and discord; but in their union, harmony, and fully developed life. The former discovers in the arrangement of subordination and co-ordination, not merely a temporary necessity, but a legitimate and ever-abiding law,—

the latter seeks perfection perhaps in the co-ordination of creatures of the same species, and recognizes in subordination, a lower and merely temporarily necessary arrangement. And though it be true that as we pass outward from our system, which holds pretty much a central position in the great system of the fixed stars, the cosmical structures gradually assume a different character as the distance increases —as though the modification commencing in our system were continued there in a similar ratio:—first, isolated stars, and then double stars, magically united, and as it would appear, indispensable to each other, which form the transition to those more remote multiple stars, and exceedingly rich astral groups—still, we look in vain here for decisive authority to pronounce upon the point at issue. For, on the one hand, that state of isolation may be praised as one of a fulness which is self-satisfying, which possesses in itself all that is to be desired, without the necessity of depending on an adjacent body; on the other hand, it may be deplored as a state of loneliness, wanting in sympathy, harmony, and happiness. Again, we have first of all, density, concentration of light, well-defined outline, and steadfastness of form, which ever become less as distance increases, and give place to transitive forms or maturing structures: still relatively near are stars and systems of stars, which, like the growing embryo in the womb, are still surrounded by the sea of light out of which they have been formed or are now being formed, and at the greatest distance, immeasurable depths of light, in which not even the faintest trace of a complete or growing star is to be discovered. Does not this, perhaps, decisively determine the point before us? By no means; for that solidity and steadfastness of form may, on the one hand, be called a rigid inaptitude to change or improvement, and evidence of a lower stage of development, in opposition to a higher and more vigorous stage, where there is mobility and capacity for improvement, where the bodies are ever taking on fresh forms of life; while on the other hand, the very same appearances may be appealed to as proofs of an ever-increasing perfection from without toward the centre.

If we compare the notion that the earth also in a *cosmical* point of view, although in appearance one of the most paltry and insignificant worlds of the universe, may still be in idea and spiritual significance, the proper and true central spot of the whole universe,—if we compare this notion with Scripture and the views therein contained, the circumstance may be adduced in favor of it, as has already been done by its[1] abettors, that the Scriptures in the pro-

---

[1] Comp. H. Steffens, in his *Anthropologie*, I., p. 264. Breslau, 1822: "It must be maintained that the Ptolemaic system, which accords to the earth a central point in the universe, can never, on this very account, assume a truly religious and Christian significance, because it takes the appearance itself absolutely." Again he says, in his *christl. Religionsphilos.*, I., p. 205: "It is a point of great significance, that the epoch in scientific knowledge which took its rise from astronomy, should have begun in our comprehending the earth in its globular form, and as involved in a common whirl of movements with all the rest of the heavenly bodies, instead of sustaining a position of repose as a centre; for the true centre can *never* (?) outwardly appear as such. As the human consciousness is justified in ceasing to seek the significance of

vince of their *ethico-religious* representations, make the same contrast between appearance and idea the basis of their views; and consider the final reconciliation of this contrast as the ultimate end of all history; so that the incongruity in the cosmical is merely a reflex image of the incongruity in the spiritual sphere, and the one serves to explain and establish the other.

There is, to a certain extent, solid ground in this argument, which should not be overlooked. The theologian cannot, may not, nor does he need to hinder or oppose the astronomer in regarding the earth as a subordinate member of our planetary system, and this system itself as the smallest of all cosmical systems, if his scientific researches force him to this conclusion,—for the astronomer has another rule whereby to measure magnitude and glory than that of the theologian. Man judges according to the outward appearance, but God looketh on the heart;[1] and the divine, and with him the christian, whether he be an astronomer or not, is bound as far as he is able, to see things as *God* sees them, which is to be done in the light of Divine revelation. The astronomer, *as such*, sees things in their outward relations. His object rests with the appearance; it is his province *here* to distinguish between truth and error, illusion and reality, what is imaginary and what is

---

morality in outward works, and in recognizing it only in the sentiments lying beyond all such outward manifestation; so the centre of the universe is to be likewise recognized not merely in its outer but in its inner secret, character."

[1] 1 Sam. 16 : 7.

actual in the appearance. He has a perfect right, from *his* stand-point, to assign the earth a subordinate position in the solar system, and the latter a similar one in respect to the whole universe. The divine is accustomed, rather, to judge the outer by the inner, the visible appearance by the hidden idea; to look for majesty in the form of a servant, and exaltation in lowliness: to him it is said, "But whosoever will be great among you, let him be your minister."[1] And he must, since he is accustomed to the incongruity and the contrast which here below are everywhere presented by the appearance to the idea, from the outset be disposed to give his assent to the heliocentric doctrine, and the results of astronomy in general. These results cannot, nor will they, take him by surprise or inopportunely; but, on the other hand, will but corroborate a truth which is the soul of his whole system of knowledge, and exhibit themselves to his mind wholly in accordance with the analogy of faith.

But still, we can give our assent to the view taken by *Steffens*, only on condition that it undergo not unessential modifications and limitations. And in the first place, we are rigidly opposed to the notion that a true but unapparent centrality now belonging to the earth in respect to the cosmos, is *never* to be manifested and rendered perceptible from outward relations. We cannot but regard the contrast between the appearance and the idea, as having merely a relative, but not an absolute necessity; and hence, as being merely of a transient, but not of a permanent character. From the philosophical, and still

[1] Matt. 20 : 26.

more decidedly from the theological stand-point, we are forced to regard an ultimate reconciliation and removal of these contrasts, an ultimate triumph of the appearance over the idea, as the necessary and final end of all history.

For as in the sphere of morals all christian effort is directed to the task of adequately represesting faith in works, and the sentiments of the heart in language, so also in respect to history, all the prophecies of the Bible touching the future completion of all things, look to the advent of the time when everything that is veiled or hidden shall be brought to light, when the outward condition shall conform to the inward *reality*, the deceptive appearance or shadow give place to the inner substance.

There must dwell uppermost in the idea, a living effort to overcome all that is inadequate, faulty, or contradictory in the manifestation, to assert its own prerogatives, and cast off the shackles which have been imposed upon it from without. For otherwise the idea were lifeless.

But if the idea contain vital energies, and assert its life in the effort to attain an adequate manifestation, this effort must ever be followed by a progressive result, however slow or deeply hidden that result may be; and the final issue must display itself in the complete triumph of the idea. For, otherwise, we should have either a *dualistic Manicheism*, which regards antagonism of principles, not as temporary and accidental only, but eternal and necessary, — or a *pantheistic* world, where to all eternity the existing is supplanted by what is coming into existence;

where the idea, circling in severe and eternal labor, produces nothing but abortions. Where an eternal passing away is opposed by a coming into being, where there is conflict without victory, and sin appears as a good in itself, a mightiest agency of development: where there is commencement without consummation, an eternal blossoming without any fruit, an onward striving before which the goal ever recedes;—truly nature were here a Sisyphean labor, in which the soul of the world is ever anxiously but abortively engaged.

If, now, our solar system, and in it, our earth, be, notwithstanding the oppositions of empirical science, the culminating point of all creation, where in the past the Lord appeared in the form of a servant, that he might come again in the future with great glory, and raise the place of his temporary humiliation to the place of his eternal glory, manifesting upon it the highest and most immediate evidences of the Omnipotent Presence in the sphere of the created,—if this be so, there must be observable more or less distinct traces, not only of a capacity and basis for this highest stage of development, but also of a development already begun and more or less advanced to that stage. If the earth be indeed the most precious germ of the whole creation, the living rudiment of the future blossom and fruit, as in the grain cast into the earth, must already be present in it.

We admit the correctness of *Steffen's* view, so far as it regards earth and man, the former in a cosmical and the latter in an ethico-religious point of view, as having attained their high significance in the history of the universe, not merely by accident, but from having been originally called thereto by a destiny in harmony with their original endowments. But we feel ourselves called upon to expose its falsity and inadequacy, so far as it either denies or altogether ignores the fact that, allowing the high distinction shown to the nature and destiny of the earth and of man *before* the nature and destiny of all other worlds and their inhabitants, the latter must at the same time be acknowledged, from another point of view, as having a decided distinction shown them in a cosmical and ethical respect, the one on the grounds of astronomical, and the other on the grounds of theological investigations.

It may be true, as *Steffens* and *Hegel* seem to have intimated in the passages referred to, that the peculiar, manifold, and complex relations and connections of our solar system, the solid, concrete forms of the bodies belonging to it, and perhaps also, many other peculiarities existing in nature, the dissimilarity of which to corresponding conditions in the celestial or stellar worlds is less conspicuous, are to be taken as evidences that our system is the only one of its kind, and unexampled in its dignity and destiny. But it must be acknowledged, on the other hand, that to all these marks of distinction, if they be regarded as such, there still belong at present, defects, inaptitudes, and incumbrances, to which the worlds of the fixed stars are not subjected. Solidity or firmness of material composition, so highly esteemed, is counterbalanced or detracted from by a hampering incapacity for change of form; and concreteness of structure brings along with it a state of isolated loneliness without sympathy. The varied and rich connections and relations of the system condition those restless and painful actings and reactings of opposites, the despotic sway of the greater over the less and the subordinate, the imperious dominion of crude naked mass over the powers of the will and the mind; the disturbing alternation of light and darkness, of heat and cold, of summer and winter, of blooming and fading, of coming into being and ceasing to be. We cannot, therefore, ascribe to the earth, and to the solar system in general, an *absolutely* higher position, in the present stage of their development. We must allow the fact, that those glorious worlds on high still have very many, diverse, and special marks of high distinction; and grant that absolute sovereignty in this respect is to be looked for, only when in the progressive development and final perfection of our solar system, there shall have been added, as it were to its present prerogatives, in full realization, the high claims of the celestial worlds.

It has already been seen, in the fourth chapter (comp. particularly § 36), that with respect to the inhabitants they contain, the same contrast, with a corresponding pre-eminence or inferior condition, obtains between the earth and the celestial worlds.

But, apart from these necessary and not unessential restrictions, the view of *Steffens* cannot vanquish and remove all doubts and difficulties, which from a cosmico-astronomical point of view, may present themselves against the occurrence of the Incarnation upon the earth. The whole scope of the question before us is by no means satisfied in showing how the *Earth*, in respect to its cosmical and ethical position, should have had, in preference to all other worlds, the nearest and the most decisive claims to such a pre-eminence, if it were to be possessed by any world in general. But, on the contrary, the real difficulty is to show *that* such a high honor was destined for the earth ALONE, and *how* this should have been; to show *that* the rest of the worlds were either not capable of, or did not stand in need of an analogous incarnation of Deity for themselves, and *why* this should have been so. We are called upon to clear up the question, whether the Incarnation upon the earth stands in *any relation* to the life and history of spiritual beings upon other worlds, and *what* there is of a necessary, essential, or decisive nature, in such relation—a question satisfactorily answered only when it is shown that in the high distinction conferred upon the earth, the rest of the worlds have in no manner been slighted, overlooked, or left behind.

§ 17. *Continuation.*

In order fully to acknowledge the claims of the point last indicated, and give it that consideration it deserves in the construction of a Biblical theory of the world, it has been strenuously and vigorously attempted even of late, to incorporate into the Christian theory of the present time, an incarnation of Deity upon *all* worlds, corresponding to the incarnation upon the earth, as something in accordance with the Bible, and as an axiom demanded alike by the results of astronomy and the admitted principles of a Christian-theistic speculation.[1]

There lies at the foundation of this idea of an incarnation of Deity upon all worlds, another notion, which, as early as the middle ages, found many defenders; and which, notwithstanding its complete overthrow by the Reformers and old Protestant divines, has again taken deep root in the theology of more modern times.[2] It is this: there was an absolute necessity supposed in the very creation itself, and not conditioned through the entrance of sin, that God should become incarnate, in order that humanity might thereby be enabled to reach the high end for which it was predestinated. God had assumed human nature, it is argued, though man had never sinned; but not in a state of humiliation, to suffer and die for humanity; rather, at once in a state of majesty and glory, in order that through the union of the Divine and human natures in the God-man, he might fill up the impassable chasm between God and man, exalt the creature of God's power to be the child and heir of God, and the co-heir with

---

[1] Comp. the article by Dr. Chr. H. Weisse: *Christus das Ebenbild des unsichtbaren Gottes*, Eine Frage an die christl. Thelogie unserer Zeit, in the *theol. Studien und Kritiken*, 1844, IV. p. 913–966.

[2] For example, it is advocated by Liebner, *die christliche Dogmatic aus christologischem Princip.:* Göttingen, 1849; Dorner, *die Lehre von der Person Christi*, p. 527 seq: Stuttgard, 1839; Martensen, *christl. Dogmatik*, p. 194: Kiel, 1850; J. P. Lange, *positive Dogmatik*, p. 212 seqq., &c. On the other hand, it is combated by Thomasius in *der Zeitschrift für Protestantism. und Kirche*, 1850 (Januarheft), and by Jul. Müller in *der deutchen Zeitschrift für christl. Wissenschaft*, 1850, No. 40–43.

Christ,[1] and make man a partaker of the Divine nature,[2] so that he might be like God.[3]

If this view be a legitimate one, it is not difficult, indeed we are impelled by a necessity which we can hardly escape, to extend it from men to angels also, and from the earth to all other habitations of created beings. But a closer examination of it will show clearly, that it is devoid of speculative necessity no less than of support from the Bible.

The highest and ultimate end of all histories and developments in connection with created life, is—according to the demands of speculation and the teachings of revelation—"*that God may be all in all;*"[4] that all creatures may, without foregoing their freedom, individuality, and independence, return again and merge themselves in the eternal source of all life, from whence they originally sprung; that *the dualism* which was constituted in the creation of free, personal beings, and which manifests itself in the independent existence of a free will, besides the free will of the Divine Being, may finally result in a permanent and undisturbed unity, without the removal or the endangering of the subsisting *duality*,—that in this consummation the movements of all created existences may be brought to rest, the longings and hopes, the aims and efforts of the rational creature, be fully met, and satisfied in the fullest possession and most complete enjoyment; that not only the real existence and manifestation of a sad opposition between the Divine freedom and the freedom of the

---

[1] Rom. 8 : 17. [2] 2 Pet. 1 : 4. [3] 1 John 3 : 2.

[4] ἵνα ᾖ ὁ θεὸς τὰ πάντα ἐν πᾶσιν, 1 Cor. 15 : 28.

creature, should be overcome, but also the abstract possibility of a revolt or lapse to such a state of opposition, should be for ever excluded.

We are bound to concede that if there be no other means whereby the last and highest destiny of all created beings can be reached, than an incarnation in all worlds where spiritual beings manifest their freedom, then is the reception of this axiom unavoidably necessary, from the clear demands of the Christian-theistic faith.

But such a supposition is an erroneous one, and hence its consequence cannot claim consideration.

We grant that all creatures are called, at the end of their development, to return to the eternal source of life from whence they originally sprung; so "*that God may be all in all.*" It is clear that we cannot (as Pantheists) regard this return as a passing away, a ceasing or an annihilation of the individuality. The *individuality* which was constituted at the creation, remains as such, even after the return of the creature to the Deity; and not till then, indeed, does it manifest its highest advancement and completeness. This return can be conceived of, only in the following manner: God placed the created individual in existence, but without Himself, by a creative act of his will. But that individual needed development, and was endowed with powers of development. There lay something more and something higher in the creative idea, than was effected at the time, by the act of creation; and the latter deposited this merely in the capacity of a potency, a tendency, and a capability. Were the created individual a free, personal, spiritual being, then must it cause the potency to be unfolded, and its destiny to be reached, by means of its own freedom. But, on the other hand, if an existence belonging purely to the sphere of nature, and not endowed with freedom, then must it attain its development through the impulse of the natural tendency (instinct) implanted in it. Here, however, the influence of the free being placed over it and for its assistance, might be either advantageous or prejudicial to its development. In the creation itself there was constituted a duality of the Creator and the creature, which was liable, through the misuse of freedom on the part of the creature, to degenerate into a complete and inwardly antagonistic dualism.

But had the creature, endowed with freedom as well as not so endowed, unfolded itself wholly in accordance with the will of God; then would both dualism have been for ever prevented, and the duality for ever preserved; then would the creature, which at the creation was placed without the Deity, have returned to Him in its own development, and thus the Divine creative idea have been realized. The commencement and the end, the potency and its evolution, the design and its fulfillment, thus unite in a harmonious and well-rounded (einheitlichen) whole.

The duality, then, constituted through the creation, is an abiding and never-ending one. Whether, therefore, it degenerate into dualism, the creature opposing itself antagonistically to the Creator; or whether the creature return to the Creator, under the sway of a third and a higher law (the complete realization in itself of the idea of the creation); in neither case is there a recurrence of the primitive absolute unity, but the duality remains in both cases; reconciled and united in the one, separated and antagonistically opposed in the other.

Pantheism, on the contrary, holds that the creature (whether it take the one or the other direction) is, at the end of its development, absorbed into the Deity—that it ceases to be a creature any longer, and again becomes God.

Theism cannot, of course, allow that the creature becomes God in *this* sense, nor in *any* sense, even though the individuality be not destroyed. Theism cannot grant that *man* at the end of his normal development, shall really become *God*, nor yet *God-man*, but merely *divine man*. For the creature can return to God only in so far, and only in the manner and the measure it has proceeded from Him. If it be *purely* a creature, the product of his *will* without the impartation of his *nature;* without personality, without freedom, without spiritual essence; it can merely return or be conducted back to his *will*, *i. e.*, be sustained and unfolded in accordance with the Divine will; so that at the end of the development the unfolded creature shall fully correspond to the idea and aim of the Divine creative will. But if the creature be a free, personal being, belonging not only to the sphere of nature, but *also* to that of spirit; if it be the offspring of God,[1] the image of God,[2] or spirit from Spirit, and hence have proceeded from the Divine will, with the *impartation* of the Divine na-

[1] Acts 17 : 28.

[2] Gen. 1 : 27.

ture—then it can and must (in respect to its spiritual aspect) return back to the Divine Being. Nor does it require for this end, in case its development be a normal one, any extraordinary assistance from Heaven. The physical chasm between the *Divine Being* and the *creature*, is, indeed, in itself, an infinite one; and God *alone* can fill it up. But this he *did* in the creation itself, as he then imbued the creature with his own being, and made it to partake of his own nature. At least, he breathed into the nostrils of *man*, His living breath of life, and made him after his own image and likeness, as the offspring of God. And something of the same kind must have occurred in the creation of the angels also; for they, too, are free, personal, spiritual beings.

The powers bestowed upon the creature through its creation, or at least, in its creation, were straightway sufficient, in case they were properly used, to conduct such being, each one after its own manner, to the predetermined goal.

True, the case were different, if these powers were misused: if instead of a normal and godly development, an abnormal and ungodly course were taken; if the creature, which through the creation was made *in* God and *for* God, should rend itself away from Him, and, taking its position without God, antagonistically oppose itself to Him. A moral chasm would then arise, which would at once become a physical one also; since the bonds which bound the divine being in man with its eternal source, should thus be torn asunder. Such a chasm were, both in its physical and moral aspects, an infinite one, which the creature of itself could never fill up or pass over. If, however, this chasm were to be filled up, and the fallen and rebellious creature led back to God, conducted to its original destination; this could not possibly be effected otherwise than by intervention on the part of God. The fallen creature has no power, in itself, to raise itself again to God; hence it becomes necessary that God must condescend to its low estate, that he may recover it from destruction, that he may renew and complete it, by raising it from the depths of sin and misery to a place before his throne.

The ground of the Incarnation is to be found *here* and *nowhere else;* in the sin of man, or rather, in the decree of Divine grace to conduct man, despite his sin and fall, to the goal for which he was destined at the creation.

Christian speculation is doubtless under the guidance of a religious motive, in ascribing to the incarnation an absolute necessity, arising from the creation itself; but this motive is founded upon an erroneous supposition. It rests entirely upon the supposition that man, through the incarnation, is to reach a higher position, and attain to a glory incomparably greater than he could have obtained without redemption, and consequently, without the entrance of sin also. It must be confessed, that the exalted terms in which the Holy Scriptures attempt to express the future great glory and blessedness of the redeemed of the earth, may easily, but none the less erroneously on that account, be held up in justification of this supposition.

It is, to our mind, altogether inconceivable from the Christian-theistic stand-point, that man, had he not sinned, and had he been true to his destiny, could not have attained any thing like the degree of advancement, any thing like the degree of glory and blessedness now made possible to him through his sin, wickedness, and rebellion against God. Why, thus, we should have cause to rejoice that we had become sinners and rebels against God; and sin should be made, by the Divine decree itself, the indispensable means of carrying out that decree;—sin itself should be the first and greatest of all blessings!

An *Augustine*, indeed, dared to utter the bold language: *O felix culpa, quæ talem meruit habere redemptorem!* and expressions similarly bold are still to be found in sacred songs of the present day. Nor would we absolutely condemn such expressions, coming from the depths of an humble and pious soul, any thing but disposed to treat sin playfully. There is a time for all things, and therefore for every thing also, an improper time; thus with paradoxes. If the Apostle were competent to call the wisdom of God folly, while the latter is indeed the fountain of all wisdom and knowledge; perhaps such an one as Augustine might be justified in calling sin, though indeed the source of all misery, the ground of the saint's blessedness. There are at times profound and genuine stirrings of the religious emotions, in which the common expressions of every-day life appear too cold, too inexpressive, and too meagre, to exhaust in an adequate degree the depths of feeling within the soul. The mind then lays hold of paradoxes, in order to bring more vividly into view the helpless inadequacy of all common forms of speech, on such sacred and privileged occasions.

This expression of Augustine's is a paradox, which, like every paradox, is a one-sided truth carried in the warmth of the feelings too far; which designedly ignores all other aspects of the truth, in order to direct the whole attention to this one; which is so wholly absorbed and affected by the one view, that it can neither think nor speak of any thing besides it.

We may, in certain moods of religious feeling, be so overcome with what we as sinners may attain to through redemption, or with what we should have come short of without it, that for the moment all other interests seem to vanish. The unspeakable blessedness derived from the grace of God, upon the occasion of our transgression, may so absorb all sense and reason, that for the time we should forget entirely from whence we have fallen by sin; what we have lost thereby, and what we might have attained to, had sin never entered the human family. But, should we attempt to raise what is only relatively true, to the level of a scientific principle, and continue what is natural and allowable only in certain frames of religious feeling and emotion, into our ordinary processes of reflection, and make it the grave judgment of the understanding,—then what had heretofore been half truth, would become wholly error; then would that which arose from the inmost soul as a high hymn of praise to the grace of God, become a slander against the Divine holiness. Were we in calm reason to say: God be praised that Adam fell into sin,—this would be the import of our words: God be praised that we are sinners, that we have sinned—which were simply, blasphemy.

There are but two ways of avoiding such a wanton impeachment of God's character. We must either give up the view, that a higher and more glorious state is to be arrived at through the medium of redemption, than could have been gained in a sinless and normal development; and acknowledge that Redemption was the only object of the Incarnation, so that the decree of the incarnation stands or falls with the decree of redemption: or we may retain that view, and then imagine that the incarnation was conditioned by the creation itself, as the necessary complement of the latter; so that not the incarnation, in itself, but merely its actual earthly character,—connection with the low estate, the misery and the condemnation of fallen human nature—was conditioned by the occurrence of sin.

It is for the Scriptures to decide between these two modes of apprehension, and it requires but little examination to see that they pronounce in favor of the first one.

It is clear upon the face of Holy Writ, that in all cases, as often and repeatedly as the subject of the incarnation is treated of, sin alone is represented as the cause, and redemption as the object of this miracle of Divine love: this must be conceded even by the abettors of the opinion we would combat. But they maintain that the Scriptures, being concerned everywhere with the concrete reality of the sinful state of man, could have had no occasion nor any motive for telling us what would have happened, had sin never entered the world; and moreover, that Christian speculation, since it feels the need of having its horizon extended in this direction, is justified in the attempt, and capable of completing the Biblical theory of the world in respect to this point, from the Christian consciousness,as begotten and fostered by Divine revelation.

But still, we cannot help viewing the matter in a different light. The question whether God should have become man, had man never been guilty of sin, is by no means one possessed of significance for speculation alone, and not affecting a practical acquaintance with the facts of salvation. If we be compelled to answer this question in the affirmative, the answer will so significantly affect the doctrine of salvation, give it so wholly different a substructure, and force it to assume from its foundation to its very summit so essentially different a coloring, that the Scriptures, notwithstanding their signally practical tendency, could not well have been wholly silent here. It will not do to reply, therefore, that they say nothing in regard to this point, because they should have had no motive for so doing. They are silent not because they do not consider the matter of sufficient importance to mention, but because they know nothing of an incarnation apart from sin; because it was not conceived why such a doctrine could be broached, since it is of itself so apparent that the incarnation must be conditioned through the existence of human guilt alone.

And our opponents appeal, moreover, to the necessary judgment of the Christian consciousness, that it is altogether inconceivable that a higher and more glorious portion awaits man, through the entrance of sin, than could have been acquired by means of a normal sinless course of development. Were the supposition here involved a correct one, its consequences also, as we have before observed, would have to be admitted. But it requires but a few words to show that it is incorrect and without foundation.

However strong the language is, and superabounding the terms, in which the New Testament describes the glory and blessedness of the redeemed in heaven; still there is nothing there said which cannot, yea, which must not be conceived as having been involved in the destinies and capacities assigned to the human race in the creation itself. The glory of man's primeval state and the glory of his future state, are related to each other as the germ to its development, as is destiny to its realization. There is nothing absolutely new to be discovered in the glory of the redeemed, nothing that we are not to suppose was already existing as a germ, capability, or a beginning, in the *image of God*, in which man was created. It was in this image that our right to be children and heirs of God was involved,[1] in it man was already made a partaker of the Divine nature,[2] and in it was already instituted man's likeness to God.[3]

Sin and Redemption are correlatives. The severer and more dangerous the disease, the more vigorous and powerful must be the remedy. The significance

---

[1] Rom. 8 : 17. [2] 2 Pet. 1 : 4. [3] 1 John 3 : 2.

of redemption is exalted, just in the degree that we swell the enormity of sin; and conversely, the greater the measures taken by God for our redemption from sin, the greater also must be the depth and extent of that destruction into which we have been cast by sin. The Christian consciousness strenuously demands that both be placed at the highest possible mark; it beholds in the one an infinite, an immeasurable, and an irremediable destruction; in the other an infinite, an immeasurable, and an adorable salvation. Our opponents, however, do violence and injury to the Christian consciousness, just in that point where it is most delicately sensitive, and can least suffer its intuitions to be called into question. For in their refusal to acknowledge that the incarnation, *as such*, was conditioned by sin; and their affirmation that merely a particular form of its manifestation—its lowly and humiliating aspect—was thus conditioned, they detract from the enormity and weight of sin, and from the value of redemption. That God should assume human nature, is the one great and infinite act of condescension and self-renunciation on the part of Deity; but this one only adorable miracle of eternal love is not, forsooth, to stand in any way connected with sin! The incomparably lesser act of self-denial only; that *the* man in whom God should otherwise have become incarnate, assumes the sorrows of humanity and suffers mortal death, this alone is to be laid to the account of sin! How much of the significance of redemption is thus lost, and how much less forbidding is the aspect of sin! But what is still far worse, not at all comporting with the

Christian consciousness, and wholly uncompatible with the teachings of Sacred Writ; redemption ceases to be the free gift of Divine mercy, and resolves itself into a necessity arising out of the creation itself. For if the creation demand for its completion and for the consummation of the destinies of humanity, an incarnation of Deity, this *must* come to pass, no more if man take an abnormal, than if he take a normal course of development. Thus sin is less to be held responsible for the sad condition of man, since the powers assigned to him in the creation still need reinforcement through a future incarnation of God; and it loses in abhorrent significance in opposition to the great plan of God, since the incarnation should have taken place without it. True, the *Incarnation* would thus still remain an adorable miracle of Divine love, a decree of free grace sufficient to exhaust all praise—but *not so* with *Redemption.* The *latter* were conditioned through the decree of the incarnation, but not through overflowing Divine compassion in view of sad and miserable estate of fallen humanity.

It may be considered as established, therefore, that the incarnation upon the earth was conditioned alone through the free grace of God,in view of overcoming and eradicating sin and its consequences; and that humanity would never have required the incarnation of Deity, to reach that high position which now indeed can be reached only by means of the incarnation, had sin, with its disturbing and destructive influences, never entered the race.

Having become possessed of this result, we return again to the question with which we set out: "Is the assumption of an Incarnation of Deity in the other worlds inhabited by reasonable beings, necessary or admissible even ?"

Such an assumption is not admissible; for there is no place for it in the Biblical theory of the world, and it is not demanded by the Christian consciousness. At least, the worlds whose inhabitants have never fallen, stand in no need of such an extraordinary aid; since in the creation itself there was given to all creatures the means and capabilities requisite, that each one might in *its own* manner reach that great and common end, comprehending all things: "*that God may be all in all.*" The question appears in a different light, however, in connection with those worlds where spiritual beings have experienced a fall similar in some respects to that of man upon the earth. It is not to be simply and immediately repelled in such cases, nor is it any more to be affirmatively answered, without due consideration. For it must first be discovered whether these beings, like man, are *capable of salvation.*

Human science is wholly unable to discover any traces of the presence of reasonable beings upon other worlds, to say nothing of the moral condition of such beings. Hence it belongs altogether to Scripture to answer our inquiry. Only two kinds of spiritual beings are known to the Bible and spoken of by it: Angels and Men. It does, indeed, acquaint us with the fact that a part of the angels at least, fell from their allegiance to God; but we are at the same time expressly told that they are incapable of salvation (Comp. chap. 4, § 21). Hence, we must sum up the following as the result of this discussion; that an incarnation of God can have occurred upon the earth only, and nowhere else; and that the inhabitants of the other worlds either do not *require* a redemption, and with it an incarnation as the procuring cause, since they *have not* been the subjects of a fall, or that they are *incapable* of redemption if they be fallen beings.

§ 18. *Continuation.*

The design of the incarnation was to conduct fallen man back to communion with God, and to advance him to that goal for which by virtue of his being made in the image of God he was destined and rendered capable. The ultimate end of redemption[1] is

[1] ["The assumption of the human by the divine nature, to say nothing of its primary consequences, supersedes a multitude of questions and speculations that might have been entertained relative to the station which man may natively be fitted to occupy. And it should not escape notice that human salvation is, with great uniformity of terms, spoken of by the inspired writers, as a restoration, a recovery; it is the bringing him back to the dignity he had lost. No expressions are employed which might seem to indicate that an alteration, or extension of the original plan of the human system had been admitted; or as if an arbitrary derangement of the ranks and orders of the intelligent system had been made, in consequence of which the family of Adam are to be promoted over the heads of others, to a place higher than their qualities should fairly warrant.

Philosophical theories of human nature are in fault, on the side both of presumption and of frigid diffidence. For too much is assumed in behalf of man in what belongs to his actual condition, and his unassisted powers; and far too little in what relates to his original destination, to the importance of his present behavior, and to his future lot. But the Scriptures, in their history of man,

no other and no higher than that of the creation; but *redemption demanded an incomparably higher species of Divine manifestation*, an infinitely greater self-abnegation on the part of God, than the creation. For the creation had to do with a mere bringing about; it originated a pure beginning, a capability, which through its own development was competent to reach its final goal. This beginning was, through the power of sin, rent away from the source of its life; the capability was destroyed, the further development was rendered impossible, and the personality sunk into a depth of destruction from which no created power could retrieve it. The mission of redemption was therefore a much greater and more comprehensive one; it demanded not merely the institution of a new, but also a negation of the old; not merely a restitution to the lost, but also an evolution to the yet unattained.

The question as to how the incarnation of Christ upon the earth is related to the spiritual inhabitants of the other worlds, hence coincides with the question as to how the creation of man is related to the same beings. The incarnation of God no more

---

set out from a point more elevated, follow him through a course that descends to the lowest depths; and again present him as emerging, and as setting out on an upward path that leads to an immeasurable height. . . . The style of the Bible, in this point, prepares us to receive whatever it may have to affirm concerning human destinies; and leave is given at once to entertain the greatest conceptions, when, in the first page of the sacred canon, it is said, and said with emphasis, that *God created man in his own image; in the image of God created he man.*"—ISAAC TAYLOR, *Saturday Evening*, pp. 316, 317, 330. — TR.]

involves the depreciating, prejudicing or neglecting of the claims of the other spiritual inhabitants of the universe, than did the creation of man in the image of God. The circumstance that man, already in the creation, was destined for a higher goal than were they, and that despite the sin of Adam this goal was to be reached, by virtue of a new and highest miracle of Divine grace, could in nowise be of disadvantage to them.

But truly an inestimable advantage might it be. A schism had been introduced into the world of the other spiritual creatures, through the fall and revolt of a part of the angels. The harmony of the universe had been destroyed. In order to restore it, man was created, yet falling himself also, was redeemed, since he was capable of redemption.

The incarnation hence results to the advantage of the whole universe. If the view, of old entertained, which regards man as the microcosm, *i. e.* as the representation of all creatures, as that product of the creative hand in which all substances and potencies, all powers and capacities of the body and the soul, of nature and the will, which are scattered at large and singly throughout the universe, are to be found in concentrated form,—if this view be correct, then may it be also true and conceivable, that God, in taking immediately upon himself human nature, thereby mediately took upon himself also the nature of all other creatures.

Speculation, empirical science, and revelation (chap. 4, § 9), all agree so clearly and decidedly in the view, that man is to be regarded as the microcosm of the *terrestrial* world, that we can dispense with all further proof in regard to this point. All terrestrial forces and substances, all potencies of animal and vegetable life, are present in man in concentrated and sublimated form. An incarnation of God results, therefore, to the advantage also of all the rest of *earthly* creatures.

The next inquiry will be, whether man also may be regarded as the microcosm of the universe, as the representative of all other and extra-mundane creatures.

Empirical science and experience, as the matter at present stands, can not be acknowledged as competent arbiters in this question. It will be readily conceded, that empirical science proves nothing of such a position in man; but no less surely must it be conceded, that there are very many things both in heaven and upon earth, of which our present empirical science neither knows nor is capable of knowing anything.

The ignorance of science, however, in regard to this point, is for *these* reasons not decisive: in the first place, because the potency of the beginning (*i. e.* the powers lent in the creation) has not yet been unfolded; but rather, disturbed in its normal evolution, as a result of the fall, has been perverted to abnormal revolution;—and in the second place, because the restoration by means of redemption, is not completed, has not yet advanced to *that* point where all revolution is overcome, and the neglected evolution adequately represented.

If in general anything decisive is to be gathered

in regard to this question, it is clear that it can be from revelation alone. And there are *three* points which may here come into consideration: the original destiny of man, assigned to him in the *beginning* through the creation, but disturbed and interrupted through the power of sin;—then the potency and fullness of the restoration, in the *mean time*, representing itself in the triumph and exaltation of the God-man;—and finally, the fulness of the *end*, which shall at length have imparted itself from the exalted Son of Man to all his people, *i. e.* those who have been born of Him, and regenerated to a new life and a new development.

Let us regard these three points of christian revelation somewhat more closely, in order to see *whether* they offer us any thing in answer to the present question, and if so, *what* it may be.

As to what concerns the *creation*, it is clear from the Bible itself, that the earth is to be regarded as the world last created, and also man as the last of all personal creatures. When man, the crown and seal of all earthly creatures, had been created, then had God finished all the works of creation; then began the rest of God, which marked the absolute cessation of all pure creative activity. The earth and man, through this their position in the scale of creation, acquired an unwonted and culminating significance in the universe: here was the goal and end of all Divine creative activity, the close and consummation of the whole idea of creation.

Still more clearly does this culminating and closing destiny and position of man in relation to the whole universe, stand forth to view, when we recognize as correct, that view, which in the fourth chapter of this work, we have sought to show as legitimately resulting from the accounts of revelation contemplated as a whole. The view is this: that the earth, being transformed into a dreary chaos through the fall of the angels, was renewed in the six days' work, and assigned to man as his dwelling-place, in order that he might do away with the disharmony in the universe, and restore all to peace and order.

If we now proceed to the consideration of the Biblical doctrine of the *God-man*, we shall here discover more clearly the admissibility of such a view.

Christ, the God-man, in whom human nature represented itself in its absolute ideality, was, according to the abundant declarations of the New Testament, exalted, after the completion of His work upon earth, above all creatures in heaven and upon earth, so that He sustains, preserves, and fills all things. This *exaltation*, however, refers not to His divine, but to His human nature: indeed, strictly taken, exclusively to the latter, since his Godhead from its very nature already possessed a like exalted position. "He took upon him," says the Apostle, Phil. 2 : 7–11, "He took upon him the form of a servant, and was made in the likeness of men, and being found in fashion as a man, he humbled himself, and became obedient unto death, even the death of the Cross. *Wherefore* God also hath highly *exalted* him, and given him a name which is above every name; that at the name of *Jesus* every knee should bow, of things in heaven, and things in earth, and things under the earth; and that every tongue should confess that Jesus Christ is Lord, to the glory of God the Father." The same Apostle speaks still more clearly in Eph. 1, 20–23: "God raised him (the *man Jesus*) from the dead, and set him at his own right hand, in heavenly places; far above all principality, and power, and might, and dominion, and every name that is named, not only in this world, but also in that which is to come: and hath put all things under his feet, and gave him to be head over all things to the Church which is his body, *the fulness of him that filleth all in all.*" And in verse 10 he says that the purpose of God consists in this: "that he may gather together in one all things in Christ (the *God-man*), both which are in heaven, and which are on earth, even in him; in whom also we have obtained an inheritance."

Here, now, the view that man, according to his original destiny, subsequently disturbed by sin, but happily to be restored through redemption, is to be regarded as the microcosm, the potential representative of the macrocosm, receives its express Biblical confirmation. For, in the first place, the *man Jesus* here evidently appears as the microcosm. But what holds good of the man Jesus, holds good also of all men redeemed by him, and conducted to their proper destiny. For the essence of redemption, in its positive aspect, consists in this: that Christ, as the Son of man, as the representative and prototype of humanity, as the second Adam, represented the idea of humanity in its full completeness: primarily in his own person, in order then as *head* of the organism, a *member* of which he became through his incarnation, to draw us after himself, and conduct us to a like perfection (comp. chap. 4, § 26); since we have entered into communion with his victorious life, in like manner as he became united with our helpless and sinful life. Besides, the Church, which is his body, and of which he is the head, is here expressly designated as the "*fullness of him who filleth all in all.*" He, the head, filleth all in all, and the Church, his body, is *his* fulness, with which, and through which, he filleth all in all.

No less clearly and distinctly is this view favored by the Biblical doctrine of the *end of the world.* The end of the development of our terrestrial world, is, according to Holy Writ, the end of *all* world-development: the judgment of man coincides with the judgment of all creatures, and the destruction, purification and renovation of the earth is also connected with the renovation of the heavens. Now, the Scriptures contain no intimation, nor do they anywhere in the least imply, that the entrance of this common end of the world, is in any measure conditioned by extra-mundane developments, unconnected with the earth: nay rather, they make it wholly dependent upon terrestrial developments; and the consummation of the celestial worlds and the inhabitants of heaven is delayed, merely because one cannot be made perfect without the other; because the consummation consists precisely in this, that all things be gathered together in one, and God be all in all,[1] Comp. chap. 4, § 24.

---

[1] Heb. 11 : 40; Eph. 1 : 10; 1 Cor. 15 : 28.

§ 19. *The Catastrophe of the End of the World.*

We have indicated above (in § 1), as the three chief points in the Biblical theory of the world, which are menaced with abandonment, as irreconcilable with the results of astronomy, the following: the Biblical doctrines of the creation of the world, of the redemption of the world, and of the judgment of the world. The two first mentioned points we have already sufficiently taken into consideration; and have clearly shown that they may be fully harmonized with the results of astronomy. It now remains to dispose of the third point in a like satisfactory manner.

According to the teachings of sacred Writ, the whole fabric of the world (not merely the earth, but also the heavens at the same time) awaits a catastrophe by which it is to be changed and renewed (as an old garment is cast off and supplied by a new). chap. 4, § 34, 35.

We have already seen, in chap. 5, § 5, that astronomy, as far as it, supported by the experience and observation of thousands of years, and sustained by the most delicate calculations, is in a condition to pronounce upon the stability of the present cosmical order and arrangements, must give this as its deliberate judgment: that our solar system at least, and in all probability the heavens of the fixed stars, bears the character of the most undisturbed and immovable harmony, order, and stable adjustment; since there is no power or accident whatever within the knowledge of this science, by which the existing order might be destroyed, altered, or endangered. Nay rather, it clearly shows that all apparent disturbances which celestial bodies exert upon each other, are so wisely and nicely designed and adjusted in the complex web of celestial movements, that they, instead of being tokens of a probable or possible destruction of this or any other system, much rather appear as the presages and pledges of the undisturbed continuance of the existing order.

It would be maintained that the Biblical doctrine of a future renovation of the world, in connection with a destruction of the same, must give way before the astronomical doctrine of the unshaken stability of the present cosmical arrangements.

The best reply to any such assumption is given by the Scriptures themselves, and in the very place where they teach most clearly and fully of a future destruction of the world: in 2 Pet. 3 : 4 seqq. It is there answered to those who say: "Where is the promise of his coming? for since the fathers fell asleep, all things continue as they were from the beginning of the creation:"—"This they are willingly ignorant of, that by the word of God the heavens were of old, and the earth standing out of the water and in the water; whereby the world that then was, being overflowed with water, perished: but the heavens and the earth, which are now, *by the same word are kept in store*, reserved unto fire," etc.

Here reference is made to the analogy of a historical fact, which may be regarded as in a certain sense a type or exponent of that general and tremendous catastrophe of the world—to the deluge. In the relations between land and sea, between the production and consumption of water, there existed of old, notwithstanding any partial disturbances, such a fixed, well-ordered, and constant proportion, that no antediluvian philosopher, even of the most learned and highly advanced order, could have suspected or foreseen any tokens of the possibility or probability of such a universal and mighty catastrophe, involving and transforming the whole surface of the earth; and yet the flood broke forth when it was least expected, and sources of destruction were opened in the fountains of the great deep and from the windows of heaven, in a manner surprising and appalling to all minds. "In the six hundredth year of Noah's life, in the second month, in the seventeenth day of the month, the same day were all the fountains of the great deep broken up, and the windows of heaven were opened, and the rain was upon the earth forty days and forty nights."[1]

"And as the days of Noah were, so shall also the coming of the Son of Man be."[2] As formerly, from the profound depths of the earth, never pentrated by the inquiring eye of man, and from the regions aloft, where the clouds are formed according to a law which no human investigation has yet discovered, there suddenly broke out floods of destruction, which in a moment silenced all sceptics and deriders with their appalling terrors,—so also there may lie hidden in the heights and depths of the universe, latent forces, which in future may leap forth at the call of the Mighty Creator and

[1] Gen. 7 : 11, 12. [2] Matt. 24 : 37.

44

Judge of the world, with an energy and universality capable of bringing about at once a transformation and renovation of the heavens and the earth.

But what is to be the outward manifestation and nature of this final catastrophe, as foretold by prophecy? The Scriptures say: "The heavens shall pass away with a great noise, and the elements shall melt with fervent *heat;* the earth also and the works that are therein shall be *burned up.* Nevertheless, we, according to his promise, look for new heavens and a new earth, wherein dwelleth righteousness."[1]

Of all the elements known to us, none is so mighty, so pervading and energetic, as fire. Of all destructive elements fire is the most destructive; but since it destroys that which is perishable, and separates dross and impurities from the genuine and the pure, it also frees the imperishable and the noble from the bonds of the perishable and the ignoble, and places the former in all its purity, its excellence and glory, in its true position. Hence fire has ever been recognized not only as the symbol of ruin and destruction; but also, with equal propriety, as the type of the most energetic and thorough purification and renovation.

If, therefore, the final catastrophe of the world is to be, not merely a ruinous and destructive, but at the same time, and signally, a purifying and renovating process, it is clear that of all the means known to us, none better adapted to secure the end in view can be imagined, than fire.

But as fire is the most energetic and mighty of all

[1] 2 Pet. 3 : 10–13.

the elements, so also is it the most universally disseminated: it lies hidden in all bodies, and may be called forth at any moment, by mechanical and dynamic means. An inextinguishable furnace of fire glows within the hidden depths of the earth; fiery bolts leap forth from the clouds of the heavens; fire is begotten by the sun; and those as it were spiritual agencies of electricity, which in all probability flit through the regions of the created everywhere, seeking ever and in vain their equilibrium, involve a signal fulness and intensity of fire-development.

Though it be further foretold by prophecy, that fearful signs in heaven and upon earth shall precede or accompany the final catastrophe; that the sun and moon shall lose their light, the stars fall from the firmament, and the sign of the coming of the Son of man be seen in the heights of the heavens; these matters are no more subjects for the judgment of astronomy, than the final catastrophe itself. This science can furnish little or nothing in explanation of such matters, or for a physical understanding of them; but still less can it presume upon the possession of any power to prove the physical impossibility of such occurrences.

That the sun and moon may become obscured, is matter of experience from year to year: portentous appearances in the heavens, which involuntarily fill the bosom of the beholder with painful or astounding forebodings, are by no means unheard of, as is proven in the sudden and remarkable advent of comets, from time to time. Stars have vanished from the heavens under the eye of the astronomer, and our November nights repeatedly display the scene of thousands of small asteroids falling from the heavens, and the like.

We would by no means maintain that the obscuration of the sun and the moon, in that great day of the future, is to be nothing else than an ordinary eclipse of the sun or moon; that the sign of the Son of man is to be identical with the appearance of a comet, or even that the falling of the stars from the heavens is to be referred to a mere shower of shooting stars; nay, rather, we believe that those points in prophecy denote something wholly different, something heretofore unseen and unheard of; but these facts of experience may certainly be taken as presages or tokens of the *possibility* of these appearances in the heavens, as foretold in prophecy.

§ 20. *The Duration of the present Course of the Earth.*

Our earth must revolve around the sun 18 million times, before the sun itself, together with the whole solar system, completes a single revolution in that wide sweep of movement in which it is involved with all the fixed stars, about that throne of cosmical power, which lies in the centre of the system of the Milky-Way. The great year of the universe, therefore, in which the heavens complete one revolution around the common centre, comprehends, if *Mädler* be correct, 18 millions of terrestrial years (chap. 5, § 9.)

How tiny and insignificant does our earth here appear, how meagre the idea and compass of mortal time, as it here sustains, limits, and controls us! How short and paltry does the period of the existence of the earth and the human race appear, when opposed to such a rule of measurement! What are six thousand in opposition to eighteen millions of years!

The present order of things upon the earth has existed, according to Scripture, almost six thousand years. How long shall it yet continue, till the great day that shall close the course of the present world; when the heavens and the earth, upon the coming of the Son of man, shall be transformed and renewed, in order that a new and ever-enduring period may be introduced?

The Scriptures distinctly reply: "It is not for us to know the times or the seasons which the Father hath put in his own power. Of that day and that hour knoweth no man: no, not the angels which are in heaven."[1]

The Apostles, and with them pious believers of all ages, have regarded the day of the future as near at hand. It was not objective prophecy which expressed itself in this lively expectation; but rather, the subjective state of the devout and religious mind, the sentiment of earnest longing and ardent desire, which was founded in reason and fully justified. Centuries have since gone by, and centuries, yea, thousands of years may yet flow on, before the subjective point of expectation shall coincide with the objective point of the fulfilment.

Yet, though it be *possible* that still hundreds and thousands of years should pass away, before the great day of the end, still it is impossible upon the ground

[1] Acts 1 : 7; Mark 13 : 32, 33.

presented by Scripture, to imagine its objective manifestation so far in the future, that there should exist a corresponding relation on the one hand, between the present course of the earth, with the close of which also the end of the heavens in their present constitution is to coincide, and the cosmical period of revolution belonging to the heavens as a whole, on the other. The position of the incarnation of Christ in the middle of the age of the world, the ever-increasing clearness of the signs of the times, the approach of the final fulfilment of the instituted conditions and the harbingers of the time of the consummation—all this forbids us most imperiously to seek for the boundaries of the developments of the earth, at such a remote and obscure distance.

Are we therefore to understand that the heavens are to be changed as an old garment, before they have reached a single year of their existence, before they have completed a single revolution?

A twofold misunderstanding, with the solution of which this question loses all significance, lies at its foundation. Those six thousand years of the Biblical chronology, as we have already seen, assuredly do not refer to the beginning of the whole universe, nor even to the first beginning of the earth, but to the restitution and new-creation of the latter; or rather, merely to the creation of man, who first appeared after this new-creation. But between the primeval creation and this new creation there lies an undefined and indeterminable period.

And *this*, further, is then overlooked: that the future age of the world, to which the judgment appears as the entrance, cannot be one *apart from time.* Time, which is a necessary correlative of the creature, shall certainly *not cease*, but shall only be absorbed into eternity; just as the creature shall not cease to be the creature, but shall be exalted to a participation in the fulness of the Divine glory (Comp. chap. 4, § 30). But if time do not cease to be time, in the eternity of the future age of the world, certainly the movements and revolutions of the worlds, which are the media and indices of time, shall also not cease. The heavens shall certainly not be annihilated through the final catastrophe, but only be renewed, consummated, and rendered glorious; and the less the heavens are affected by that destruction, which in the purifying fires of the judgment day shall separate the dross as hell, so much the less also shall they be altered from their present constitution.

§ 21. *The Cosmical Consummation.*

Finally, let us cast a glance at the *cosmical state of consummation* of the future world.

Here at length must the whole dignity and worth of the earth, and of its inhabitants, men, have arrived at full and open manifestation. All devastation and ruin brought upon the earth by the twofold catastrophe of the fall of angels and men, must now be overcome, and rooted out; and all destinies attached to the earth through the counsel of divine wisdom, both in its original creation and also in its new-creation for man, have arrived at the highest and fullest unfolding and manifestation of themselves.

As we have previously been compelled to concede cosmical distinction to the heavenly worlds as the dwelling-place of the angels, over the earth in its present condition; so also must we expect that in the consummated state of our at present so lowly and imperfect abode, it will after its own manner, have arrived at a level with the angelic abodes, in respect to those points in which it now falls below them, as they manifest a superior development and perfection; and on the other hand, that the distinguishing features which now, in harmony with its destiny, belong to the earth in contrast to the rest of the celestial worlds, but still as unfolded germs, veiled in the form of lowliness, disturbed and perverted through the curse of sin, shall have appeared in their complete fulness and perfection.

Hence we expect that in the cosmical regions of the earth in the future, at least an equally vigorous co-operation of the now antagonistic contrasts shall take place, with that of now more favored regions: that sin and death, together with all their shadows and fruits, shall be removed: that an equally vital harmony, an equally close communion and reciprocal influence, equally intimate bonds of sympathy and love, shall be found between the members of our solar system, now isolated and existing in their individual capacities. Perhaps this shall take place in a similar manner to that observed in the heavens: perhaps these worlds, so completely separated, yet related and belonging together, attuned to the harmony of a higher music of the spheres, shall celebrate a like sacred and holy jubilee of assembled celestial hosts: perhaps also they shall then stand in the most vital and immediate communication with each other — similar also in this to the celestial worlds — perhaps then that dark, unilluminated, and unilluminable sea of ether belonging to our system, most thoroughly pervaded with light, shall also afford us an "eternal sunshine," and the very same sea of ether which now so rudely dissociates world from world, then most intimately unite them, as the light-atmosphere of the heavens of the fixed stars binds together all the worlds that float within it.

But wherein is the greater glory, the pre-eminence which our earth in future is to possess over all other worlds, to consist? In this: that redeemed, glorified humanity, originally created in the image of God, and again restored to that image, is to dwell there; that the Lord of glory, who has taken upon himself our nature to all eternity, shall there dwell among his people, whom he is not ashamed to call brethren;[1] that he shall bring with himself upon the glorious earth, the immaculate, unfading and imperishable inheritance of His sonship, of which they shall be co-heirs with Him; that He shall there establish among them the most glorious throne of His grace and power, of His glory and majesty; that He himself, the Uncreated Light, shall there shed around their souls the beams of a sacred and holy light which no mortal eye could endure.

As to the conditions and changes that shall be hereby produced, in the *physical* condition of the earth and the wide system to which it belongs, and

[1] Heb. 2 : 11.

in the *cosmical* position of both to the whole universe, it here behooves us to stop short in mute silence and profound adoration, hoping only in the future to arrive at the beyond all measure glorious reply to such a significant inquiry.

We have previously seen that the earth is alone in its manifested lowliness: in like manner it shall be alone, only in an opposite sense, in its future exaltation. As man is made a little lower than the angels, and still is "the embryo of the highest of creatures," so also is the earth made below the worlds of the angels, and yet is it "the most noble germ of the whole creation:"—as Judea is the least and most despised of all lands upon the earth, and still is the *glorious land;*[1] as Bethlehem was little among the thousands of Judah,[2] and yet the Sun of Righteousness there arose with healing in his wings;[3] so also is our region the Judea of the universe, our poor earth the Bethlehem of this holy land, small and lowly, yet precious above all:—and as in the prophetic dream, the sun, moon, and stars, made obeisance to Joseph, the least of all his brethren, so also in future shall the same bow down before the earth, the least of worlds in the universe.

Formerly, when Jehovah laid the foundations of the earth, the morning stars, beholding with adoring wonder, sang together in choral songs of praise; and as the Eternal Word, full of grace and truth, left the throne of glory to clothe Himself in flesh and blood, then swelled in higher and fuller notes the chorus of the heavenly host: *Glory to God in the highest, and*

[1] Dan. 11 : 16–41. [2] Mic. 5 : 2. [3] Mal. 4 : 2.

on earth peace, good will toward men. In the future also, when the Son of Man shall come again in the clouds, surrounded by all the glory of his eternal Godhead, to renew the heavens and the earth, and consummate all things, then shall those sacred messengers of His might and goodness, whose bosoms are thrilled with unspeakable joy at every new token of the spread of God's kingdom upon the earth,[1] behold with adoring wonder the development of those heaven-born mysteries they now desire to look into, and sing in purer tones and loftier chorus their eternal hallelujahs.[2]

[1] Luke 15 : 7.

[2] Acts 5 : 12, 13.

THE END.

Notices by the Press of Kurtz's Sacred History,

PUBLISHED BY LINDSAY AND BLAKISTON, PHILADELPHIA.

"* * * It would seem that the author of this work is one of that class of individuals on whom God has bestowed ten talents. * * * The author is a methodical thinker; he narrates in the most beautiful language and in great clearness, though in a condensed form. * * * The translator has placed in the sacred historical library a work of rare merit."—*Easton Whig, of Jan.* 17, 1855.

"The author's remarkable genius and vast attainments have already given him a place among the greatest lights in theology and history on the continent of Europe. The present work * * * requires to be thoroughly examined in order to a full appreciation of its highly evangelical type, of its lucid arrangement, of its felicitous selection of historical events, of the harmony of the various parts, and the bearing of the whole upon one glorious consummation. * * There are few minds, if any, that have thought so extensively or so profoundly on the subjects of which it treats, that they may not be instructed by it. Dr. Schaeffer has performed his work as translator in a manner that fully satisfies those who are most competent to judge of the merits of the translation."—*Albany Argus, of Jan.* 24, 1855.

"We cannot but regard this work as a valuable aid to our own students and instructors, from its clear and pregnant summary of facts, its lively and original suggestions, and its constant exhibition of unity in all God's plans and dispensations, of which even the most pious and attentive readers of the Bible are too much accustomed to lose sight.

"This book is, according to the Lutheran standard, thoroughly orthodox in matters of doctrine, and is more thoroughly religious in spirit than any similar German work with which we are acquainted.

"The English translation is, in our opinion, highly creditable to its author; not only accurate, so far as we have yet had time to judge it, but less disfigured by undue adherence to German idiom, by awkward stiffness, and by weak verbosity, than any version we have recently examined."—*Biblical Repertory and Princeton Review, of Jan.* 1855.

"It is a work of great value, not only on account of its literary excellence, and the profound theological knowledge displayed in it, but especially as supplying a great want in a clear, simple, and thorough explanation of all the difficult points and obscure questions both as to doctrine and ecclesiastical polity in the Bible.

"All who are desirous of a thorough understanding of Bible history should possess themselves of this learned and interesting work."—*Eastonian, of Jan.* 27, 1855.

Notices by the Press of Kurtz's Sacred History,

PUBLISHED BY LINDSAY AND BLAKISTON, PHILADELPHIA.

"This volume deserves to be in every family; all may read and study it with profit. It is well adapted for schools and seminaries of learning and theology. * * We are pleased to learn that arrangements have been already made for its immediate introduction into Esther Female Institute and Capital University. We know of no work in any language, in all the bounds of sacred literature, calculated to exert a more wholesome and beneficial influence in the cause of Christ, than this work."—*Lutheran Standard, (Columbus, O.) of Jan.* 26, 1855.

"* * * The present volume treats of the subject of Sacred History on a novel plan. It furnishes a suggestive comment on the incidents recorded in the Bible, considered as illustrations of the divine purpose in the salvation of man. The style is clear, compact, and forcible, presenting a mass of weighty thoughts, in simple and appropriate language."—*N. Y. Tribune, of Jan.* 5, 1855.

"* * An important addition to the line of text-books. The plan of the work is as novel as it is happy. * * * Like all other of the recent German theological and metaphysical works, the analytical arrangement is exquisitely delicate and minute, perhaps too much so; and the amount of valuable historical material as well as of doctrinal exposition it contains, bears a proportion to the amount of space which those who are accustomed to our own looser method of composition may well welcome."—*Episcopal Recorder.*

"The arrangement is admirable, the explanatory remarks are instructive, and the whole work one of marked ability." * * —*Baltimore (Baptist) True Union.*

"All classes of readers may study it with advantage."—*N. Y. Commercial.*

"An admirable volume. Its literary and theological merits are of a high order, and entitle it to a wide circulation among the lovers of a religious literature. The translator has faithfully executed his task."—*Christian Chronicle.*

"It is a work of great value as a text-book for Bible classes and schools, and which may be made extensively useful in a family."—*(Boston) Daily Evening Traveller.*

"We have perused this volume with great satisfaction. It is a succinct yet comprehensive sacred history, narrated in a style of great purity and attractiveness; and though its subject is ancient, and hundreds of volumes have been written upon it, yet the book is as full of freshness and charm as if it were a romance."—*New York Observer, of April* 19, 1855.

www.ingramcontent.com/pod-product-compliance
Lightning Source LLC
LaVergne TN
LVHW021305110826
845150LV00003B/483

* 9 7 8 1 4 2 5 5 5 9 5 5 7 *